FRUIT SCIENCE

FRUIT SCIENCE

Editor

Detlev Wechsler

scitus
academics

Fruit Science

Edited by **Detlev Wechsler**

This Book is Printed on Acid-Free Paper. ♾

ISBN: 9781681179377

Library of Congress Control Number: 2017949076

SCITUS Academics LLC,
616, Corporate Way, Suite 2, 4766,
Valley Cottage, NY 10989

www.scitusacademics.com

Printed in United States of America.

Preface

Fruit farming deals with the growing of fruit crops, including nuts, primarily for use as human food. The subject of fruit and nut production deals with intensive culture of perennial plants, the fruits of which have economic significance. Horticulture is a boon of nature which is refined by human skill as a science to obtain more and more benefits. It involves rigorous cropping expertise, including the improvement, production, distribution and use of vegetables, fruits, woody landscape and greenhouse plants. Horticulture is now one of the fastest growing industries with striking professional opportunities. An increasing proportion of the world's population is living in metropolitan environments where their understanding of farming, and therefore of food production, is becoming progressively more poor.

This volume 'Fruit Science' offers PRACTICAL KNOWLEDGE of the production of tree fruits and nuts, grapes, small fruits and vegetables, and other row crops, and forages. The concentration details factors influencing the growth, development, and productivity of these crops (e.g., site selection, cultivar selection, field and plant establishment, pest management, harvesting, and postharvest handling). The concentration also focuses on ongoing and newly emerging specialty industries and concerns such postharvest technology, plant breeding and biotechnology, integrated pest management, and precision agriculture. It also provides an overview of the health benefits associated with fruit consumption, including brief discussions of underlying protective mechanisms, identifies key scientific findings regarding the health benefits of fruit consumption. Also included in this volume are other essential features of fruit processing operations, such as: the food additives used, microbiology, quality assurance, packaging, grades and standards of fruits, and waste management.

Table of Contents

CHAPTER 1

Trait-Based Characterisation of Soil Exploitation Strategies of Banana, Weeds and Cover Plant Species

Florence Tardy[1*], GaeÈ lle Damour[1*], Marc Dorel[1], Delphine Moreau[2]

[1]*CIRAD, UPR GECO, Capesterre-Belle-Eau, Guadeloupe, France,*

[2]*AgroeÂcologie, AgroSup Dijon, INRA, Univ. Bourgogne Franche-ComteÂ, Dijon, France*

ABSTRACT

Cover plants can be introduced in cropping systems to provide agroecosystem services, including weed control via competition for resources. There is currently no consensus on how to identify the best cover plant species, while trait-based approaches are promising for screening plant species due to their agroecosystem service provision potential. This study was carried out to characterize soil exploitation strategies of cover plant species in banana agroecosystems using a trait-based approach, and in turn identify cover plant species with a high weed control potential via competition for soil resources in banana cropping systems. A field experiment was conducted on 17 cover plant species, two weed species and two banana cultivars grown individually. Four functional traits were measured. Two of them (i.e., the size of the zone explored by roots and the root impact density) were used to characterize root system soil exploration patterns. Two other traits (i.e., specific root length and root diameter) were used to characterize resource acquisition within the soil zone explored by the roots. All studied traits exhibited marked variations among species. The findings suggested a trade-off between the abilities of species to develop a limited number of large diameter roots exploring a large soil zone versus many thin roots exploring a smaller soil zone. Three soil-resource exploitation strategies were identified among species: (i) with large diameter roots that explore a large soil zone; (ii) with small diameter roots and a high specific length that explore a smaller soil zone; and (iii) with a high total root-impact density and an intermediate specific root length that explore the uppermost soil layers. Interestingly, in our panel of species, no correlations with regard to belowground and aboveground strategies were noted: species with an acquisitive belowground strategy could display an acquisitive or a conservative aboveground strategy. The findings of this study illustrated that a trait-based approach could be used to identify plant species with potential for competing with weeds, while minimising competition with banana. Six of the 17 studied cover crop species were identified as having this potential. The next step will be to assess them for their weed control performances in banana cropping systems with low reliance on herbicides.

INTRODUCTION

Cover plants can provide agroecosystem services. For instance, they may improve resource use efficiency or control soil erosion and pests [1–3]. They can be grown during the fallow period or in association with the cash crop to control weeds through resource competition, thereby reducing reliance on herbicides. The effectiveness of biological control of pests such as weeds varies with the cover plant species [3–5]. However, to date, no tools are available to identify the most suitable cover plant species according to the targeted services.

Weed control by cover plants relies mainly on competition for available resources between the cover plant and the weed species. When choosing a cover plant for an association with the cash crop, the challenge is to maximise competition with weeds while minimising competition with the cash crop so as to avoid negative impacts on crop growth and yield. There can be aboveground competition between species for light and/or belowground competition for water and nutrients. In agroecosystems with a tall cash crop such as banana, the cash crop:weed competition for light is limited to a short period after crop planting, when the cash crop plants are still small. Belowground competition occurs throughout the crop cycle, probably because cash crop and weed plants share a similar spatial niche in the soil. This suggests that cover plants cannot be used to compete with weeds for soil resources without also generating competition with banana for the same resources. Weed control using cover plants should aim to maximise competition with weeds while minimising competition with banana. This could be achieved using cover plants with above- and belowground spatial niches that differ from those of banana [6], potentially leading to more complete use of resources. Aboveground complementarity between cover plants and cash crops could be achieved by choosing cover plants that are smaller than banana plants. Belowground complementarity could be achieved if the roots of both crop species explore different soil zones [7].

Trait-based approaches, originally developed in the field of comparative functional ecology, are a promising way to screen plant species for their ability to provide ecosystem services [8–10]. They rely on functional traits, i.e. the morphological, physiological and phenological features of individual plants that affect their performance [11]. Traits are related to plant-driven processes and can be used to compare a large number of species (e.g. [12]). Each species can be characterised by a combination of trait values to define its strategy, e.g. the manner in which a species secures carbon profit during vegetative growth and ensures future gene transmission [13]. Plant strategies can be characterized for specific processes (e.g. reproductive strategies in [14]) or at a more global scale via the so-called "leaf economic spectrum", a fundamental trade-off between leaf traits associated with resource acquisition or conservation revealed by ecological studies [15]. The leaf economic spectrum has revealed two contrasted plant ecological strategies, acquisitive to conservative strategies, showing that a plant cannot maximise both the relative growth rate and nutrient retention [16]. A gradient of variation in trait values between resource acquisitive and conservative strategies also exists for roots, i.e. the so-called "root economic spectrum" [17,18]. In both leaf and root economic spectra, species with an acquisitive strategy are defined as making rapid use of nutrients, resulting in a high relative growth rate. Aboveground, they are characterised by a high specific leaf area (leaf area per biomass unit) and a low leaf dry-matter content [15] and, belowground, by a high specific root length (root length per biomass unit) and a small root diameter [17–19]. Conversely, species with a conservative strategy are defined as conserving nutrients, resulting in a low relative growth rate. They are characterised by the opposite combination of trait values. We thus assumed that assessing plant traits in order to characterize the resource exploitation strategies of a range of plant species would provide a basis for assessing their potential ability to compete for resources.

In a previous study [20], we hypothesised that traits related to exploration of an aboveground niche (e.g. plant height and width) and those related to light acquisition capacity (e.g. aboveground dry biomass) reflected plant light exploitation strategies. We assumed that characterising the light exploitation strategy of a range of plant species could generate information on their ability to compete for light and, therefore, to control weeds. On this basis, the light exploitation strategy was characterized in the French West Indies for 21 species, including potential cover plant species, weed species and banana cultivars. In line with this previous paper on light resources, the present study was focused on soil resource exploitation strategies. The overall objective was to characterize the soil exploitation strategies of these species using a trait-based approach, and thus to identify cover plant species with good potential to favour weed control via soil resource competition in banana cropping systems.

Plant root systems may have different architectures, which dictates their ability to grow depthwise and widthwise and to produce many roots. Plants thus have different soil exploration capacities [21]. Spatiotemporal soil exploration is known to play a key role in a plant's access to both mobile resources, such as water and nitrogen, and relatively immobile resources, such as phosphorus ([22] and references therein, [23–27]). The ability of a plant species to explore a soil can be assessed both by the size of the zone explored by its root system and by the total root-impact density within the zone, representing the root occupation within this zone (Table 1). Besides, the ability of a plant to acquire soil resources within this soil zone can be assessed through its capacity to take up soil resources and transport them efficiently. Two root economic spectrum traits reflect these processes (Table 1): specific root length (with a high value reflecting an acquisitive strategy) and root diameter (with a high value reflecting a conservative strategy).

Table 1. Root traits used to characterize the ability of a plant species to exploit soil resources.

	Abbreviation	Description	Unit	Associated function	References	Mean (CV)
Traits related to spatial exploration of soil niches	Z	Median zone explored by the roots[1]	cm^2	Soil interception zone (+)	[28]	764.8 (0.6)
	DI	Total root-impact density[2]	Number of root intersections.dm^{-2} of soil	Interception efficiency (+)	[10]	8.3 (0.6)
Traits related to resource acquisition	SRL	Specific root length	$m.g^{-1}$ of root	Uptake capacity (+)	[29,30]	33.5 (0.9)
	D	Root diameter	Mm	Transport efficiency (+)	[17–19]	0.9 (0.8)
Markers of competition	DI 0–20 cm	Root-impact density in the 0–20 cm soil layer	Number of root intersections.dm^{-2} of soil	Niche complementarity		
	DI 40–80 cm	Root-impact density in the 40–80 cm soil layer				

(+) indicates a positive relationship between the trait value and the function. CV = coefficient of variation corresponding to the mean/standard deviation.

[1] The median zone explored by the roots was calculated as the product of the root depth and root width measured in a vertical soil profile.

[2] Total root impact density was the number of root intersections divided the median zone explored by the roots.

Using the same species panel as in our previous study [20], the present paper characterised traits linked to the exploitation of soil resources (i.e. soil exploration by the root system and resource acquisition within the explored soil zone). Our aim was fourfold: (i) to identify root traits that differentiate plant species, (ii) to analyse correlations among root traits in order to determine whether soil-resource exploitation strategies could be identified in our panel of plant species, (iii) to analyse links between plant exploitation for soil resources (using data from the present study) vs. plant exploitation for light resources (using data presented in [20]), and (iv) to illustrate how the information gained in the present study could be used to facilitate the choice of potential cover plant species in cropping systems. The study was based on a 6-month field experiment, which is the maximum cycle duration of the annual cover crops studied, where plants were cultivated individually in classical soil and climate conditions for banana production.

MATERIALS AND METHODS

Plant species and experimental conditions

A field experiment was conducted to measure traits related to the exploration of both aboveground and belowground spatial niches and resource acquisition in 17 cover plant species and two weed species commonly found in banana cropping systems (Table 2 and S1 File). We also characterised two dessert banana cultivars: Cavendish, the cultivar grown in banana plantations worldwide, and a new cultivar created by CIRAD (cultivar 925), which is resistant to black Sigatoka (*Mycosphaerella fijiensis*), a major banana disease in the French West Indies. The experiment was conducted at the CIRAD Neufchateau experimental station in Guadeloupe (French West Indies, 16°04'48"N, 61°35'53"W, 263 m a.s.l.) from 24 April to 6 November 2013. It was carried out in a 0.4-ha field previously under fallow whose vegetation was destroyed with herbicides and spading before the beginning of the experiment. Four individual plants per species, corresponding to four repetitions, were sown by hand with a 16-m^2 plot per plant (for *B. decumbens*, only three repetitions could have been processed). The plots were randomly distributed in the field. The soil was an andosol (FAO World reference base for soil resources). at the beginning of the experiment, the soil contained, 59.4 g organic matter kg^{-1}, 34.4 g C kg^{-1}, 3.5 g N kg^{-1} and 0.6 g P kg^{-1} on average, with a pH of 5.6. Fifty grams of urea fertilizer (46% of nitrogen) were applied at the base of each plant at the beginning of the experiment to ensure non-limiting nitrogen nutrition. Weeds that grew spontaneously around the plants were regularly removed by hand to ensure non-limiting growth conditions and to assess the growth potential of the tested species. Plants were grown under a rainfed regime with cumulated precipitation of 2,829 mm. The mean daily temperature was 25.6°C, ranging from 22.8°C to 29.7°C. Mean total solar radiation was 462 +/- 40 $MJ.m^{-2}.month^{-1}$.

Table 2. Species name, abbreviation and status in the banana cropping system. See photos of these plants in S1 File.

Species name	Abbreviation	Status	Plant type
Arachis pintoi	AP	Cover plant	Perennial herbaceous plant
Brachiaria decumbens	BD	Cover plant	Perennial grass
Bidens pilosa	BP	Weed	Annual herbaceous plant
Brachiaria ruziziensis	BR	Cover plant	Perennial grass
Cajanus cajun var. Guadeloupe	CCG	Cover plant	Pluri-annual shrub
Centrosema pascuorum	CP	Cover plant	Annual herbaceous plant
Crotalaria spectabilis	CS	Cover plant	Annual herbaceous plant
Crotalaria zanzibarica	CZ	Cover plant	Pluri-annual herbaceous plant
Dolichos lablab	DL	Cover plant	Pluri-annual vine
Gliricidia sepium	GS	Cover plant	Perennial tree
Momordica charantia	MC	Weed	Annual vine
Mucuna pruriens var. deeringiana	MD	Cover plant	Annual vine
Musa spp. var. CIRAD925	B925	Banana cultivar	Perennial herbaceous plant
Musa spp. var. Cavendish	Bcav	Banana cultivar	Perennial herbaceous plant
Vigna unguiculata var. David	N	Cover plant	Short annual vine
Neonotonia wightii	NW	Cover plant	Perennial vine
Paspalum notatum	PN	Cover plant	Perennial grass
Pueraria phaseolides	PP	Cover plant	Perennial vine
Ricinus communis	RC	Cover plant	Perennial shrub
Stylosanthes guianensis	SG	Cover plant	Pluri-annual
Tagetes patula	TP	Cover plant	Short annual herbaceous plant

Plant traits related to belowground water and nutrient competition

Plant species were characterised with traits that accounted for soil exploration and resource acquisition. The traits are listed in Table 1 along with the main associated functions. We considered the four traits listed in the Introduction: the median zone explored by the roots (Z), total root-impact density (DI), specific root length (SRL) and root diameter (D). In addition, we described root distribution in the spatial niche with two markers, i.e. the DI values in two soil layers (0–20 cm and 40–80 cm).

Measured and calculated plant traits

Aboveground traits [20] were measured as a complement to belowground traits to study possible correlations between aboveground and belowground traits and strategies. In species with a short life cycle, traits were measured at mid-flowering, i.e. when half the twigs had started to flower; in species with longer cycles, they were measured 6 months after sowing.

For belowground traits, a 1-m deep and 1-m wide trench was dug 20 cm from the base of the plant. The root intersections in the vertical soil profile were counted on a 5 cm × 5 cm mesh. The median root depth was calculated as the depth at which 50% of the root intersections were visible, and the median root width as the width at which 50% of the root intersections were visible. The median zone explored by the roots (Z) was calculated as the product of the median depth and median width. Median values were chosen to describe the zone where half of the roots—and presumably half of the acquisition—were present irrespective of the form of the root system. The total root-impact density (DI) was calculated as the number of visible root intersections divided by the surface area of the soil explored by the roots in a vertical soil profile. Like DI, the root-impact density per soil layer was calculated independently for the two soil layers (0–20 and 40–80 cm). For instance, in the 0–20 cm layer, DI was the number of visible root intersections between 0 and 20 cm soil depth divided by the surface area of the soil explored by the roots between 0 and 20 cm. For each cover plant species, the root-impact density was expressed relative to that of the Cavendish banana cultivar (Bcav) to assess the complementarity of the spatial niche in the two species. In the 0–20 soil layer: $\Delta 0\text{–}20 = DI_{Bcav\ 0\text{–}20} - DI_{Cover\ plant\ 0\text{–}20}$. In the 40–80 soil layer: $\Delta\ 40\text{–}80 = DI_{Bcav\ 40\text{–}80} - DI_{Cover\ plant\ 40\text{–}80}$. The root-impact densities were analysed (i) in the 0–20 cm layer, where resources were plentiful due to fertilisation and (ii) in the 40–80 cm layer where nitrates were leached by rain.

In addition, for each plant, three 1000-cm^3 soil samples were removed in the form of 10*10*10 cm cubes. The first sample was taken at the base of the plant at a depth of 0 to 10 cm, the second sample was taken under the base of the plant at a depth corresponding to half the maximum depth, and the third sample was taken at half the maximum width and depth. Each sample was washed to collect fine and coarse roots. The roots were scanned at 400 dots per inch (Epson expression 10000XL Pro scanner). The length and diameter of each root sample were measured with WinRHIZO Pro 2009a software (Regent Instruments, Quebec, Canada). The root diameter (D) was calculated by averaging the root diameters in three cubes per plant. The root contents of the three soil cubes were then pooled and weighed after drying for 72 h at 70°C to evaluate root dry biomass. The specific root length (SRL) was calculated by dividing the total length of the roots in the three samples by their root dry biomass.

For aboveground traits, the specific leaf area (SLA) was measured on three green leaves taken from the top of the plants using standard protocols [31]. The whole plant was used to measure aboveground dry biomass (Bma) and leaf dry biomass. In addition to Bma and SLA, the aboveground traits used to describe the competition abilities for light exploitation were height (H), crown width (CW) and leaf area ratio (LAR), with the latter being the ratio of total plant leaf area (product of leaf biomass and SLA) to Bma. The aboveground trait measurements are described in detail in [20].

Statistical analysis

All statistical analyses were performed with R 3.1.1 software (alpha = 0.05) [32]. The Fligner test was conducted to assess the homogeneity in the variance of traits among species. The Kruskal-Wallis test (non-parametric test) was used to test differences between species due to the small number of replicates; we used the "Kruskal" function in the "*agricolae*" package [33] with false discovery rate correction. Correlations between traits were assessed using Spearman's rank order correlation coefficient. Principal component analyses (PCA) were performed with the traits that explained most of the variability between species on the two first axes (with four replicates per species, except for two species with only three replicates, n = 82).

RESULTS

Soil-resource exploitation traits

Mean values and coefficients of variation of functional traits across species are listed in Table 1and the details for each species in the panel are listed in S1 Table. Traits exhibited marked variations among species. The trait with the largest (25-fold) difference among species was the median root exploration zone (Z) which ranged from 52 cm^2 (*T. patula*) to 1323 cm^2 (*P. phaseolides*). Root diameter (D) was the trait with the smallest (9-fold) difference among species, ranging from 0.22 (*B. pilosa*) to 1.99 mm (*D. lablab).*

Correlations between soil resource-exploitation traits

Z was positively correlated with D and negatively correlated with the total root-impact density (DI) and specific root length (SRL) (Table 3 and Fig 1). D was negatively correlated with DI and SRL (Fig 2). DI and SRL were positively correlated.

Table 3. Correlations between root traits related to the ability of the species to exploit soil resources.

	Median zone explored by the roots (Z)	Total root-impact density (DI)	Specific root length (SRL)	Root diameter (D)
Median zone explored by the roots (Z)		**0.007**	**<0.001**	**<0.001**
Total root-impact density (DI)	**-0.30**		**<0.001**	**<0.001**
Specific root length (SRL)	**-0.54**	**0.39**		**<0.001**
Root diameter (D)	**0.45**	**-0.46**	**-0.89**	

The section below the diagonal presents the Spearman's ranked correlation coefficient (n = 82). The section above the diagonal presents the P-value associated with the correlations. Significant correlations are in bold.

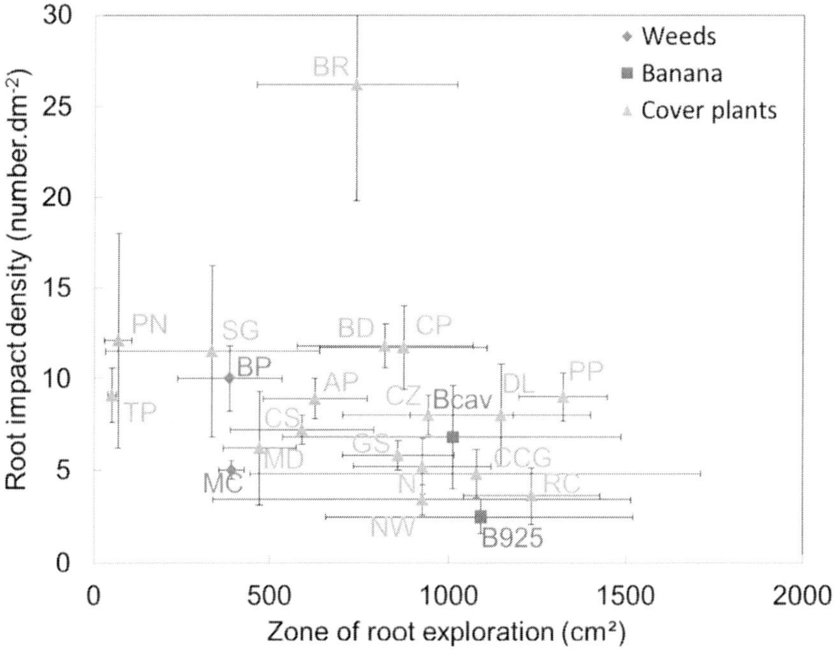

Figure 1. Species distributions regarding the relationship between total root-impact density and root exploration zone. See Table 2 for the species names. (n = 4 replicates for all species except BD (n = 3)).

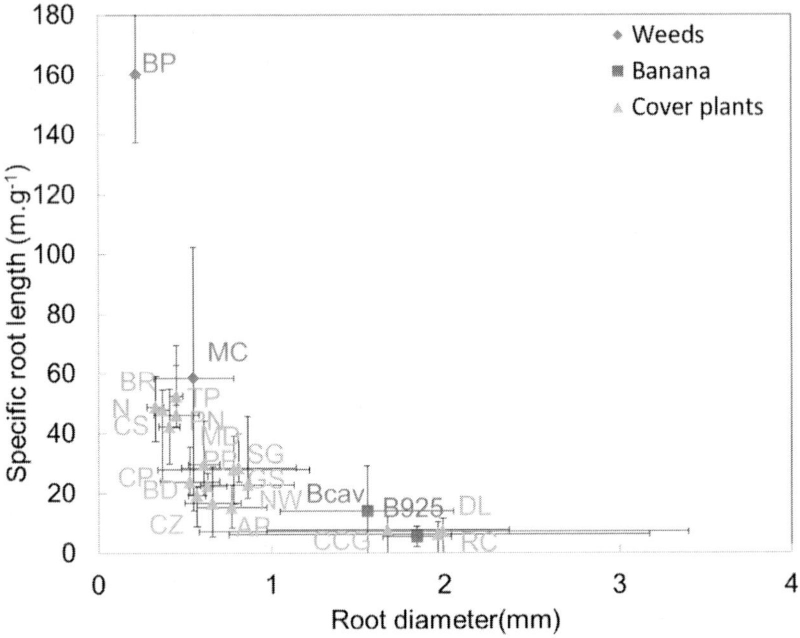

Figure 2. Species distributions regarding the relationship between specific root length and root diameter. See Table 2 for the species names. (n = 4 replicates for all species except BD (n = 3)).

The weed species BP had a smaller Z and a higher DI than the banana cultivars (Fig 1). The cover plant species had a wide range of Z, and *B. ruziziensis* was unique as it had a very high DI and an intermediate Z.

As shown in Fig 2, the two weed species had smaller root diameters, D, than the banana cultivars and a much higher SRL. The roots of the cover plant species had intermediate D and SRL values compared to that of weed species and banana cultivars.

Soil-resource exploitation strategies

A principal component analysis of the 21 species, with four traits related to soil resource exploitation strategies, explained 73% of total variability on two axes (Fig 3a). Axes 1 and 2 explained 51% and 22% of variability, respectively. SRL, Z and D contributed (30%, 23% and 30%, respectively) to axis 1 while DI contributed (62%) to axis 2.

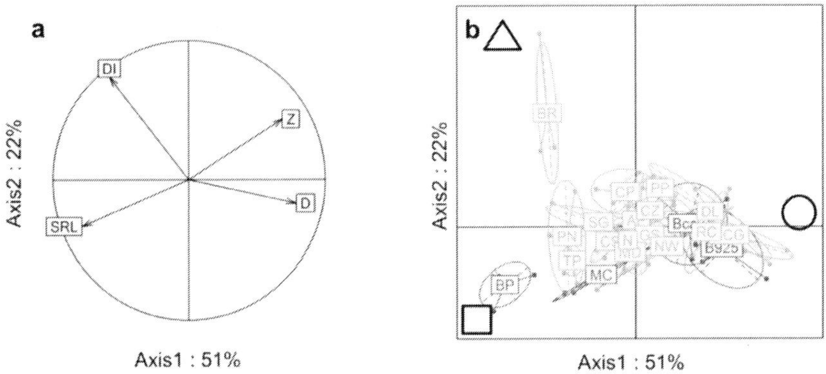

Figure 3. Principal component analysis on four soil-resource exploitation traits and 21 species (n = 4 replicates for all species except BD (n = 3)). a Correlation circle between traits; b Representation of weed species (in blue), banana cultivars (in red) and cover plant species (in green) on the first two axes (see Table 2 for the species names). The soil-resource exploitation traits are: SRL for specific root length, DI for total root-impact density, Z for median root exploration zone and D for root diameter. The symbols represent contrasted plant strategies, with a triangle for species with a high DI (total root-impact density), a square for species with a high SRL (specific root length) and a circle for species with high Z (median root exploration zone) and D (root diameter).

Axis 1 differentiates plant species with high D and high Z (on the right) from plant species with high SRL and high DI (on the left). Axis 2 differentiates plant species according to their DI values.

Fig 3b shows the projection of individual plants in the two-dimensional principal component analysis. Species were located along a continuum from high D and high Z (strategy represented by a circle in Fig 3b) to high SRL (strategy represented by a square in Fig 3b) or high DI (strategy represented by a triangle in Fig 3b). The two banana cultivars (Cavendish, Bcav; and Cirad 925, B925) were close and characterised by high Z, high D and low SRL. At the opposite end of the scale, the two weed species *B. pilosa* (BP) and *M. charantia* (MC) were characterised by high SRL and low Z values. The cover plant species were on a continuum from species with a high Z and high D (e.g. *C. cajan* var. Guadeloupe, CCG) to species with either a high SRL (e.g. *T. patula*, TP) or a high DI (e.g. *B. ruziziensis*, BR) (Fig 3b).

Links between soil-resource and light exploitation traits

We characterised species for soil-resource exploitation traits (this study) and light exploitation traits [20]. We analysed correlations between the four traits explaining soil-resource exploitation and the five traits explaining light resource exploitation (Table 4). Among the correlations, Z was positively correlated with aboveground dry biomass (Bma), height (H) and crown width (CW). Bma was positively correlated with D and negatively correlated with SRL. H and CW were positively correlated with D and negatively correlated with SRL.

Table 4. Correlations between traits related to the ability to exploit soil resources (vertically) and traits related to the ability to exploit light (horizontally).

	Aboveground dry biomass (Bma)	Specific leaf area (SLA)	Leaf area ratio (LARa)	Height (H)	Plant crown width (CW)
Median root exploration zone (Z)	**0.63 <0.001**	-0.07 0.536	-0.10 0.388	**0.42 <0.001**	**0.52 <0.001**
Total root-impact density (DI)	0.02 0.836	0.18 0.154	-0.06 0.610	**-0.23 0.036**	-0.04 0.747
Specific root length (SRL)	**-0.55 <0.001**	0.19 0.085	0.16 0.137	**-0.45 <0.001**	**-0.34 <0.01**
Root diameter (D)	**0.47 <0.001**	**-0.26 0.020**	-0.16 0.153	**0.47 <0.001**	**0.27 0.013**

The first number in each box is the Spearman's ranked correlation coefficient ($n = 82$); the second value is the P-value associated with the correlations; significant correlations ($P<0.05$) are in bold.

To analyse links between strategies for the exploitation of soil resources on the one hand and light on the other, PCA was performed on the 21 species with the four traits explaining soil-resource exploitation and the five traits explaining light exploitation (Fig 4). PCA showed that two axes explained 60% of total variability (Fig 4a). Axes 1 and 2 explained 33% and 27% of the variability, respectively. SRL, Z, D and H contributed significantly (19%, 18%, 17% and 18%, respectively) to axis 1. SLA, Bma, LARa and CW contributed significantly (25%, 22%, 12% and 28%, respectively) to axis 2.

Axis 1 mainly revealed a contrast between soil-resource exploitation, ranging from species with high D, high H and high Z (on the right) to species with high SRL (on the left). Axis 2 mainly differentiated plant species according to their light exploitation traits with high SLA and LARa values at the bottom and low values at the top.

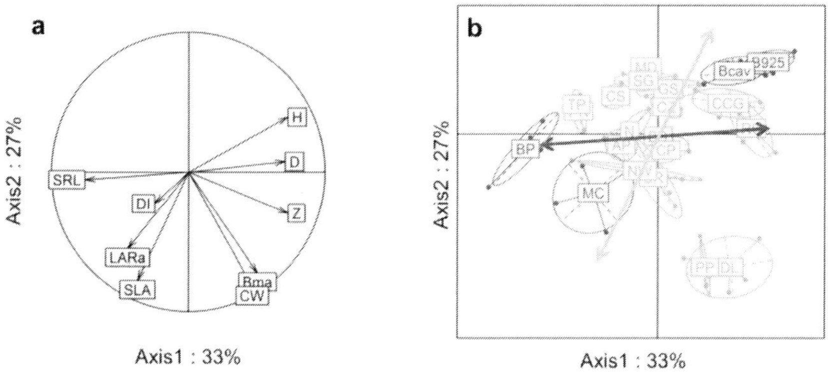

Figure 4. Principal component analysis on four soil-resource exploitation traits and six light exploitation traits and 21 species. a Correlation circle between traits; b Representation of weed species (in blue), banana cultivars (in red) and cover plant species (in green) on the first two axes (see Table 2 for the species names). The light exploitation traits are: Bma for aboveground dry biomass, SLA for specific leaf area, LARa for aboveground leaf area ratio, H for height and CW for plant crown width. The soil-resource exploitation traits are: SRL for specific root length, DI for total root-impact density, Z for the root exploration zone and D for root diameter. The yellow arrow represents the leaf economic spectrum gradient (according to the SLA direction) and the red arrow represents the root economic spectrum gradient (according to the SRL direction).

Fig 4b shows projections of the two-dimensional PCA for individual plants. Species were located along a continuum from high D, high Z and low SLA (the two banana cultivars Bcav and B925), to high SRL, high SLA and high LARa, (the two weed species, *B. pilosa*–BP—and *M. charantia*—MC). The cover plant species were located between the two extremes and scattered according to Bma. Fig 4b shows the gradient in the leaf economic spectrum according to the SLA direction (acquisitive strategy marker; yellow arrow), and the gradient representing the root economic spectrum according to the direction of SRL and D (respectively markers of acquisitive and conservative strategies; red arrow). The two banana cultivars (Bcav and B925) displayed a conservative strategy for both light (low SLA) and soil resources (low SRL), while the two weed species displayed acquisitive strategies for both light (high SLA) and soil resources (high SRL). Some species, including *P. phaseolides* (PP) and *D. lablab* (DL), displayed a light acquisitive strategy and a soil-resource conservative strategy. Species such as *C. spectabilis* (CS) displayed a light conservative strategy and a soil-resource acquisitive strategy.

Prospection zone for soil resources

The root-impact density revealed the soil occupation profile in each soil layer. The soil occupation profile of each cover plant and weed species was compared to that of the Cavendish banana cultivar (see example in Fig 5a). In two distinct soil layers (0–20 cm and 40–80 cm), a Δ was calculated for each cover plant species as the root impact density of banana minus the root impact density of the cover plant species. In the 0–20 cm soil layer, the root-impact density of species with a positive Δ0–20 was lower than that of banana, e.g. *T. patula*(TP) or *R. communis* (RC) in Fig 5b. In the 40–80 cm soil layer, the root-impact density of species with a negative Δ40–80 was higher than that of banana, e.g. *A. pintoï* (AP) or *B. ruziziensis* (BR) in Fig 5b. In banana, the roots that contributed most to soil-resource acquisition were located in the 0–20 cm soil layer. Consequently, species with a lower root density than banana in the 0–20 soil layer (positive Δ0–20 value) and a higher root density than banana in the deeper soil layer (negative Δ40–80) could be considered as species with a preferential root occupation zone complementary to the banana preferential zone, and therefore of potential interest for associations with banana. These species are at the bottom right in Fig 5b: *N. wightii*, *C. zanzibarica*, *C. spectabilis*, *V. unguiculata*, *M. deeringiana*, and *G. sepium* (NW, CZ, CS, N, MD and GS, respectively).

DISCUSSION

Trade-offs between traits linked to spatial-niche exploration and soil-resource acquisition

We characterised a range of plant species according to their soil-resource exploitation strategies by assessing root traits involved in soil exploration by the root system and in the acquisition of resources within the explored soil zone. We observed trade-offs between traits.

In accordance with previous studies ([19] and references therein), we found a negative correlation between root diameter (D) and specific root length (SRL). This supports the hypothesis of a trade-off between processes involved in soil resource acquisition. Species with a high SRL are characterised by rapid nutrient acquisition and turn-over [17–19]. Large diameter roots have more mechanical strength to penetrate tough soil layers, which helps to access deeper soil resources while increasing the uptake of water and leached nutrients [34]. This is consistent with the root economic spectrum opposing conservation species (high D) and acquisition species (high SRL), regardless of botanical families [17,18].

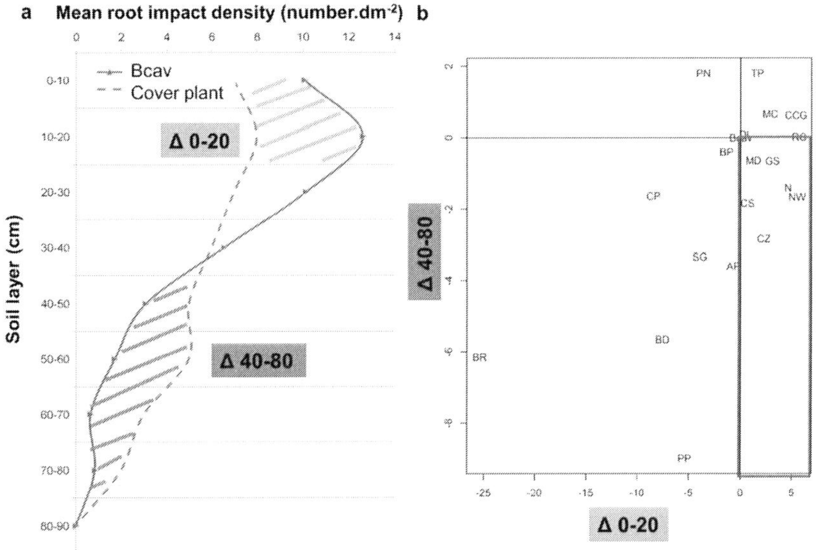

Figure 5. Species distribution according to differences in root-impact density between the cover plant species and the Cavendish banana cultivar (Bcav) in two soil layers (0–20 cm and 40–80 cm). a Banana root-impact density profile (red line) and root-impact density profile of an ideal cover plant species (blue dashed line) which is complementary to the banana profile, i.e. with a positive Δ0–20 (lower root-impact density than banana) and negative Δ40–80 (higher root-impact density than banana). **b** Species distribution according Δ0–20 and Δ40–80. The area outlined in red corresponds to the root-impact density profile required for associations with banana. See Table 2 for the species names.

Interestingly, we obtained a series of correlations showing that species exploring a large median soil zone (Z) had large diameter roots (D) but a low total root-impact density (DI). Conversely, species exploring a small soil zone had small diameter roots with a high total root-impact density. These findings suggest a trade-off between the species abilities to develop a limited number of large diameter roots exploring a large soil zone versus many thin roots limited to a smaller soil zone. They illustrated the scale versus precision foraging theory [29] opposing: (i) plant species that forage slowly for resources at a large scale but with limited spatial colonisation due to a small number of roots per soil unit (e.g. *R. communis*), and (ii) plant species that forage rapidly at a smaller scale but with more precision due to a large number of roots per soil unit, thereby increasing their chance of reaching nutrient patches (e.g. *S. guianensis*). These results are consistent with the two trade-offs found in the literature, i.e. a trade-off between the "exploitation potential" (Z in our study) and "exploitation efficiency" (DI in our study) shown in [28], and a trade-off of carbon allocation between producing thick roots (which are more "carbon-expensive" than thin roots) and fast growing roots [35].

Plant species soil resource exploitation strategies

Characterising strategies of the panel of species with respect to soil resource exploitation revealed three contrasted strategies (Fig 3b), with species located along a continuum between these three strategies. (i) We identified species, such as banana, exploring a large soil volume (Z > 1,000 cm^2) with large diameter roots (> 1.5 mm) able to penetrate tough soil layers (this strategy is represented by a circle in Fig 3b). As water is generally found in deep soil layers, plant species with this strategy may have the advantage of being able to acquire water and leached mobile resources. According to the root economic spectrum [17,18], these species could be considered as conservative. (ii) We identified species, such as *T. patula*, with a very high SRL (> 45 $m.g^{-1}$) (this strategy is represented by a square in Fig 3b). The specific root length is known to be greatest in species with high nitrogen [36] and phosphorus [37] absorption abilities. Moreover, high SRL reflects a high relative growth rate [17,18,38]. We hypothesise that species with this strategy rapidly increase their root length to absorb nutrients efficiently. According to the root economic spectrum, these species could be considered as acquisitive. The strategies of both weed species were close to this strategy. Their rapid expected growth and nutrient acquisition could explain their abundance in the studied banana agroecosystem. (iii) We identified species, like *B. ruziziensis*, with a high total root-impact density (more than 11 root $intersections.dm^{-2}$) and a small median soil zone (Z < 800 cm^2) (this strategy is represented by a triangle in Fig 3b). We hypothesised that these species favour the production of many thin roots to explore the superficial soil layers. This strategy may favour the acquisition of immobile nutrients, such as phosphorus, in the uppermost soil layer [25,39,40].

Links between aboveground and belowground traits and strategies

Until now, only a few papers have analysed the links between aboveground and belowground strategies using a range of plant species [41]. In our study, a few positive correlations were found between traits characterising the size of the aboveground and belowground spatial niches (Table 4): the median zone explored by roots (Z) was positively correlated with aboveground biomass (Bma), height (H) and crown width (CW). Although weak, a significant negative correlation was also found between the root diameter (a high D value is an indicator of a conservation strategy in the root economic spectrum) and the specific leaf area (a high SLA value is an indicator of an acquisition strategy in the leaf economic spectrum), in accordance with the findings of [42]. However, in contrast with [19], no significant correlation was found between the specific root length (SRL) and the specific leaf area (SLA), which are indicators of acquisitive strategies in the root and leaf economic spectra, respectively. Consequently, the gradient of the leaf economic spectrum according to the specific leaf area (SLA, marker of the acquisitive strategy; yellow arrow on Fig 4b) was not superimposed on the gradient representing the root economic spectrum positioned according to the specific root length and root diameter (SRL and D, respectively markers of acquisitive and conservative strategies; red arrow on Fig 4b). Overall, these results did not confirm the "plant economic spectrum" hypothesis put forward by Reich (2014), which presumes a similar strategy for aboveground and belowground compartments. Accordingly, some species displayed a similar ecological strategy aboveground and belowground—this was the case for banana cultivars with a conservative strategy for both light (low SLA) and soil resources (low SRL), and for the two weed species showing acquisitive strategies both aboveground and belowground. Moreover, some plant species (cover plant species in our study) displayed a light acquisitive strategy and a soil-resource conservative strategy, while others displayed a light conservative strategy and a soil-resource acquisitive strategy. Our findings thus indicated that all combinations between aboveground and belowground strategies are possible in plants.

Practical application for identifying potential cover plant species to control weeds

We assessed the potential of cover crop species to compete with weeds and banana by growing a large panel of cover plant species in the absence of competition with banana. Our results provided insight for identifying cover plant species with a good potential for controlling weeds in banana cropping systems. The objective when choosing a cover plant species for associations with banana is to simultaneously minimize competition with banana while maximising competition with weeds. Cover plant species are thus required to preferentially explore different soil layers from those explored by banana and, within these layers, to maximise soil exploration and resource acquisition. We compared the root-impact densities of different cover plant species in shallow and deep soil layers relative to that of banana (cv. Cavendish). We identified six cover plant species that preferentially explored deep soil layers, while banana explored shallow soil layers (Fig 5b). These species were therefore be assumed to minimize soil

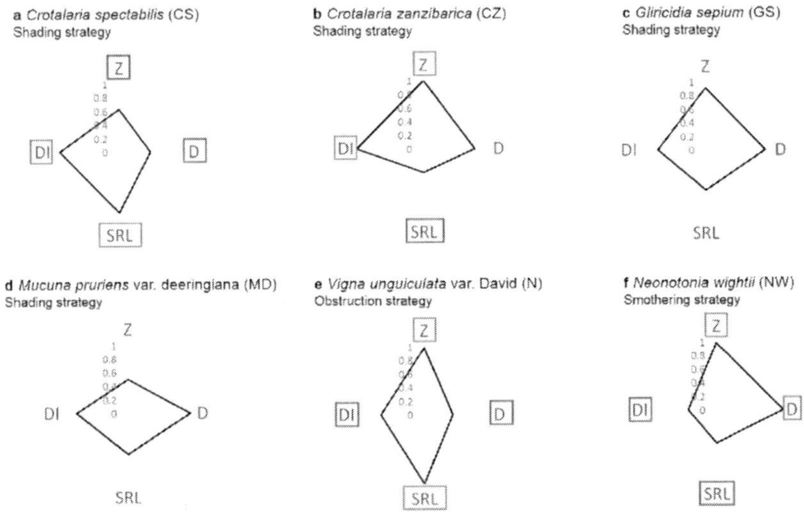

Figure 6. Comparison of the six potential cover crop species with regard to the four studied root traits: Root exploration zone (Z), root diameter (D), Specific Root Length (SRL) and total root impact Density (DI). a *Crotalaria spectabilis* (CS). b *Crotalaria zanzibarica* (CZ). c *Gliricidia sepium* (GS). d*Mucuna pruriens* var. *deeringiana*. e *Vigna unguiculata* var. David (N). f *Neonotonia wightii* (NW). For each trait, values are expressed in proportion to the highest value per trait observed for that trait in the panel of species studied. For each species, when the name of a trait is surrounded by green, then the species is one of the two species (among the six species) with the highest values for this trait. For each species, when the name of a trait is surrounded by red, then the species is one of the two species (among the six species) with the lowest values for this trait. Light exploitation strategy [see 20] of each species is mentioned under its name.

resource competition with banana. We also characterized each species by four traits related to spatial exploration of soil niches and acquisition of resources within the explored soil zone. A cover plant species with a good weed control potential could be expected to combine high values for these four traits. However, due to trade-offs between traits, none of the strategies identified and none of the studied species displayed this characteristic (Fig 3b). Therefore, we could not determine which of the six species, identified as minimizing competition with banana, had the best potential for maximising competition with weeds. Further studies should be focused on determining whether it is better to choose a species with intermediary values for each root trait (*Gliricidia sepium* in Fig 6c), or a species with high values for some traits and low values for others (five other species in Fig 6a, 6b, 6e, 6f and 6g). Indeed, at the current state of knowledge, it remains unclear which traits, among the four studied root traits, have the greatest impact for maximising competition with weeds. Moreover, in line with our previous study [20], the six potential cover plant species displayed different light exploitation strategies (Fig 6). Therefore, the ability of the six species (differing in both root traits and light exploitation strategies) to maximise competition with the most problematic weed species will have to be assessed in the field. Growing each of the six potential cover crop species, in competition with banana and/or weeds, would also allow us to search for any modifications in their rooting patterns in response to competition.

CONCLUSIONS

With increasing interest in trait-based approaches to help in identifying potential cover plant species [10,43], the present study offers a new test of this approach. Of particular interest is the focus on root traits, an important yet understudied aspect of plant functional trait research. Our study generated answers to the four questions outlined in the Introduction. (i) All of the studied root traits varied among species and were therefore identified as relevant for analysing interspecific differences. (ii) Correlations among root traits indicated trade-offs between traits linked to spatial-niche exploration and soil-resource acquisition. Interestingly, they allowed us to position the different species on a scale versus precision foraging gradient and, on this basis, to identify three contrasted soil-resource exploitation strategies among species. (iii) We did not confirm the hypothesis of a plant spectrum due to the discrepancy between the belowground strategies (characterized in the present study) and the aboveground strategies (characterized in our previous paper [20]). (iv) The information gained in the present study illustrated how a trait-based approach could be used to identify potential cover plant species that could be introduced in cropping systems, according to their ability to compete for resources with weeds and banana. Among the 17 studied cover plant species, six were identified as potential good candidates. The next step will be to assess their performance in the field to biologically control weed species without affecting banana, in cropping systems with lower reliance on herbicides.

ACKNOWLEDGMENTS

We thank Charles Meynard, Mylène Ramassamy, Steewy Lakhia, Christina Racel, and Jean-Luc Jean-Louis for their technical support. We would also like to thank David Manley for improving English writing, and Nathalie Colbach and the anonymous reviewers who helped greatly improve this manuscript.

REFERENCES

1. Altieri MA (1999) The ecological role of biodiversity in agroecosystems. Agriculture Ecosystems & Environment 74: 19–31.
2. Koohafkan P, Altieri MA, Gimenez EH (2012) Green Agriculture: foundations for biodiverse, resilient and productive agricultural systems. International Journal of Agricultural Sustainability 10: 61–75.
3. Lu YC, Watkins KB, Teasdale JR, Abdul-Baki AA (2000) Cover crops in sustainable food production. Food Reviews International 16: 121–157.
4. Malezieux E, Crozat Y, Dupraz C, Laurans M, Makowski D, Ozier-Lafontaine H, et al. (2009) Mixing plant species in cropping systems: concepts, tools and models. A review. Agronomy for Sustainable Development 29: 43–62.
5. Vandermeer JH (1989) The Ecology of Intercropping. Cambridge: Cambridge University Press. 237 p.
6. Hunt MA, Battaglia M, Davidson NJ, Unwin GL (2006) Competition between plantation Eucalyptus nitens and Acacia dealbata weeds in northeastern Tasmania. Forest Ecology and Management 233: 260–274.
7. Ehrmann J, Ritz K (2014) Plant: soil interactions in temperate multi-cropping production systems. Plant and Soil 376: 1–29.
8. Martin AR, Isaac ME (2015) Plant functional traits in agroecosystems: a blueprint for research. Journal of Applied Ecology 52: 1425–1435.
9. Wood SA, Karp DS, DeClerck F, Kremen C, Naeem S, Palm CA (2015) Functional traits in agriculture: agrobiodiversity and ecosystem services. Trends in Ecology & Evolution 30: 531–539.
10. Damour G, Garnier E, Navas ML, Dorel M, Risede JM (2015) Using Functional Traits to Assess the Services Provided by Cover Plants: A Review of Potentialities in Banana Cropping Systems. In: Sparks DL, editor. Advances in Agronomy, Vol 134. pp. 81–133.
11. Violle C, Navas M-L, Vile D, Kazakou E, Fortunel C, Hummel I, et al. (2007) Let the concept of trait be functional! Oikos 116: 882–892.
12. Gardarin A, Daurr C, Colbach N (2011) Prediction of germination rates of weed species: Relationships between germination speed parameters and species traits. Ecological Modelling 222: 626–636.
13. Westoby M, Falster DS, Moles AT, Vesk PA, Wright IJ (2002) Plant ecological strategies: Some leading dimensions of variation between species. Annual Review of Ecology and Systematics 33: 125–159.
14. Stewart A, Thompson K (1982) Reproductive strategies of six herbaceous perennial species in relation to a successional sequence. Oecologia 52: 269–272.
15. Wright IJ, Reich PB, Westoby M, Ackerly DD, Baruch Z, Bongers F, et al. (2004) The worldwide leaf economics spectrum. Nature 428: 821–827. pmid:15103368
16. Aerts R (1999) Interspecific competition in natural plant communities: mechanisms, trade-offs and plant-soil feedbacks. Journal of Experimental Botany 50: 29–37.

17. Roumet C, Birouste M, Picon-Cochard C, Ghestem M, Osman N, Vrignon-Brenas S, et al. (2016) Root structure-function relationships in 74 species: evidence of a root economics spectrum related to carbon economy. New Phytologist 210: 815–826. pmid:26765311
18. Roumet C, Urcelay C, Diaz S (2006) Suites of root traits differ between annual and perennial species growing in the field. New Phytologist 170: 357–368. pmid:16608460
19. Fort F, Jouany C, Cruz P (2012) Root and leaf functional trait relations in Poaceae species: implications of differing resource-acquisition strategies. Journal of Plant Ecology 6: 211–219.
20. Tardy F, Moreau D, Dorel M, Damour G (2015) Trait-based characterisation of cover plants' light competition strategies for weed control in banana cropping systems in the French West Indies. European Journal of Agronomy 71: 10–18.
21. Pages L, Becel C, Boukcim H, Moreau D, Nguyen C, Voisin AS (2014) Calibration and evaluation of ArchiSimple, a simple model of root system architecture. Ecological Modelling 290: 76–84.
22. Dunbabin VM, Postma JA, Schnepf A, Pages L, Javaux M, Wu L, et al. (2013) Modelling root-soil interactions using three-dimensional models of root growth, architecture and function. Plant and Soil 372: 93–124.
23. Chapman N, Miller AJ, Lindsey K, Whalley WR (2012) Roots, water, and nutrient acquisition: let's get physical. Trends in Plant Science 17: 701–710. pmid:22947614
24. Jungk AO (2002) Dynamics of nutrient movement at the soil-root interface; Waisel Y, Eshel A, Kafkafi U, editors. New York, USA: Marcel Dekker Inc.
25. Lynch JP, Brown KM (2001) Topsoil foraging—an architectural adaptation of plants to low phosphorus availability. Plant and Soil 237: 225–237.
26. Zhu JM, Lynch JP (2004) The contribution of lateral rooting to phosphorus acquisition efficiency in maize (Zea mays) seedlings. Functional Plant Biology 31: 949–958.
27. Hodge A, Robinson D, Griffiths BS, Fitter AH (1999) Why plants bother: root proliferation results in increased nitrogen capture from an organic patch when two grasses compete. Plant Cell and Environment 22: 811–820.
28. Berntson GM (1994) Modeling root architecture—Are there trade-offs between efficiency and potential of resource acquisition? New Phytologist 127: 483–493.
29. Maire V, Gross N, Pontes LdS, Picon-Cochard C, Soussana J-F (2009) Trade-off between root nitrogen acquisition and shoot nitrogen utilization across 13 co-occurring pasture grass species. Functional Ecology 23: 668–679.
30. Mokany K, Ash J (2008) Are traits measured on pot grown plants representative of those in natural communities? Journal of Vegetation Science 19: 119–126.
31. Perez-Harguindeguy N, Diaz S, Garnier E, Lavorel S, Poorter H, Jaureguiberry P, et al. (2013) New handbook for standardised measurement of plant functional traits worldwide. Australian Journal of Botany 61: 167–234.
32. R Core Team (2014) R: A language and environment for statistical computing. R Foundation for Statistical Computing. Vienna, Austria.
33. de Mendiburu F (2014) agricolae: Statistical Procedures for Agricultural Research. R package version 1.2–0. ed.
34. Eissenstat D (1992) Costs and Benefits of Constructing Roots of Small Diameter. Journal of Plant Nutrition 15: 763–782.
35. Hishi T (2007) Heterogeneity of individual roots within the fine root architecture: causal links between physiological and ecosystem functions. Journal of Forest Research 12: 126–133.
36. Osone Y, Tateno M (2005) Nitrogen absorption by roots as a cause of interspecific variations in leaf nitrogen concentration and photosynthetic capacity. Functional Ecology 19: 460–470.
37. Gahoonia TS, Nielsen NE (2004) Barley genotypes with long root hairs sustain high grain yields in low-P field. Plant and Soil 262: 55–62.

38. Cornelissen JHC, Werger MJA, CastroDiez P, vanRheenen JWA, Rowland AP (1997) Foliar nutrients in relation to growth, allocation and leaf traits in seedlings of a wide range of woody plant species and types. Oecologia 111: 460–469.
39. Ge ZY, Rubio G, Lynch JP (2000) The importance of root gravitropism for inter-root competition and phosphorus acquisition efficiency: results from a geometric simulation model. Plant and Soil 218: 159–171. pmid:11543364
40. Lynch JP (2007) Roots of the second green revolution. Australian Journal of Botany 55: 493–512.
41. Reich PB (2014) The world-wide 'fast-slow' plant economics spectrum: a traits manifesto. Journal of Ecology 102: 275–301.
42. Chanteloup P (2013) Patrons d'organisation des traits aériens et racinaires en prairies humides: liens avec la production primaire (quantité et qualité). Université Rennes 1; Université européenne de Bretagne. 196 pp. PhD Thesis.
43. Tribouillois H, Fort F, Cruz P, Charles R, Flores O, Garnier E, et al. (2015) A Functional Characterisation of a Wide Range of Cover Crop Species: Growth and Nitrogen Acquisition Rates, Leaf Traits and Ecological Strategies. Plos One 10: e0122156. pmid:25789485

CHAPTER 2

Combined Effects of Fertilizer, Irrigation, and Paclobutrazol on Yield and Fruit Quality of Mango

Babul C. Sarker [1,*], Mohammad A. Rahim[2] and Douglas D. Archbold[3]

[1] *Principal Scientific Officer, Pomology Division, Horticulture Research Centre, Bangladesh Agricultural Research Institute, Joydebpur, Gazipur 1701, Bangladesh*

[2] *Department of Horticulture, Bangladesh Agricultural University, Mymensingh 2202, Bangladesh*

[3] *Department of Horticulture, University of Kentucky, Lexington, KY 40546-0091, USA*

ABSTRACT

Combinations of fertilizer rates, foliar N sprays, irrigation practices, and paclobutrazol were studied to determine how much they could alter and/or improve mango (*Mangifera indica* L.) growth, flowering, and yield. Two treatment combinations derived from several years of prior studies of individual practices were compared: one combination was comprised of the best (BT) individual practices from the prior studies and included three applications of fertilizer, a 4% KNO_3 spray application before flowering, paclobutrazol at 7.5 g/L, and weekly irrigation, and the other combination was comprised of the next best (NB) individual practices including two applications of the same amount of fertilizer, a 4% urea spray before flowering, paclobutrazol at 10.0 g/L, and biweekly irrigation. Both combinations significantly reduced terminal shoot growth and leaves per terminal shoot, advanced the date of flowering and harvest, increased panicle number, length and secondary branching, increased fruit set, fruit number at harvest, fruit size, and yield, with BT producing larger fruit and a greater yield than NB. Although both combinations produced fruit with higher quality than the control, the BT combination produced fruit with the higher total soluble solids, reducing, non-reducing, and total sugar content, and vitamin C content than the NB combination. Both BT and NB combinations of the optimums identified in the prior studies were successful at advancing bloom and harvest and increasing yield more than any of the optimum individual components alone, by 14-fold more than untreated trees for the BT combination, suggesting there were additive, if not synergistic, effects on mango. Further studies are warranted to assess the sustainability of these effects over longer periods of time, and to ascertain if the effects occur across mango cultivars and production environments.

Keywords: paclobutrazol; soil drenching; panicle emergence; flowering; peel–pulp ratio

1. INTRODUCTION

Mango (*Mangifera indica* L.) is a delicious fruit which belongs to the family Anacardiaceae. The principal mango producing countries are India, China, Thailand, Indonesia, and Pakistan [1]. In Bangladesh, which ranks ninth, about 242,605 tons of fruit are produced from an area of 51,012 ha, with an average yield of 4.75 tons per ha [2], although it falls short of fulfilling national demand. Irregular flowering, low fruit set, as well as fruit retention leading to low yield, fruit of poor quality, and a short harvest period are the main hindrances to increasing mango availability. Optimizing and integrating various key management practices such as fertilizer application, irrigation, foliar sprays of KNO_3 or urea, and soil drench application of paclobutrazol might extend the availability period and increase yield and quality. Many mango growers in Bangladesh do not employ these management practices on an annual basis, or have few guidelines for optimizing them, and need a set of recommendations to increase and sustain production.

Yearly soil application of N, P, and K has markedly increased the number of mango fruit per plant, fruit weight, yield, and fruit quality [3,4,5,6,7]. Effective rates of N, P and K have ranged from 0.4 to 0.8 kg per tree. Foliar KNO_3 applications have also advanced flowering and harvest date, increased yield, and reduced alternate bearing in mango [8,9]. Maximum yield was found by foliar urea application [10]. Potassium nitrate, especially in combination with urea, produced good results on flowering and yield parameters [11]. Irrigation of mango trees can be beneficial, as it has been shown to reduce fruit drop resulting in satisfactory production [12]. However, in Bangladesh, bearing trees are often irrigated only after fruit set and thereafter at fortnight intervals, but not all growers practice this.

Soil application of the growth retardant paclobutrazol was reported to reduce tree growth and induce precocious flowering, fruit set, fruit retention, and increase yield in bearing mango trees [13,14]. In addition, improvements in fruit quality parameters, such as ascorbic acid, total sugar, reducing sugar, and total soluble solids content, have also been observed [14,15,16,17].

A series of studies assessing strategies of soil and foliar fertilization, irrigation, and soil application of paclobutrazol were carried out to identify the best treatments for optimizing mango yield, fruit quality, and harvest period, and to be used to develop management recommendations. Cow manure, widely available to mango growers in Bangladesh, was studied as a soil fertilizer. It comprised 13.8 g N/kg, 3.2 g P/kg, 0.36 g K/kg, 0.13 g S/kg, and 0.02 g Zn/kg, and was applied at 12.5 to 37.5 kg per tree in 1, 2, or 3 separate installments over several months [18]. The highest rate, split into three applications, produced the highest number of fruit, total yield, and fruit quality, and delayed the start of harvest by 11 days. To determine if foliar nutrient application was beneficial, KNO_3 at 4%, 6% and 8% (w/v), and urea at 2% and 4% (w/v), were applied before flowering on 15 November 2007. KNO_3 at 4% exhibited the highest yield, and fruit vitamin C and total soluble solids (TSS) content, followed by urea at 4% [19], with both advancing the start of harvest by 5 days. Regular irrigation starting at fruit set prevented fruit drop and improved fruit in size and quality [20]. To determine which irrigation strategies had significant impacts on fruit yield and quality, trees were irrigated on 15 October and 15 November 2007 and then from fruit set to maturity at 7, 14, or 21 day intervals versus no irrigation. Plants irrigated twice on 15 October and 15 November 2007 and then from fruit set to maturity at 7 and 14 day intervals produced about two-fold higher yields (22.2 and 21.6 kg/plant, respectively) and the best quality [21], and delayed the start of harvest by 7 days. Paclobutrazol applied at 7.5 g/L on 15 October 2007 produced the highest number of fruits as well as greatest yield per plant fruit weight, and had the best quality, compared to lower and higher rates and a December application, with harvest starting about 15 days earlier. A subsequent study indicated that a July application of paclobutrazol at 7.5 g/L or 10.0 g/L significantly advanced harvest by 22 days over control and produced 4.7-fold higher yield and bigger fruit over control.

After having assessed the cultural practices individually, combinations of the optimum individual practices were assessed to determine if they were additive or synergistic in their effects on tree growth and components of yield. Thus, an integrated experiment with split applications of fertilizer,

KNO_3 and urea sprays, periodic irrigation, and soil drench applications of paclobutrazol was performed, combining the best individual practices from the previous studies as one treatment, and the second best practices as another treatment, to determine effects on mango yield and fruit quality.

2. MATERIALS AND METHODS

2.1. Location

The experiment was carried out at the Bangladesh Agricultural University (BAU) Germplasm Centre, Department of Horticulture, Mymensingh, Bangladesh which is located at 24°26′ latitude and 90°15′ longitude with an altitude of 8.3 m above sea level. Biochemical analyses were carried out in the Department of Biochemistry of Bangladesh Agricultural University (BAU), Mymensingh, Bangladesh.

2.2. Soil and Climate

The soil at the Germplasm Centre is sandy loam which belongs to the Old Brahmaputra Flood Plain Alluvial Tract [22]. The selected soil samples of the experimental area were analyzed at the Laboratory of the Soil Science Division, Bangladesh Agricultural Research Institute, Joydebpur, Gazipur. Soil pH was 6.5 and 6.3 at a depth of 0–15 cm and 15–30 cm, respectively. The total %N, available P, and available K were 0.056, 38 μg/mL and 0.10 meq/100 mL at a soil depth of 0–15 cm, respectively, and 0.036, 13 μg/mL and 0.07 meq/100 mL at a soil depth of 15–30 cm, respectively. The experimental site enjoys a sub-tropical climate characterized by heavy rainfall (means of 361 mm/month and 2166 mm total), high temperature (mean 28.34 °C), a mean of 161 h/month of sun during April to September, and scant rainfall (mean 79 mm/month and 555 mm total), low temperature (22.6 °C) and a mean of 187 h/month of sun during October to March (Source: Weather Yard, Department of Irrigation and Water Management, BAU, Mymensingh, Bangladesh, 2005).

2.3. Experimental Material

The cultivar used in the study, Amrapali (BARI Mango-3), is precocious, medium dwarf, annual, and prolific in bearing and has good fruit quality [23]. The age of the plants was 11 years at the initiation of experiment. Plant spacing was 5 m × 5 m.

2.4. Experimental Design

The experiment was laid out in a Randomized Complete Block Design (RCBD) with 3 replications. The treatments were: (1) the best (BT) individual treatments of fertilizer (37.5 kg cow dung with 518.43 g N, 120 g P, 187.5 g K, 67.45 g S, and 8.09 g Zn per plant) applied in three installments, a 4% KNO_3 spray application on 15 November 2007 approximately 1.5 to 2 months prior to anticipated flowering, irrigation on 15 October and 15 November 2007 and then continued from fruit set (6 March 2008) to maturity at 7 day intervals, soil drench application of paclobutrazol at 7.5 g/L of active ingredient on 15 July 2007 (all of the best treatments from prior studies); (2) the 2nd or next best (NB) individual treatments of fertilizer (same as above) applied in two installments, a 4% urea spray application on 15 November 2007 approximately 1.5 to 2 months prior to anticipated flowering, irrigation on 15 October and 15 November 2007 and then continued from fruit set to maturity at 14 day intervals, and a soil drench application of paclobutrazol at 10 g/L of active ingredient on 15 July 2007; and (3) untreated control.

2.5. Application of Fertilizer and Paclobutrazol

Fertilizer was applied in two or three applications. The first application was the total amount of cow dung, triple superphosphate (TSP), gypsum, and zinc sulfate applied on 15 September. If in two applications, ½ of the urea and muriate of potash (MoP) were applied on 15 September and the remainder was applied on 15 March. If in three applications, ¼ of the total amounts of urea and MoP were applied on 15 March and again at the 3rd application on 15 May. Very light irrigation was provided at each time of fertilizer application.

Paclobutrazol (Syngenta, India) was prepared from a 25% (w/v) stock solution for final concentrations of 7.5 and 10 g/L of active ingredient. Treatments were applied as a soil drench in which 10 small holes (10–15 cm depth) were prepared in the soil around the collar region of the trees just inside the fertilizer ring [24]. One liter of the prepared solutions was poured into the holes of each tree, and the soil was reworked after application. Only water was applied to the control plants.

2.6. Tree Growth and Yield Measurements

The length and number leaves of 10 randomly-selected terminal shoots at flowering on 20 December 2007 were measured. The individual leaf area of 50 leaves was measured, taking 5 from each of the 10 above-selected shoots, using a leaf area meter and was expressed as cm^2. Starting at the first appearance of a panicle, the number of panicles per plant was counted at 10 day intervals up to completion of panicle emergence. Ten panicles were randomly selected from each treatment on 18 February 2008, and the length and number of secondary branches per panicle was recorded. The initial number of fruit of each panicle, and the fruit retained per panicle were counted at 10 day intervals starting from the pea stage up to harvest starting from 6 March 2008. The average date of harvest (when one or two ripe fruit dropped from the tree, all the fruit were harvested at one time), number of fruit/tree, fruit weight, and total yield/tree was recorded.

2.7. Fruit Quality Measurements

After harvest, 10 randomly-selected fruit per tree were ripened at room temperature, and fruit quality was determined. Each fruit was peeled, the stone removed, and each was weighed along with the remaining pulp. The edible (pulp) portion, stone-to-pulp ratio, peel-to-pulp ratio, shelf-life (the difference between the harvest date and the date up to which the fruit remained edible was considered as shelf life), total soluble solids (TSS), titratable acidity, vitamin C, dry matter, and reducing, non-reducing, and total sugar content. TSS of the pulp was measured using a hand refractometer. The titratable acidity of mango pulp was determined [25] using 0.1 N NaOH solution. Vitamin C content was determined [26]. Reducing sugar content of mango pulp was determined by the dinitrosalicylic acid method [27]. The total sugar content of mango pulp was colorimetrically determined by the anthrone method [28].

2.8. Statistical Analysis

Treatment means were analyzed by analysis of variance and separated by Fisher's Least Significant Difference (LSD) test at $p = 0.05$ [29].

3. RESULTS

3.1. Tree Growth

Both BT and NB had similar effects of reducing terminal shoot length and increasing panicle length and the number of secondary branches per panicle (Table 1). The NB treatment had the fewest leaves per terminal shoot, although the BT treatment also reduced terminal shoot length. The

NB treatment only had the smallest leaf dimensions and area per leaf. Interestingly, the BT combination had the widest panicles.

Table 1. Effects of combinations of growth regulator and management practices on leaf, shoot and panicle characters of mango. Treatments were the combinations of the best and next best treatments from prior studies, as defined in the text.

Treatment Combination	Length of Terminal Shoot (cm)	Leaves per Terminal Shoot	Leaf			Length of Panicle (cm)	Width of Panicle (cm)	Secondary Branches per Panicle
			Length (cm)	Width (cm)	Area (cm^2)			
Control	16.3 a [z]	12.5 a	22.6 a	4.5 a	63.1 a	22.9 b	17.1 b	24.8 b
Best	14.2 b	9.3 c	21.1 a	4.4 a	60.9 a	28.1 a	19.1 a	32.7 a
Next Best	12.8 b	8.4 b	19.2 c	3.8 b	52.3 b	26.9 a	17.5 b	31.7 a
LSD (0.05)	1.5	0.8	1.89	0.4	5.8	1.6	1.3	3.3

[z] Means followed by different letters within columns significantly differ by Fisher's LSD at $p = 0.05$.

3.2. Emergence, Number of Panicles, Fruit Set, and Fruit Retention

The first panicle emerged 49 and 44 days earlier for the BT and NB treatment combinations, respectively, than the control (Table 2). The BT combination produced more panicles per plant from emergence until equaled by the NB combination on the final measurement date.

Table 2. Effects of combinations of growth regulator and management practices on emergence and number of panicles per plant of mango. Treatments were the combinations of the best and next best treatments from prior studies, as defined in the text.

Treatment Combination	Date of Appearance of First Panicle	Number of Panicles per Plant at					
		30 December 2007	9 January 2008	19 January 2008	29 January 2008	8 February 2008	18 February 2008
Control	7 February 2008	0 c [z]	0 c	0 c	0 c	6.8 b	32.0 b
Best	20 December 2007	15.0 a	48.3 a	96.0 a	148.8 a	189.5 a	197.7 a
Next Best	25 December 2007	9.4 b	30.0 b	88.0 b	134.3 b	180.2 a	191.6 a
LSD (0.05)	-	1.4	3.9	7.1	10.1	12.9	16.0

[z] Means followed by different letters within columns significantly differ by Fisher's LSD at $p = 0.05$.

Initial fruit set was greater for both treatments than the control (Table 3). From 26 March to harvest, the BT combination had 2-fold greater fruit retained per panicle followed by the NB combination than the control.

Table 3. Effects of combinations of growth regulator and management practices on fruit set and fruit retention per panicle. Treatments were the combinations of the best and next best treatments from prior studies, as defined in the text.

Treatments	Fruit Set per Panicle	Number of Fruit Retained per Panicle in 2008 at								
		16 March	26 March	5 April	15 April	25 April	5 May	15 May	25 May	Harvest
Control	6.54 b [z]	5.87 b	2.65 c	1.75 b	1.45 c	1.17 c	1.10 c	1.00 c	0.90 c	0.82 c
Best	10.83 a	10.42 a	5.25 a	4.50 a	3.53 a	2.43 a	2.22 a	2.15 a	2.10 a	2.05 a
Next Best	10.17 a	9.62 a	3.42 b	2.87 ab	1.97 b	1.67 b	1.67 b	1.65 b	1.62 b	1.62 b
LSD (0.05)	1.34	1.09	0.62	1.64	0.32	0.34	0.20	0.13	0.17	0.18

[z] Means followed by different letters within columns significantly differ by Fisher's LSD at $p = 0.05$.

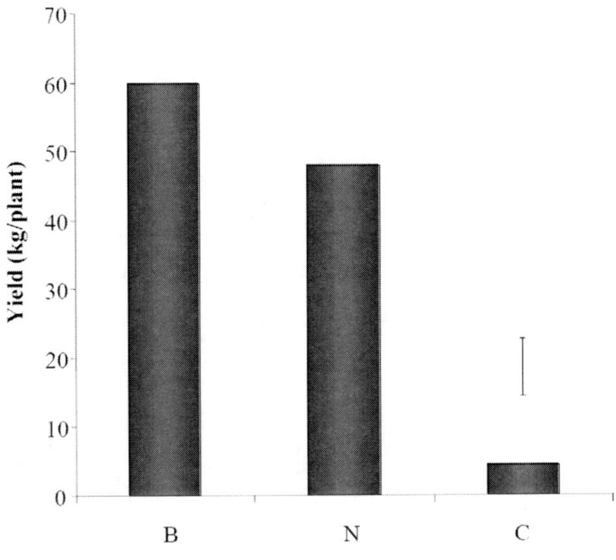

Figure 1. Effects of combinations of growth regulator and management practices on yield per plant of mango. Treatments were the combinations of the best (BT) and next best (NB) treatments from prior studies, as defined in the text. Vertical bar represents LSD at $p = 0.05$.

3.3. Harvest, Fruit Yield, and Fruit Characters

The harvest date was three or more weeks earlier for the treatment combinations than the control (Table 4). The BT combination produced the most and largest fruit, and total yield (Figure 1), followed by the NB combination, and both were significantly greater than the control. In fact, the BT combination produced 8-fold more fruit, and fruit weight was 61% greater, than the control. Fruit dimensions for both combinations were greater than the control but did not differ. The BT combination produced a higher edible portion than the control, with the NB combination intermediate between them. The stone:pulp ratio and peel:pulp ratio was lowest in the BT combination, followed by the NB combination, and was highest in the control. Both treatment combinations increased shelf life compared to controls.

Table 4. Effects of combinations of growth regulator and management practices on number of fruit and fruit characters of mango. Treatments were the combinations of the best and next best treatments from prior studies, as defined in the text.

Treatment	Date of Harvest	Number of Fruit per Plant	Fruit				Edible Portion (%)	Stone:pulp Ratio	Peel:pulp Ratio	Shelf Life (days)
			Weight (g)	Length (cm)	Breadth (cm)	Thickness (cm)				
Control	1 July 2008	23 c [z]	191.0 c	9.1 b	6.4 b	5.9 b	63.9 b	0.3 a	0.3 a	8.7 b
Best	4 June 2008	197 a	307.8 a	10.7 a	7.5 a	6.6 a	69.5 a	0.2 b	0.2 b	9.8 a
Next Best	6 June 2008	175 b	278.8 b	10.3 a	7.4 a	6.5 a	67.0 ab	0.2 b	0.2 b	9.6 a
LSD (0.05)		16	23.2	1.0	0.4	0.4	4.0	<0.1	<0.1	0.7

[z] Means followed by different letters within columns significantly differ by Fisher's LSD at $p = 0.05$.

3.4. Fruit Quality

The BT combination exhibited the highest TSS, but NB did not differ from the control (Table 5). The NB combination had a higher pulp pH than the control, with the BT combination similar to both. Both combination treatments had lower titratable acidity and moisture content, and higher vitamin C content, than the control, but did not differ from one another. The BT combination produced higher dry matter content, and reducing sugar, non-reducing sugar, and total sugar content than the NB combination, which was greater than the control.

Table 5. Effects of combinations of growth regulator and management practices on quality attributes of mango. Treatments were the combinations of the best and next best treatments from prior studies, as defined in the text.

Treatments	TSS (%)	pH	Titratable Acidity (%)	Vitamin C (mg/100 g pulp)	Moisture Content (%)	Dry Matter Content (%)	Reducing Sugars (%)	Non-Reducing Sugars (%)	Total Sugars (%)
Control	24.8 b [z]	5.74 b	0.23 a	29.4 b	81.9 a	17.9 c	4.8 c	12.8 c	17.5 c
Best	28.7 a	5.89 ab	0.18 b	35.9 a	77.7 b	22.3 a	5.4 a	14.5 a	19.9 a
Next Best	26.2 b	5.95 a	0.18 b	34.8 a	79.0 b	21.0 b	5.2 b	13.7 b	18.9 b
LSD (0.05)	1.9	0.16	0.01	1.2	1.4	1.1	0.2	0.6	0.7

[z] Means followed by different letters within columns significantly differ by Fisher's LSD at $p = 0.05$.

4. DISCUSSION

Both BT and NB combinations significantly impacted mango vegetative growth, flowering, and fruiting. While both were expected to differ from untreated trees, the primary question was whether they differed from one another, as one was comprised of the best individual treatments from prior studies while the other was comprised of the next best treatments, as the names indicate. The treatments differed in their components in some important ways. Both had equal levels of fertilizer applied, but it was applied in three installments in BT and only two in NB. KNO_3 was foliar-applied in BT, while urea was used at the same concentration in NB. For the NB combination, trees were less frequently irrigated from fruit set to maturity and had a higher rate of paclobutrazol included than the BT combination. While each component had effects on vegetative growth, flowering, and/or fruiting individually, it was not known how they might additively and/or synergistically interact to affect these traits.

Both combination treatments reduced vegetative growth compared to the control (Table 1). Our prior study had shown that paclobutrazol, regardless of concentration, caused a marked reduction in terminal shoot length, leaf number per terminal shoot, and leaf area compared to a control, but that reductions were at a maximum at a high concentration (10.0 g/L) [30]. However, the NB combination in this study had a higher paclobutrazol concentration than the BT, but it did not reduce terminal shoot growth more; although leaves per shoot, and leaf length, width, and area, were reduced more. The reduction in mango growth by paclobutrazol may be due to enhanced total phenolic content of terminal buds and alteration of the phloem-to-xylem ratio of the stem, altering assimilate partitioning and patterns of nutrient supply [31]. Soil-applied paclobutrazol reduced gibberellin production and growth in new mango shoots [32,33]. Less frequent irrigation reduced mango terminal shoot growth in our prior work [21] but both irrigation levels produced the same growth in the present study, and both grew less than the unirrigated control, so irrigation level was not a factor affecting growth.

Earlier flowering and greater panicle production were promoted by both BT and NB treatments (Table 2), although BT flowered earlier and increased panicle number in the first weeks of flowering more than NB. The reduction in shoot growth may have increased total available carbohydrates, and a higher C:N ratio in the shoots favors flower bud initiation in mango [34]. The flowering response was also likely due to the foliar KNO_3 and urea applications [19]. Others have shown similar responses to KNO_3 treatment [8,35]. The response to KNO_3 may be mediated, not by increasing N in the tissue, but by promoting ethylene biosynthesis which affects floral induction in mango [36]. Paclobutrazol also induced earlier flowering and greater panicle production in our earlier study [30]. In contrast, more frequent irrigation caused a slight delay in date of flowering [21].

Initial fruit set per panicle was similar in both treatments, but the number of fruit retained by BT was greater than by NB from early in the fruiting period until harvest (Table 4), as previously observed for paclobutrazol treatment [30], KNO_3 and urea treatments [19], and in response to irrigation [21]. The same response was noticed to paclobutrazol, and found that treated trees had higher reserves which enhanced fruit set compared to low reserves and low fruit set in untreated trees [17].

Both combination treatments advanced harvest date by 24–26 or more days in the present work (Table 4). Paclobutrazol alone can cause this, as found by [37] and in our prior work, although only by 10 to 15 days [30]. KNO_3 and urea treatment also advanced harvest date although only by 5 days [19], but irrigation delayed it by 7 days [21]. Interestingly, the differences in harvest dates were less, 27–29 days, than the differences in the first bloom dates, 44–49 days.

The BT combination increased yield by 14-fold and the NB combination by 11-fold over control trees, with BT yield significantly greater than NB yield (Figure 1). These increases far exceeded the response to each treatment component alone [18,19,21,30], when yield increases ranged from 2-fold to 7-fold greater than control trees. The present results are a strong indication that the combinations provided additive impacts on yield.

The increased total yield was the result of effects on both fruit per plant and mean fruit weight. All treatments alone also increased fruit number and yield [18,19,21,30]. The increased fruit weight in the combination treatments might be due to an increased rate of photosynthesis and an increase in chlorophyll content. Paclobutrazol increased leaf water use efficiency by increasing the rate of photosynthesis for a given level of leaf stomatal conductance or transpiration [38]. An increase in fruit weight and yield per tree from paclobutrazol has been reported in several studies with different mango varieties [13,17,39,40,41], from KNO_3 [42] and urea application [10], and from increased fertilizer application [43].

Both treatments exhibited better physico-chemical characteristics (pulp percentage, stone percentage, peel percentage, total soluble solids, titratable acidity, ascorbic acid, total reducing and non-reducing sugar) compared to the control, similar to the reports by others with the individual and/or combined treatments [4,14,15,44].

Table 6. Summary of major effects of the individual factors and their combination on key yield traits. The top individual treatment of each pair was considered the best and was included in the Best combination, and the next best treatment was included in the Next Best combination.

Treatment	Advance (−) or Delay (+) Relative to Control in Date of First Panicle	Date of First Harvest	Final Fruit per Panicle	Fruit Weight (g)	Yield vs. Control
Fertilizer in 3 applications [z]	−3	+11	1.87	217	5.6X more
Fertilizer in 2 applications	−3	+5	1.74	201	4.6X more
Control	0	0	0.62	161	-
4% KNO_3 [y]	−18	−5	1.63	193	2.5X more
4% Urea	−14	−5	1.37	203	2.2X more
Control	0	0	0.73	175	-
Irrigation at 7 day intervals [x]	+2	+10	1.47	183	1.7X more
Irrigation at 14 day intervals	+2	+10	1.17	187	1.7X more
Control	0	0	0.57	169	-
Paclobutrazol 7.5 g/L [w]	−21	−22	2.00	304	4.7X more
Paclobutrazol 10.0 g/L	−21	−22	1.70	261	3.4X more
Control	0	0	0.63	178	-
Best	−44	−25	2.05	308	14X more
Next Best	−44	−25	1.62	279	11X more
Control	0	0	0.82	191	-

[z] From [18]; [y] From [19]; [x] From [21]; [w] From [45].

5. CONCLUSIONS

This paper represents the culmination of a set of studies on the effects of individual management practices in fertilization, foliar N application, irrigation, and paclobutrazol application to identify the optimum for each alone. Both combinations significantly reduced shoot growth, likely due to the presence of the growth retardant paclobutrazol, and shifted the bloom and harvest dates to earlier times. Although both combinations produced more fruit that also had higher quality than the control, the BT combination produced a greater yield and fruit with a higher total soluble solids, reducing, non-reducing, and total sugar content, and vitamin C content than the NB combination. The BT and NB combinations of the optimums identified in the prior studies were successful at advancing bloom and harvest and increasing yield more than any of the optimum individual componentsalone (Table 6), suggesting there were additive, if not synergistic, effects on mango. Although the present results are clear, indicating that implementation of these combinations of treatments could increase yield and ensuing economic returns very significantly over use of a single treatment, there are important considerations that need to be addressed. Given the low replication number and single year of data with a single cultivar, it is not known if similar responses will occur on broader plantings of "Amrapali" (BARI Mango-3) or other mango cultivars, or under different edaphic or environmentalconditions. In addition, the long-term sustainability of the responses to the treatment combinations should be studied, as well as the possible carryover of effects of the treatments, especially to soil-applied nutrients and to paclobutrazol, into the year(s) after application. This might be a concern if there was a cumulative effect from several years of application. Nonetheless, the "packages" of treatments appeared superior to any one treatment alone, and warrant further study to answer these additional issues.

ACKNOWLEDGMENTS

The author is grateful to the authorities at the Bangladesh Agricultural Research Institute (BARI) for awarding the scholarship and deputation to complete the research work which led to the Ph.D. The author extends his special thanks to the authorities of the Bangladesh Agricultural University (BAU) Germplasm Centre, Department of Horticulture, Bangladesh Agricultural University, Mymensingh, for providing logistical support during the entire period of the field experiments.

REFERENCES

1. FAOSTAT. Food and Agriculture Organization of the United Nations, Statistics Division. 2016. Available online: http://faostat3.fao.org/home/E (accessed on 20 June 2016)
2. Bangladesh Bureau of Statistics, Statistics Division, Planning Division, Ministry of Planning. *Yearbook of Agricultural Statistics of Bangladesh*; Government of the People's Republic of Bangladesh: Dhaka, Bangladesh, 2005; p. 90
3. Reddy, Y.T.N.; Kurian, R.M.; Kohli, R.R.; Gorakh, S.; Singh, G. Effect of nitrogen, phosphorus and potassium on growth, yield and fruit quality of "Totapuri" mango (*Mangifera indica*). *Indian J. Agric. Sci.* **2000**, *70*, 475–478
4. Satapathy, S.K.; Banik, B.C. Studies on nutritional requirement of mango cv. Amrapali. *Orissa J. Hortic.* **2002**, *30*, 59–63
5. Sharma, R.C.; Mahajan, B.V.C.; Dhillon, B.S.; Azad, A.S. Studies on the fertilizer requirements of mango cv. Dashehari in sub-mountainous region of Punjab. *Indian J. Agric. Res.* **2000**, *34*, 209–210
6. Suriyapananont, V.; Subhadrabandhu, S. Fertilizer trials on mangoes (*Mangifera indica* L.) var. Nam Dok Mai in Thailand. *Acta Hortic.* **1992**, *321*, 529–534
7. Zhou, X.C.; Liu, G.J.; Yao, J.W.; Ai, S.Y.; Yao, L.X.; Zhou, X.C.; Liu, G.J.; Yao, J.W.; Ai, S.Y.; Yao, L.X. Balanced fertilization on mango in Southern China. *Better Crops Int.* **2001**, *15*, 16–20
8. Khattab, M.M.; Haseeb, G.M.; Shaban, A.E.; Arafa, M.A. Effect of paclobutrazol and potassium nitrate on flowering and fruiting of Ewais and Sidik mango trees. *Bull. Fac. Agric.* **2006**, *57*, 107–123
9. Sergent, E.; Ferrari, D.; Leal, F. Effects of potassium nitrate and Paclobutrazol on flowering induction and yield of mango (*Mangifera indica* L.) cv. Haden. *Acta Hortic.* **1997**, *455*, 180–187
10. Gupta, R.K.; Brahmachari, V.S. Effect of foliar application of urea, potassium nitrate and NAA on fruit retention, yield and quality of mango cv. Bombai. *Orissa J. Hortic.* **2004**, *32*, 7–9
11. Yeshitela, T.; Robbertse, P.J.; Stassen, P.J.C. Potassium nitrate and urea sprays affect flowering and yields of "Tommy Atkins" (*Mangifera indica*) mango in Ethiopia. *S. Afr. J. Plant Soil* **2005**, *22*, 28–32

12. Hossain, A.K.M.A. *Production Technology of Mango*; Horticulture Research Centre, BARI: Gazipur, Bangladesh, 1994; p. 122
13. Kulkarni, V.J. Chemical control of tree and promotion of flowering and fruiting in mango (*Mangifera indica* L.) using paclobutrazol. *J. Hortic. Sci.* **1988**, *63*, 557–566
14. Singh, S.; Singh, A.K. Regulation of shoot growth and flowering in mango cv. Gulab Khas by paclobutrazol. *Ann. Agric. Res.* **2006**, *27*, 4–8
15. Karuna, K.; Mankar, A.; Singh, J. Effect of urea and growth substances on yield and physico-chemical characteristics of mango. *Hortic. J.* **2005**, *18*, 131–133
16. Vijayalakshmi, D.; Srinivasan, P.S. Improving the quality attributes of "off" year Alphonso mango through chemicals and growth regulators. *Orissa J. Hortic.* **2000**, *28*, 31–33
17. Yeshitela, T.; Robbertse, P.J.; Stassen, P.J.C. Paclobutrazol suppressed vegetative growth and improved yield as well as fruit quality of "Tommy Atkins" mango (*Mangifera indica*) in Ethiopia. *N. Z. J. Crop Hortic. Sci.* **2004**, *32*, 281–293
18. Sarker, B.C.; Rahim, M.A. Effects of doses and splits of fertilizer application on harvesting time, yield and quality of mango cv. Amrapali. *Bangladesh J. Agric. Res.* **2012**, *37*, 279–293
19. Sarker, B.C.; Rahim, M.A. Yield and quality of mango (*Mangifera indica* L.) as influenced by foliar application of potassium nitrate and urea. *Bangladesh J. Agric. Res.* **2013**, *38*, 145–154
20. Singh, L.B. *The Mango Botany, Cultivation and Utilization*; Leonard Hill Ltd.: London, UK, 1968; p. 230
21. Sarker, B.C.; Rahim, M.A. Effect of irrigation frequency on harvesting time, yield and quality of mango. *J. Bangladesh Soc. Agric. Sci. Technol.* **2010**, *7*, 11–16
22. United Nations Development Programme (UNDP). *Land Resource Appraisal of Bangladesh for Agricultural Development Report 2: Agroecological Regions of Bangladesh*; Food And Agriculture Organization (FAO): Rome, Italy, 1988; p. 577
23. Singh, R.N. *Mango*; Indian Council of Agricultural Research: New Delhi, India, 1996; p. 134
24. Burondkar, M.M.; Gunjate, R.T. Control of vegetative growth and induction of regular and early cropping in "Alphonso" mango with paclobutrazol. *Acta Hortic.* **1993**, *341*, 206–215
25. Rangana, S. *Manual of Analysis of Fruit and Vegetable Products*; Tata McGraw-Hill Pub. Co. Ltd.: New Delhi, India, 1979; p. 634
26. Plummer, D.T. *An Introduction to Practical Biochemistry*; Tata McGraw-Hill Pub. Co. Ltd.: New Delhi, India, 1971; p. 229
27. Miller, G.L. Use of dinitro salicylic acid reagent for determination of reducing sugar. *Anal. Chem.* **1972**, *31*, 426–428
28. Jayaraman, J. *Laboratory Manual in Biochemistry*; Wiley Eastern Ltd.: New Delhi, India, 1981; p. 62
29. Gomez, K.A.; Gomez, A.A. *Statistical Procedures for Agricultural Research*; John Wiley and Sons: New York, NY, USA, 1984; p. 680
30. Sarker, B.C.; Rahim, M.A. Vegetative growth, harvesting time, yield and quality of mango (*Mangifera indica* L.) as influenced by soil drench application of paclobutrazol. *Bangladesh J. Agric. Res.* **2012**, *37*, 335–348
31. Kurian, R.M.; Iyer, C.P.A. Stem anatomical characteristics in relation to tree vigour in mango (*Mangifera indica* L.). *Sci. Hortic.* **1992**, *50*, 245–253
32. Cardenas, K.; Rojas, E. Effect of paclobutrazol and nitrates of potassium and calcium on the development of the mango "Tommy Atkins". *Bioagro* **2003**, *15*, 83–90
33. Ram, S. Hormonal physiology of flowering in "Dashehari" mango. *J. Appl. Hortic.* **1999**, *1*, 84–88
34. Jogdande, N.D.; Choudhari, K.G. Seasonal changes in auxin content and its role in flowering of mango (*Mangifera indica* L.). *Orissa J. Hortic.* **2001**, *29*, 10–12
35. Dalal, S.R.; Gonge, V.S.; Jadhao, B.J.; Jogdande, N.D. Effect of chemical on flowering and fruit yield of mango cv. Pairy. *Int. J. Agric. Sci.* **2005**, *1*, 24–25

36. Mosqueda-Vazquez, R.; Avila-Resendiz, C. Floral induction of mango with KNO_3 applications and its inhibition by $AgNO_3$ or $CoCl_2$ application. *Hortic. Mex.* **1985**, *1*, 93–101
37. Singh, D.B.; Ranganath, H.R. Induction of regular and early fruiting in mango by paclobutrazol under tropical humid climate. *Indian J. Hortic.* **2006**, *63*, 248–250
38. Quinlan, J.D. New approaches to the control of fruit tree forms and size. *Acta Hortic.* **1981**, *120*, 95–105
39. Singh, Z.; Dhillon, B.S. Effect of paclobutrazol on floral malformation, yield and quality of mango (*Mangifera indica*L.). *Acta Hortic.* **1992**, *296*, 51–53
40. Shinde, A.K.; Waghmare, G.M.; Wagh, R.G.; Burondkar, M.M. Effect of dose and time of paclobutrazol application on flowering and yield of mango. *Indian J. Plant Physiol.* **2000**, *5*, 82–84
41. Anbu, S.; Parthiban, S.; Rajangam, J.; Thangaraj, T. Induction of off-season flowering in mango (*Mangifera indica* L.) using paclobutrazol. *South Indian Hortic.* **2002**, *49*, 384–385
42. Oosthuyse, S.A. Effect of KNO_3 sprays to flowering mango trees on fruit retention, fruit size, tree yield and fruit quality. *Acta Hortic.* **1997**, *455*, 359–366
43. Feungchan, S.; Yimsawat, T.; Chindaprasert, S.; Hongsbhanich, N.; Daito, H. The effect of the fertilizer application interval on the mango. *Kaen Kaset Khon Kaen Agric. J.* **1989**, *17*, 100–105
44. Ghosh, S.N.; Chattopadhyay, N. Foliar application of urea on yield and physico-chemical composition of mango fruits cv. Himsagar under rainfed condition. *Hortic. J.* **1999**, *12*, 21–24
45. Sarker, B.C.; Rahim, M.A.; Pomology Division, Horticulture Research Centre, Bangladesh Agricultural Research Institute, Joydebpur, Gazipur 1701, Bangladesh. Unpublished Data. 2012

CHAPTER 3

Morphological and Pomological Diversity of Fig (Ficuscarica L.) Cultivars in Northwest of Tunisia

Badii Gaaliche,[1] Olfa Saddoud,[2] and Messaoud Mars[1]

[1]*U.R. Agrobiodiversity, High Agronomic Institute, IRESA, University of Sousse, ChottMariem 4042, Tunisia*

[2]*National Gene Bank, Boulevard YesserArafet, Charguia 1, Tunis 1080, Tunisia*

ABSTRACT

The fig (Ficuscarica L.) is one of the oldest fruit trees cultivated in Tunisia. Djebba region is located in the northwest of Tunisia. It is very famous by fig culture. Many specific fig genotypes are very appreciated locally and nationally. Taking into account these considerations, Djebba fig cultivars are subject of label products, namely, "Djebba figs." This study was focused on fig germplasm characterization of 17 cultivars in Djebba region based on morphological and pomological traits. Results revealed a large variability within the local fig germplasms. The comprehensive analyses of all the data permitted to distinguish some particular genotypes as distinct cultivars, and groups of cultivars as polyclone varieties. It was possible to discriminate six distinct cultivars and two groups of multiclone varieties (Soltani and Thgagli) with different degrees of polymorphism. Hypotheses of homonymy and synonymy were suggested for some cultivars. The diversity is currently threatened by genetic erosion. Measure of conservation is necessary to be undertaken.

1. INTRODUCTION

The common fig (Ficuscarica L., 2n = 26) belongs to the family Moraceae, with over 1400 species classified into about 40 genera. The genus Ficus contains about 700 species, mainly found in the tropics and currently classified into six subgenera [1]. Nowadays, the common fig grows wild in the Mediterranean Basin where it has been cultivated for its edible fruits for millennia in close association with olive and grapevine [2]. Three types of figs are grown commercially [3]: the common type that develops fruit parthenocarpically, the Smyrna type that requires pollination with pollen from caprifigs (caprification), and the San Pedro type that produces a first crop parthenocarpically and a second crop only after pollination. Total world fig production is over 1 million tons and 70% is produced in Turkey, Egypt, Greece, Iran, Morocco, Spain, and USA. Turkey produces nearly 25% of the total production [4].

In Tunisia, fig biodiversity is very high, and many niches are well suited for fig production. Despite its socioeconomic and historical importance, fig tree was considered as a secondary crop. Fig trees are grown all over the country, with more than 2,500,000 trees occupying about 33800 ha. The total annual production is about 27.000 mile tons [5]. Figs are mainly consumed as fresh fruits. A small portion is sun dried and little quantities are used for jam and alcoholic beverage production [6]. Tunisian fig cultivars are numerous and well adapted to local agroecological conditions [7]. Some are of the common type. Many others are of Smyrna type [8]. It should be stressed that several studies have reported the use of pomological traits in a number of Tunisian ecotypes collected from different zones [9]. Thus, 65 cultivars were described by Minangoin [10], 22 cultivars by Valdeyron and Crossa-Raynaud [11], and 28 by Lahbib [12]. It is noteworthy that a number of these cultivars have been destroyed as a consequence of the intensive urbanization in recent decades. Moreover, severe genetic erosion is also threatening the remaining germplasm. Thus, available information about the ancient landraces is very limited [6].

In order to preserve fig genetic resources, several prospections and alternative strategies for genetic resources management were considered. Pomological and morphological traits as well as molecular markers were used to analyse genetic diversity and characterize local cultivars [8, 13–19]. Recently, various studies on fig genetic resources are continuing across the country. But no research results have been reported so far on northwest fig germplasm. This study is part of a project of the "Union Tunisienne de l'Agriculture et de la Pêche" aiming to establish a label to Djebba's fig. The objectives are (i) identification of different fig cultivars grown in Djebba area and (ii) analysis of genetic diversity within the germplasm.

2. MATERIALS AND METHODS

2.1. Plant Material and Study Area. This study concerned the region of Djebba (altitude, 700 m; latitude, 36° 40′N; longitude 9°0′E) located at the northwest of Tunisia. The climate is subhumid with mild winter and hot summer. Annual average temperature is around 20°C. Thermal amplitude is about 16.5°C in summer and 8°C in winter. Average annual rainfall is about 600 mm. The morphological and pomological variability was studied on adult fig trees of 17 local cultivars (one tree per cultivar) grown in this region (Table 1).

Table 1: Studied fig cultivars.

N°	Cultivar	Code	Type
1	Zidi	ZD1	Smyrna
2	Zidi Artab	ZD2	Smyrna
3	Soltani Abiadh	SAB	Smyrna
4	Soltani Ahmar	SAH	Smyrna
5	Thgagli Abiadh	TAB	Smyrna
6	Thgagli Akhdher	TAK	Smyrna
7	Nemri	NMR	Smyrna
8	Hemri	HMR	Smyrna
9	Bouhouli	BHL	San Pedro
10	Wahchi	WAH	San Pedro
11	Garaï	GRI	San Pedro
12	Khenziri	KNZ	San Pedro
13	Khartoumi	KRT	Smyrna
14	Boukhobza	BKB	Smyrna
15	Zergui	ZRG	Smyrna
16	Bouharrag	BHR	Smyrna
17	Fawari	FAW	Smyrna

San Pedro produces first and second crops and Smyrna produces only second crop.

2.2. Morphological and Pomological Traits

Morphological and pomological parameters were chosen according to the fig descriptors [20]. On branches, 13 parameters related to the dimensions of shoots and internodes were measured (Table 2). For each cultivar, measurements were made after fall of leaves on six shoots per tree. For leaves, 16 parameters related to the lamina and petiole were measured on 20 mature leaves per tree and cultivar (Table 2).

Table 2: Morphological and pomological traits used for fig cultivar characterization.

Traits	Code
Branch	
Terminal bud length (mm)	TBL
Terminal bud diameter (mm)	TBD
TBL/TBD ratio	TBL/TBD
1-year-old shoot length (mm)	S1Y
Length of the 1st internode on S1Y (mm)	L1N1Y
Diameter of the 1st internode on S1Y (mm)	D1N1Y
Length of the 2nd internode on S1Y (mm)	L2N1Y
Diameter of the 2nd internode on S1Y (mm)	D2N1Y
2-year-old shoot length (mm)	S2Y
Length of the 1st internode on S2Y (mm)	L1N2Y
Diameter of the 1st internode on S2Y (mm)	D1N2Y
Length of the 2nd internode on S2Y (mm)	L2N2Y
Diameter of the 2nd internode on S2Y (mm)	D2N2Y
Leaf	
Leaf shape (1: calcarate; 14: cordate)	LS
Shape of lobes (1: lyrate; 4: spatulate)	SL
Leaf color (1: light green; 12: dark green)	LC
Leaf margin dentation (1: undulate; 5: crenate)	LMD
Apex shape (1: truncate; 3: acute)	AS
Density of hairs (1: none; 7: pubescent/dense)	DH
Petiole color (1: light green; 6: pinkish)	PC
Leaf length (cm)	LL
Leaf width (cm)	LW
Petiole length (mm)	PL

Petiole diameter (mm)	PD
Depth of basal sinus (mm)	DBS
Depth of lateral sinus 1 (mm)	DLS1
Depth of lateral sinus 2 (mm)	DLS2
Depth of lateral sinus 3 (mm)	DLS3
Depth of lateral sinus 4 (mm)	DLS4
Fruit	
Fruit weight (g)	FW
Fruit length (mm)	FL
Fruit diameter (mm)	FD
Neck length (mm)	NL
Neck diameter (mm)	ND
Ostiole diameter (mm)	OD
Fruit shape (1: spheroid; 10: pyriform)	FS
External color (1: violet black; 13: yellowish green)	EC
Skin cracks (1: cracked skin; 4: not cracked skin)	SC
Internal color (1: greenish white; 11: red)	IC
Fruit cavity (3: very small; 9: large)	FC
Skin thickness (mm)	ST
Flesh thickness (mm)	FT
pH	pH
Total soluble solids (°Brix)	TSS
Titratable acidity (%)	TA

On fruits (second crop), 16 parameters related to the size, external and internal appearance, and juice quality were taken into account (Table 2). During the harvesting period, pomological measurements and chemical analysis carried out on samples of 20 mature fruits per tree and cultivar were randomly collected. Total soluble solids (°Brix) were determined with a digital refractometer, PR-101 ATAGO, and Norfolk, and titratable acidity (citric acid %) was determined by titrating fig juice with 0.1 M NaOH.

2.3. Statistical Analysis

For all parameters, analysis of variance (one-way ANOVA) was used to determine differences between cultivars. Comparison of the mean values was made using the Duncan's multiple range test ($P < 0.05$). Multivariate relationships among cultivars were revealed through a principal component analysis (PCA) using a correlation matrix derived from the significant characters. The squared Euclidean distance was used to perform cluster analysis. Statistical analyses were computed using SPSS 13.0 statistical software.

3. RESULTS

3.1. One-Way ANOVA.

The average values of parameters measured on branches are shown in Table 3. Analysis of variance showed highly significant differences between cultivars for all parameters except terminal bud diameter (TBD). The study of morphological characteristics of fig tree shoots allowed the differentiation in biology of some cultivars such as Zidi (ZD1), ZidiArtab (ZD2), and Nemri (NMR). Moreover, the most discriminating variables were terminal bud length (TBL), 1-year-old shoot length (S1Y), and 2-year-old shoot length (S2Y).

Table 3: Mean values and significance degree of differences between fig cultivars for branch characteristics.

Cultivar	TBL (mm)	TBD (mm)	TBL/TBD	S1Y (mm)	L1N1Y (mm)	D1N1Y (mm)	L2N1Y (mm)	D2N1Y (mm)	S2Y (mm)	L1N2Y (mm)	D1N2Y (mm)	L2N2Y (mm)	D2N2Y (mm)
ZD1	9.70def	5.27	1.82bcde	240.30a	38.13a	11.17bc	39.59a	10.35abc	264.90a	36.05	14.38ab	39.06b	13.74ab
ZD2	7.73f	5.33	1.46e	166.65cd	24.27cde	13.61ab	21.67defgh	12.08a	164.42bcd	30.28	15.62a	24.79bcd	15.41a
SAB	8.57ef	4.69	1.82bcde	125.29d	15.19e	7.69d	16.33fgh	6.51d	131.00bcd	23.12	8.27h	18.52d	8.55fg
SAH	8.30ef	5.31	1.58de	135.36cd	18.18cde	9.68cd	15.32gh	9.23bcd	144.83bcd	20.17	11.43cde	20.58bcd	11.29cde
TAB	11.29bcde	4.86	2.32a	157.08cd	22.76cde	8.32cd	19.01efgh	7.36de	152.32bcd	22.00	10.11efgh	17.05d	10.47def
TAK	10.77cdef	5.15	2.12abc	169.80cd	18.64cde	9.03cd	19.94defgh	9.04bcd	122.24cd	18.61	10.50efg	18.55d	10.22defg
NMR	11.82bcd	5.92	1.98abcd	234.74ab	35.19ab	14.63a	35.99ab	10.76ab	192.88b	27.08	13.45bc	30.70abc	13.17bc
HMR	13.07abc	5.95	2.29ab	164.16cd	23.68cde	9.13cd	23.65defgh	9.13bcd	172.96bc	24.20	10.96def	27.44bcd	10.30defg
BHL	15.15a	6.48	2.33a	181.04cd	20.44cde	10.44cd	25.70cdef	9.86bc	162.56bcd	28.07	12.11cde	31.11abc	12.07bcd
WAH	9.19def	5.09	1.80cde	155.29cd	25.03cde	8.75cd	26.73bcde	8.45cd	138.06bcd	30.07	8.99fgh	32.02ab	8.77fg
GRI	11.15cde	5.18	2.14abc	169.08cd	24.32cde	9.82cd	24.28cdefg	9.90bc	145.97bcd	20.15	10.14efgh	23.69bcd	10.37def
KNZ	9.40def	5.25	1.80cde	187.30bc	26.62bcd	8.63cd	29.87bcd	9.12bcd	157.47bcd	26.47	8.38h	41.50a	9.37efg
KRT	9.72def	5.94	1.67cde	133.84cd	22.42cde	11.52bc	28.63bcde	10.41abc	127.76bcd	27.67	12.65bcd	27.40bcd	11.86bcd
BKB	10.03def	5.89	1.89abcde	147.82cd	17.46de	11.11bc	13.81h	10.03bc	122.07cd	22.61	13.30bc	19.82cd	13.08bc
ZRG	14.08ab	6.13	2.33a	154.10cd	25.23cde	11.11bc	25.71cdef	10.72ab	147.05bcd	23.23	13.07bcd	23.64bcd	13.04bc
BHR	8.30ef	5.31	1.58de	161.59cd	18.18cde	9.68cd	15.32gh	9.23bcd	144.83bcd	20.70	12.67bcd	19.64cd	11.94bcd
FAW	9.51def	4.54	2.10abc	151.43cd	27.78bc	7.73d	33.77abc	6.62d	104.72d	22.59	8.52gh	21.21bcd	8.08g
F-value	5.04**	1.22NS	3.93**	3.12**	4.03**	3.82**	6.15**	5.76**	3.42**	1.72NS	11.09**	4.26**	8.40**

Means in each column followed by the same letters are not significantly different at $P < 0.05$ according to Duncan's multiple range test.
* Significant at $P < 0.05$; ** significant at $P < 0.01$; NS: nonsignificant; TBL: terminal bud length; TBD: terminal bud diameter; TBL/TBD: TBL/TBD ratio; S1Y: 1-year-old shoot length; L1N1Y: length of the 1st internode on S1Y; D1N1Y: diameter of the 1st internode on S1Y; L2N1Y: length of the 2nd internode on S1Y; D2N1Y: diameter of the 2nd internode on S1Y; S2Y: 2-year-old shoot length; L1N2Y: length of the 1st internode on S2Y; D1N2Y: diameter of the 1st internode on S2Y; L2N2Y: length of the 2nd internode on S2Y; D2N2Y: diameter of the 2nd internode on S1Y.

The averages of characters related to leaf are shown in Table 4. Analysis of variance showed highly significant differences between cultivars for all parameters. The variability of leaf characteristics was very important and allowed the distinction between some cultivars for several parameters such as leaf shape, leaf margin dentation, and leaf dimensions (LS, LMD, LL, and LW), petiole color (PC), and the depth of basal and lateral sinus (DBS and DLS1, DLS2, DLS3, DLS4).

Table 4: Mean values and significance degree of differences between fig cultivars for leaf characteristics.

Cultivar	LS	SL	LC	LMD	AS	DH	PC	LL (cm)	LW (cm)	PL (mm)	PD (mm)	DBS (mm)	DLS1 (mm)	DLS2 (mm)	DLS3 (mm)	DLS4 (mm)
ZD1	12.84a	2.35e	6.65de	5.00a	2.79abc	4.79cdef	3.00fg	29.45a	25.42a	99.05ab	7.67a	46.37a	88.25ab	89.43a	---	---
ZD2	12.76ab	3.80a	6.43def	4.05b	2.74abc	3.86ghi	3.48cdef	30.05a	26.10a	94.58abcd	7.20ab	53.18a	77.72bc	76.64b	---	---
SAB	11.85c	3.45abc	3.05h	2.20d	2.95a	2.60k	3.95bcd	22.57cd	19.60cd	75.12fg	6.15c	11.00gh	70.07cd	71.10bc	---	---
SAH	8.00d	3.00cde	3.30h	2.95cd	2.65abcd	4.70cdefg	5.05a	21.97cde	19.33cde	86.95bcdef	5.75cd	17.28fgh	86.23ab	88.36a	35.95bc	37.65bc
TAB	7.25ef	3.25bcd	9.05ab	2.95cd	2.60bcd	4.60defg	3.85bcd	21.06cdef	19.23cdef	102.78a	5.63cd	25.24cdef	78.02bc	73.44bc	34.16bc	31.92cd
TAK	12.25bc	3.85a	7.60bcd	2.70cd	2.65abcd	5.90ab	4.10bc	16.71h	16.10g	61.32h	5.08de	13.34gh	38.41e	39.97f	---	---
NMR	7.70de	3.20bcd	7.30cd	3.25bc	2.95a	3.60hij	2.25h	25.71b	22.34b	76.30fg	7.00b	22.78cdef	90.46a	93.01a	40.54ab	41.02ab
HMR	12.35abc	2.90cde	2.60h	2.45cd	2.85ab	5.30bcd	3.70bcde	20.08efg	18.07def	93.57abcde	5.64cd	11.10gh	69.87cd	67.80bcd	---	---
BHL	12.85a	2.95cde	2.95h	3.05cd	2.85ab	3.20ijk	3.50cdef	20.34defg	17.53efg	82.76def	5.57cd	18.88efg	59.02d	57.73de	---	---
WAH	7.55def	2.90cde	2.90h	2.85cd	2.90ab	5.50abc	4.10bc	21.21cde	19.52cde	85.44cdef	5.65cd	9.49h	70.30cd	71.89bc	28.58cd	28.58de
GRI	7.30ef	2.80de	3.55gh	2.35cd	2.85ab	3.00ijk	2.50gh	21.36cde	19.04cdef	84.70def	5.46cd	23.57cdef	67.44cd	68.47bcd	28.61cd	27.12de
KNZ	11.90c	3.10cde	4.85fg	2.85cd	2.95a	5.10bcde	3.95bcd	18.93fg	17.29fg	81.21ef	4.68e	----	48.17e	47.94ef	---	---
KRT	6.40g	2.70de	8.85bc	2.10d	2.40d	3.20ijk	2.60gh	21.58cde	19.82cd	102.05a	5.86c	21.06def	92.12a	88.76a	46.41a	45.92a
BKB	2.95h	2.95cde	5.65ef	2.25cd	2.80abc	6.30a	3.20ef	20.63def	18.86cdef	89.26bcde	6.12c	29.80bc	70.11cd	76.51b	38.81b	37.73bc
ZBG	7.55def	3.75ab	5.15ef	3.00cd	2.60bcd	4.20fgh	3.35def	23.23c	20.03cd	65.44gh	5.93c	26.31bcde	65.03d	63.47cd	23.81d	21.82e
BHR	7.10f	3.70ab	8.65bc	5.00a	2.50cd	4.30efgh	4.20b	18.20gh	17.26fg	80.92ef	5.81c	28.12bcd	71.07cd	70.33bc	31.13cd	31.04cd
FAW	7.35ef	3.05cde	10.45a	5.00a	2.70abc	2.90jk	5.20a	26.18b	20.89bc	97.32abc	7.43ab	33.57b	95.03a	91.55a	38.94b	40.49ab
F-value	267.18**	5.16**	23.52**	10.62**	3.12**	16.00**	16.24**	25.79**	18.64**	9.34**	14.60**	24.76**	16.58**	14.23**	7.43**	10.33**

Means in each column followed by the same letters are not significantly different at $P < 0.05$ according to Duncan's multiple range test.
* Significant at $P < 0.05$; ** significant at $P < 0.01$; NS: nonsignificant; ----: absence of basal sinus; ---: absence of lateral sinus 3 and 4.
LS: leaf shape; SL: shape of lobes; LC: leaf color; LMD: leaf margin dentation; AS: apex shape; DH: density of hairs; PC: petiole color; LL: leaf length; LW: leaf width; PL: petiole length; PD: petiole diameter; DBS: depth of basal sinus; DLS1: depth of lateral sinus 1; DLS2: depth of lateral sinus 2; DLS3: depth of lateral sinus 3; DLS4: depth of lateral sinus 4.

The averages of characters related to fruit are given in Table 5. Analysis of variance showed highly significant differences between cultivars for all parameters. The most discriminating variables were fruit weight (FW) and fruit dimensions (FL, FD).

Table 5: Means values and significance degree of differences between fig cultivars for fruit characteristics.

Cultivar	FW (g)	FL (mm)	FD (mm)	NL (mm)	ND (mm)	OD (mm)	FS	EC	SC	IC	FC	ST (mm)	FT (mm)	pH	TSS (°Brix)	TA (%)
ZD1	96.45a	60.27a	60.52a	5.31cd	14.70a	12.74a	3.79cdef	1.79hi	4.00a	8.84abc	4.37bcdef	1.18de	12.13g	4.65ef	13.23g	0.21fg
ZD2	51.83efg	47.03de	43.90gh	7.71bc	11.80bc	5.81defg	3.90cde	1.86h	1.76f	9.95a	3.10efgh	1.25cd	17.56abc	4.88cd	15.95ef	0.17h
SAB	59.10cde	61.50a	45.82fgh	14.43a	12.06bc	6.45cdef	4.50bc	11.95a	1.55f	9.00abc	1.90h	1.53ab	18.31ab	4.52gh	15.46f	0.25a
SAH	44.45ghi	51.08bc	39.93i	14.89a	13.54ab	5.50efg	5.65a	3.25f	3.45bcd	9.35ab	3.70defg	1.05e	16.01cde	5.10a	15.58f	0.18h
TAB	62.20cd	42.75ef	53.65c	----	----	8.22b	2.70g	12.00a	3.75abc	7.85c	6.40a	1.26cd	9.05h	4.89cd	15.56f	0.20g
TAK	51.95efg	42.17f	50.26de	5.95cd	9.66cde	5.18fg	3.15efg	10.95c	2.90e	8.95abc	6.00ab	1.56ab	14.86de	4.45h	15.96ef	0.22bcdef
NMR	82.36b	46.57def	58.62ab	----	----	7.34bc	3.00fg	4.50d	3.75abc	8.25bc	4.60bcde	1.48b	19.07a	4.82d	16.48de	0.23abcde
HMR	57.83cde	58.12a	44.95fgh	10.90b	11.45bc	4.84g	5.00ab	3.70e	3.45bcd	8.80abc	2.20gh	1.43bc	18.90ab	4.59fg	17.56c	0.25a
BHL	64.53c	48.38cd	56.03bc	3.44d	10.58cd	5.08fg	3.15efg	2.65g	3.90ab	4.65e	5.70abc	1.05e	16.00cde	4.91bcd	15.36f	0.21cdef
WAH	41.67hij	43.50ef	44.27gh	5.92cd	10.97bcd	6.36cdef	3.40efg	11.40b	3.70abc	6.30d	2.80fgh	1.16de	17.85abc	4.95bc	19.02a	0.23abc
GRI	54.23def	45.86def	49.99de	3.91cd	7.55e	7.12bcd	2.70g	10.00c	3.60abc	3.05f	4.70abcde	1.05e	19.24a	4.29i	16.32de	0.21defg
KNZ	44.27ghi	44.31def	47.34efg	5.21cd	10.42cd	5.78defg	2.85g	11.30bc	3.35cd	8.50bc	5.00abcd	1.54ab	12.13g	4.97bc	18.32b	0.23abcd
KRT	52.77defg	46.36def	47.99ef	5.71cd	11.31bc	5.81defg	4.25bcd	10.35c	3.80abc	8.55bc	4.20cdef	1.19de	14.24ef	4.46h	17.20c	0.24ab
BKB	34.54j	35.20g	43.81h	4.26cd	8.34de	5.59efg	3.55defg	11.90a	3.70abc	9.00abc	3.10efgh	1.72a	11.34g	4.71e	15.48f	0.22bcdef
ZBG	37.38ij	43.06ef	45.49fgh	----	----	6.85cde	2.70g	1.65h	3.10de	6.65d	5.20abcd	1.63ab	15.45de	5.00ab	16.84cd	0.19gh
BHR	46.87fgh	42.76ef	43.98gh	----	----	6.44cdef	3.15efg	1.40i	3.80abc	2.00f	2.70fgh	1.55ab	12.40fg	4.91bcd	17.00cd	0.21efg
FAW	53.98defg	53.64b	52.99cd	6.34cd	11.78bc	6.41cdef	3.85cdef	2.15h	3.60abc	5.45de	4.10cdef	1.58ab	16.89bcd	4.55fgh	17.26c	0.25a
F-value	26.00**	26.30**	28.16**	21.32**	5.50**	23.00**	10.46**	977.45**	23.24**	29.54**	6.12**	13.13**	21.22**	46.17**	31.75**	10.00**

Means in each column followed by the same letters are not significantly different at $P < 0.05$ according to Duncan's multiple range test.
* Significant at $P < 0.05$; ** significant at $P < 0.01$; NS: nonsignificant; ----: neck less fruit.
FW: fruit weight; FL: fruit length; FD: fruit diameter; NL: neck length; ND: neck diameter; OD: ostiole diameter; FS: fruit shape; EC: external color; SC: skin cracks; IC: internal color; FC: fruit cavity; ST: skin thickness FT: flesh thickness; TSS: total soluble solids; TA: titratable acidity.

3.2. Principal Component Analysis

Principal component analysis(PCA) was performed taking into account shoots, leaves, and fruits. The eigenvalues obtained by PCA indicate that the first three components provide a good summary of the data. They explained 51.91% of the total variability. The first component, PC1, had large positive loading for shoot length of the two years (S1Y and S2Y), lengths of the 1st, and 2nd internode on S1Y (L1N1Y and L2N1Y), diameters of the 1st, and

2nd internode on SY2 (D1N2Y and D2N2Y), depths of basal and lateral sinus (DBS, DLS1 and DLS2), leaf dimensions (LL and LW), fruit weight and diameter (FW and FD), and ostiole diameter (OD) and negative loadings for petiole color (PC), fruit external color (EC), and total soluble solids (TSS). It represented 26.46% of the total variation (Table 6). This component separated the cultivars Zidi (ZD1) and Nemri (NMR) from Soltani Abiadh (SAB), ThgagliAkhder (TAK), and Khenziri (KNZ) based on shoot length, fruit size, depth of sinus, and ostiole diameter. It allowed the distinction of cultivars Bouharrag (BHR) and Wahchi (WAH) from ZD1 based on petiole color, fruit size, and total soluble solids (Figure 1). The second component, PC2, explained 13.67% of the variability observed. It is positively correlated with fruit length and shape (FL and FS), fruit neck dimensions (NL and ND) and negatively correlated with terminal bud length (TBL), 1-year-old shoot length (S1Y), and fruit cavity (FC). This component allowed the distinction of cultivars Zergui (ZRG), Bouhouli (BHL), Garaï (GRI), KNZ, and TAK from SAB, SoltaniAhmar (SAH), and Hemri(HMR) on the basis of fruit shape, fruit neck, and cavity (Figure 1). The third component, PC3, accounted for 11.78% of the total variability estimated from shoots, leaf, and fruit traits. It is positively correlated to the apex shape (AS), fruit length (FL), and titratable acidity (TA), and negatively correlated to the internode diameter of shoots of the two years (D1N1Y, D2N1Y and D1N2Y, D2N2Y), shape of lobes (SL), depth of basal sinus (DBS), and pH. This component allowed the differentiation of cultivars Fawari (FAW), SAB, and HMR with thin shoots and acid fruits from cultivars Boukhobza (BKB), BHR, ZRG, and SAH presenting thick shoots and low acid fruits (Figure 1). The projection of cultivars in the 1–3 plot exhibited the distinction of some cultivars with particular traits such as ZD1 and ZidiArtab (ZD2) and others which are grouped together and having similar characters such as (SAB, WAH, and KNZ) and (BKB, BHR, and ZRG) (Figure 1).

3.3. Cluster Analysis

Morphological and pomological analysis based on different characters showed high polymorphism with 17 fig cultivars. The dendrogram based on squared Euclidian distance clustered cultivars into five major groups (Figure 2). The first cluster was constituted with three cultivars: Zidi, ZidiArtab (ZD2), and Nemri (NMR). ZD2 was the most divergent from the other cultivars ($d = 20$).It is characterized by long internodes, spatulate, and large leaves, deep sinus, and acid fruit juice. The cultivars ZD1 and NMR meet at d = 15.9. They shared the long shoots, large fruits size, and leaves with serrate margin dentation. The cultivar SoltaniAbiadh (SAB) formed the second cluster at d = 16.9. This

Table 6: Definition the first three components of ACP made on the basis of morphological and pomological traits of fig cultivars.

Principal components	PC1	PC2	PC3
Cumulative, %	26.46	40.13	51.91
Proportion, %	26.46	13.67	11.78
Character*		Eigenvalue	
TBL	0.01	**−0.56**	0.23
SY1	**0.71**	**−0.47**	0.28
L1N1Y	**0.82**	−0.22	0.34
D1N1Y	0.61	−0.24	**−0.46**
L2N1Y	**0.69**	−0.19	0.53
D2N1Y	0.47	−0.38	**−0.50**
SY2	**0.80**	−0.16	0.20
D1N2Y	**0.63**	−0.16	**−0.65**
L2N2Y	0.45	−0.28	0.48
D2N2Y	0.61	−0.24	**−0.66**
SL	−0.29	−0.08	**−0.58**
AS	0.02	−0.05	**0.57**
PC	**−0.41**	0.50	0.00
LL	**0.84**	0.32	−0.11
LW	**0.85**	0.27	−0.19
PD	**0.80**	0.38	−0.12
DBS	**0.74**	0.06	**−0.50**
DLS1	**0.60**	0.55	−0.05
DLS2	**0.60**	0.54	−0.07
FW	**0.75**	−0.04	0.43
FL	0.35	**0.62**	**0.52**
FD	**0.62**	−0.40	0.48
NL	−0.18	**0.74**	−0.09
ND	0.41	**0.67**	0.18
OD	**0.74**	−0.09	0.20
FS	−0.01	**0.85**	0.00
EC	**−0.57**	−0.19	0.25
FC	0.10	**−0.70**	0.07
pH	0.04	−0.05	**−0.35**
TA	−0.28	0.17	**0.71**
TSS	**−0.50**	−0.02	0.16

* See Table 2.

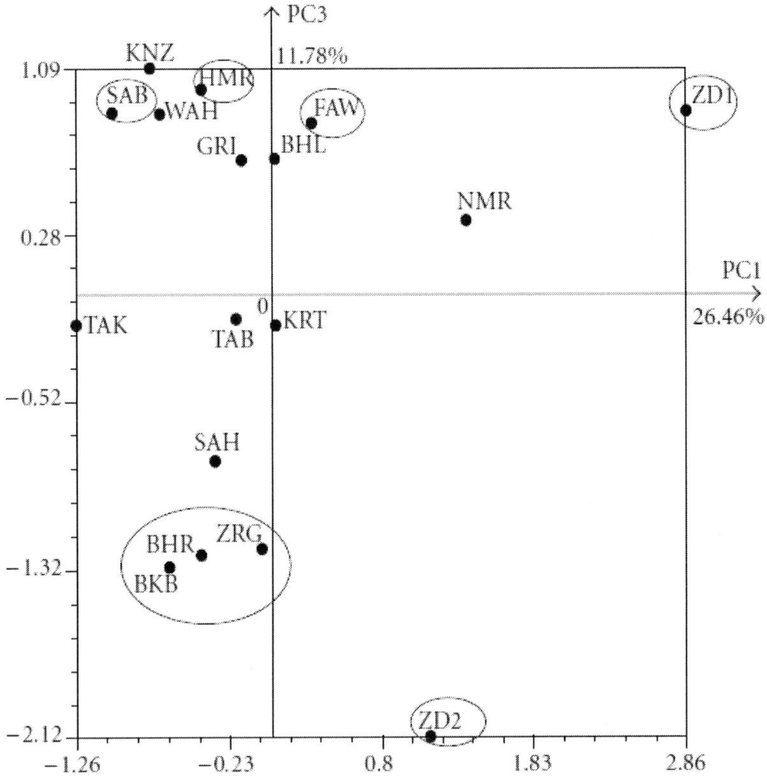

Figure 1: Plot of the first and third principal components resulting from a PCA of the fig cultivars using morphological and pomological characters.

cultivar had the shortest shoots, flattened buds, and cordate leaves with apex acute. The fruits were very long, crackled with high acid juice. The third group, detached at d = 15.9, was constituted with the unique cultivar Fawari (FAW). It is distinguishable, particularly, by short vegetative shoots and internodes, flattened buds, and dark green, large and crenate leaves, with deep sinus and yellow petiole. Fruit shape was oblate producing high acid juice. The cultivar SoltaniAhmar (SAH) formed the fourth cluster at d = 12.85 of the remaining genotypes. It has short and thin shoots with flattened buds. The fruits were relatively small, purple with very thin skin, and give low acid juice with high pH. The leaves were pubescent, serrate with widened lobes and rounded apex. The fifth group consisted of 11 cultivars with a maximum internal dissimilarity level of 10. This cluster was divided into three subgroups. The first subgroup, detached at d = 10, comprised the cultivar ThgagliAkhder (TAK). It is distinguished essentially from the remaining genotypes by small leaves with spatulate lobes, shallow sinus, and short petioles. The second subgroup contained the cultivar Bouharrag (BHR) associated with a level d = 7.85 to the three cultivars ThgagliAbiadh(TAB), Boukhobza (BKB), and Khartoumi (KRT). They are characterized particularly by green leaves with widened lobes, fruits oblate, and sweetened juice. The third subgroup consisted of six cultivars with a maximum internal dissimilarity level of 7.61. Two subsets were identified in this third subgroup. Subset 1 composed of cultivars Garaï (GRI) and Bouhouli (BHL) which associated between them at high level of similarity (d = 2.37) and rejoined the cultivar Zergui (ZRG) at d = 7.61. These cultivars presented conical buds, light green cordate leaves, and oblate fruits without neck. Subset 2 composed of cultivars Khenziri (KNZ), Wahchi (WAH), and Hemri (HMR) having cordate leaves with widened lobes and serrate margin dentation, acute apex, and yellowish green petiole. The fruits give a sweetened juice. WAH and KNZ have the highest level of similarity observed (d = 0.71). The averages values of their main characteristics were similar.

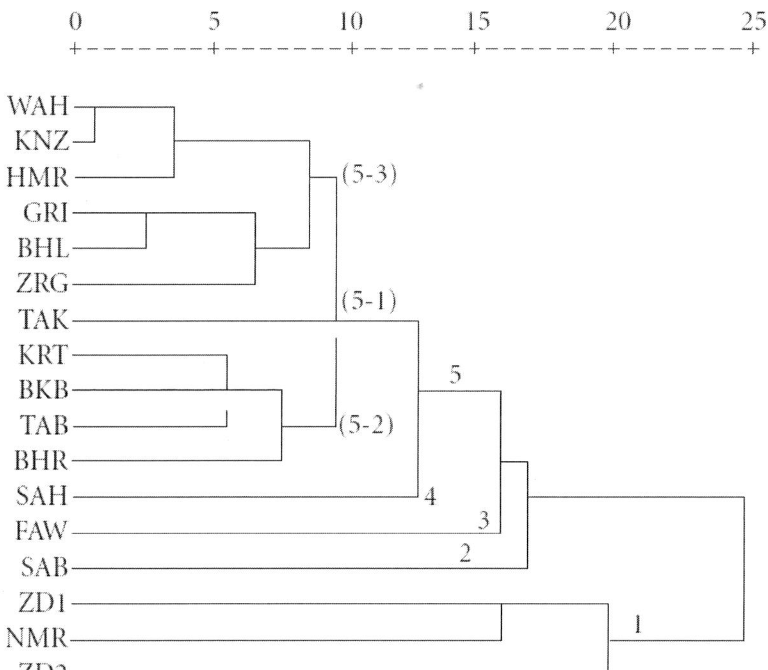

Figure 2: UPGMA dendrogram (based on squared Euclidian distance) of fig cultivars performed using morphological and pomological characters.

4. DISCUSSION AND CONCLUSIONS

Morphological and pomological traits considered in this study showed a large variability for Djebba figs. Among the 45 variables analyzed, those of high discriminating level were leaf dimensions, shoot dimensions, petiole color and dimensions, depth of sinus, fruit shape and color, fruit weight and dimensions, ostiole diameter, and juice acidity. Results of the PCA based on shoots, leaf, and fruits traits showed that 51.91% of the total variability is accounted by the three PCs. Studies based on morphological and pomological traits conducted for pomegranate and peach founded, respectively, 49.29% [21] and 63% [22] of the total variability. Characters related to shoots, leaf, and fruits were powerful for studying the genetic diversity of domestic fig in Djebba. Results showed that, among these characters, some were good criteria for discriminating between cultivars. Similar results were reported by Salah et al. [9] in fig collection in the oasis of Nefzaoua and by Chatti et al. [13] in fig collection in ChottMariem. In addition, morphological study based on the characteristics of fig trees has shown that the first three axes of the PCA amounted to 71.7% and 81.9% of the total variability, respectively, for leaves and fruit traits [15].

The cluster analysis showed high degree of diversity in the germplasm of Djebba fig. Among the cultivars with the same names, there is a similarity between SoltaniAbiadh (SAB) and SoltaniAhmar (SAH) and between ThgagliAbiadh (TAB) and ThgagliAkhder (TAK). However, the levels of similarity observed were not always high enough to believe that they are synonymous. Both cultivars Soltani have almost the same characteristics of shoots and leaf and differed mainly by the fruit external color. Both cultivars Thgagli have a significant similarity in terms of shoots and fruit, while they have a divergence in leaf and petiole. Thus, for Soltani and Thgagli, it seems to be "polyclones cultivars" or "multiclones cultivars." Both cultivars Zidi and ZidiArtab have many differences that were focused on shoots strength developed during the two years. Also, differences were noted for leaves and fruits. The hypothesis of homonymy between these two cultivars could be proposed.

Significant similarities were observed among the cultivars Wahchi (WAH), Khenziri (KNZ), and Hemri (HMR) and between Garaï (GRI) and Bouhouli (BHL) for all morphological and pomological traits. Thus, it seems to be a case of synonymy between these cultivars.

The characters adopted in this study could be used to establish a catalog of local fig cultivars. The concordance between the results of PCA and cluster analysis showed that morphological and pomological analysis can provide reliable information on the variability in fig tree. The overall analysis of all traits brings out a wide diversity in plant material that may have important implications for genetic resources management. This diversity could be due to the antiquity of the culture in this mountainous region and particular cultural practices [7]. More interest has been focused on the diversity since it was known that the domestication of fig tree occurred independently in different areas especially around the edge of the Mediterranean. Thus, it is very interesting to conduct the proper management of these genetic resources. This can be addressed by different tools such as the establishment of ex situ collections. The on-farm conservation can ensure the sustainability of these resources. It is also possible to explore the techniques of tissue culture as an alternative as future protocols for in vitro micropropagation and even cell-cultured fig are already developed [23–26]. Further studies are needed involving chemical, biochemical, and molecular markers [14, 16, 17, 19]. They would clarify the genetic variation at the molecular level in these cultivars.

ACKNOWLEDGMENTS

This study is part of research program of the Research Unit on Agrobiodiversity (UR03AGR04) financed by the Ministry of Higher Education, Scientific Research and Technology (Tunisia). Authors would like to thank Mrs. W. Brini for data analysis and technicians and farmers for their efficient collaboration.

REFERENCES

1. C. C. Berg, "Flora malesiana precursor for the treatment of Moraceae 1: The main subdivision of Ficus: the subgenera," Blumea, vol. 48, no. 1, pp. 167–178, 2003.
2. D. Zohary and M. Hopf, Domestication of Plants in the Old World, Oxford University Press, Oxford, UK, 1988.
3. W. B. Storey, "Fig Ficuscarica (Moraceae)," in Evolution of Crop Plants, N. W. Simmonds, Ed., pp. 205–208, Longman, New York, NY, USA, 1976.
4. Faostat, "FAO statistics database on the World Wide Web," 2009, http: //apps. fao.org/.
5. MAE, 2010, Ministère de l'Agricultureet de l'Environnement. Budget économique 2011, Agriculture etPêche, Novembre-Décembre 2011.
6. M. Mars, K. Chatti, O. Saddoud, A. Salhi-Hannachi, M. Trifi, and M. Marrakchi, "Fig cultivation and genetic resources in Tunisia, an overview," ActaHorticulturae, vol. 798, pp. 27–32, 2008.
7. M. Mars, B. Gaaliche, I. Ouerfelli, and S. Chouat, "Systèmes de production et ressourcesgénétiques du figuier (Ficuscarica L.) àDjebba et Kesra, deux villages de montagne au nordouest de la Tunisie," Revue des RégionsArides, vol. 22, no. 1, pp. 33–45.
8. M. Mars, M. Marrakchi, and T. Chebli, "Multivariate analysis of fig (Ficuscarica L.) germplasm in southern Tunisia," in ActaHorticulturae, vol. 480, pp. 75–81, 1998.

9. M. B. Salah, N. Kadri, M. Ben Mimoun, and R. Hellali, "Répertoire et description de six variétés populations de figuier (Ficuscarica L.)dans les oasis de Nefzaoua," Revue des RégionsArides, pp. 139–144, 2004.
10. N. Minangoin, "Monographie des variétés de figuestunisiennes," in Congrèsd' Agronomie du Cinquantenaire, vol. 1, pp. 336–364, Baconnier, Alger, Algeria, 1931.
11. G. Valdeyron and P. Crossa-Raynaud, Les Fruits de Tunisie, Annales du service botaniqueetagronomique de Tunisie, 1950.
12. T. Lahbib, "Etude pomologique des variétés de figuier (Ficuscarica L.) répertoriées au Sahel tunisien," in Mémoire de Fin d'Études du Cycle de Spécialisation, INAT, Tunis, Tunisie, 1984.
13. K. Chatti, A. Hannachi-Salhi, M. Mars, M. Marrakchi, and M. Trifi, "Analyse de la diversitégénétique de cultivars tunisiens de figuier (Ficuscarica L.) àl'aide de caractèresmorphologiques," Fruits, vol. 59, pp. 49–61, 2004.
14. Salhi-Hannachi, K. Chatti, O. Saddoud et al., "Genetic diversity of different Tunisian fig (Ficuscarica L.) collections revealed by RAPD fingerprints," Hereditas, vol. 143, no. 2006, pp. 15–22, 2006.
15. O. Saddoud, G. Baraket, K. Chatti et al., "Morphological variability of fig (Ficuscarica L.) cultivars," International Journal of Fruit Science, vol. 8, no. 1-2, pp. 35–51, 2008.
16. O. Saddoud, G. Baraket, K. Chatti et al., "Using morphological characters and simple sequence repeat (SSR) Markers to characterize Tunisian fig (Ficuscarica L.) cultivars," ActaBiologicaCracoviensia, vol. 53, no. 2, pp. 7–14, 2011.
17. K. Chatti, G. Baraket, A. B. Abdelkrim et al., "Development of molecular tools for characterization and genetic diversity analysis in Tunisian fig (Ficuscarica) cultivars," Biochemical Genetics, vol. 48, no. 9-10, pp. 789–806, 2010.
18. G. Baraket, K. Chatti, O. Saddoud et al., "Genetic analysis of Tunisian fig (Ficuscarica L.) cultivars using amplified fragment length polymorphism (AFLP) markers," ScientiaHorticulturae, vol. 120, no. 4, pp. 487–492, 2009.
19. G. Baraket, K. Chatti, O. Saddoud et al., "Comparative assessment of SSR and AFLP markers for evaluation of genetic diversity and conservation of Fig, Ficuscarica L., genetic resources in Tunisia," Plant Molecular Biology Reporter, vol. 29, no. 1, pp. 171–184, 2011.
20. IPGRI and CIHEAM, Descriptors for Fig. International Plant Genetic RessourcesInstitue, Rome, Italy, and International Centre for Advanced Mediterranean Agronomic Studies, Paris, France, 2003.
21. M. Mars, Ressourcesgénétiques du grenadier (Punicagranatum L.) en Tunisie : prospection, conservation et analyse de la diversité [ThèseDoctoratd'EtatEs Sciences Naturelles], Faculté des Sciences, Université Tunis EL Manar, 2001.
22. S. Perez, S. Montes, and C. Mejia, "Analysis of peach germplasm in Mexico," Journal of the American Society for Horticultural Science, vol. 118, pp. 519–524, 1993.
23. G. Günver and E. Ertan, "A study on the propagation of figs by the tissue culture techniques," ActaHorticulturae, vol. 480, pp. 169–172, 1998.
24. Demiralay, Y. Yalçin-Mendi, Y. Aka-Kaçar, and S. Çetiner, "In vitro cloning of Ficuscarica L. var. Bursa Siyahi through Meristem culture," ActaHorticulturae, vol. 480, pp. 165–167, 1998.
25. S. Hepaksoy and U. Aksoy, "Propagation of Ficuscarica L. clones by in vitro culture," BiologiaPlantarum, vol. 50, no. 3, pp. 433–436, 2006.
26. S. Hepaksoy and U. Aksoy, "In vitro propagation of Ficuscarica cv. Sarilop clone selected for its high performance," ActaHorticulturae, vol. 798, pp. 199–204, 2008.

CHAPYTER 4

Evaluating the Efficiency of Wicking Bed Irrigation Systems for Small-Scale Urban Agriculture

Niranjani P. K. Semananda*, James D. Ward and Baden R. Myers

School of Natural and Built Environments, University of South Australia, Mawson Lakes Campus, Adelaide 5095, Australia

ABSTRACT

A wicking bed (WB) is a plant driven system where plants receive water through capillary rise from a self-contained coarse material-filled subsoil reservoir. WBs have been widely promoted as a water-efficient irrigation solution for small-scale and urban food gardens. However, little published research exists to support popular claims about their effectiveness. In this study, the performance of WBs was compared with best-practice, precision surface irrigation in terms of water use efficiency (WUE), fruit yield, fruit quality and labour input, using tomato (*Solanum lycopersicum*) as the experimental crop. The influence of WB design variables (reservoir depths and soil bed depths) was tested. Results showed that WBs performed as well or better than precision surface irrigated pots, showing statistically significant improvement in WUE, yield and fruit quality. The results also suggest an optimum design exists for soil depth (where 300 mm outperformed 600 mm) but not reservoir depth (no difference between 150 and 300 mm). The WBs were more labour efficient, requiring significantly less frequent watering to achieve the same or better WUE. WBs are inherently low-tech and scalable and appear well-suited to a variety of urban agriculture settings.

Keywords: capillary; irrigation; tomato; urban agriculture; water use efficiency; wicking

1. INTRODUCTION

More than a half of the world's population lives in urban and peri-urban areas, and this proportion can be up to 90% in developed countries [1,2]. However, land available for crop production within cities is considerable [3]. In the UK, about 87% of households have access to a private garden and these contribute between 22% and 36% of the total urban area [4]. In Australia, more than half of households are growing some of their own food [5]. A wide acceptance and expansion of urban agriculture can be seen even in densely populated cities like New York, USA, as a measure towards sustainable use of land, and in the pursuit of economic and food security [6,7]. There is an increasing trend towards growing crops (especially organic products) in household and community gardens in urban areas. This is due to many factors, including concerns over the side effects of herbicides, insecticides and fertilizers in commercial agriculture [8,9]. Urban gardens also provide a significant contribution to urban green infrastructure, providing benefits to the urban ecosystem. Such benefits include reduced urban heat island impacts, improved human health and well-being, and improved aesthetics [2,10,11]. Urban agriculture can also provide a source of food or additional income for households in developing countries [12].

The growing interest in urban food production adds pressure to municipal water supplies. Municipal treated water is the main water supply used to irrigate urban gardens, and in Australia, about 34% of household water consumption is used to irrigate residential gardens [13]. In addition to supply constraints, the water used for irrigation in urban settings is extremely expensive compared to rural water supplies for commercial horticulture [14,15]. There is therefore a great interest in the development of efficient irrigation systems which can secure urban water resources and provide urban gardeners with an economically efficient means of irrigation that is not technologically complex [16,17,18].

Although improved practices exist for efficient crop irrigation in commercial farm management, there has been limited adoption of these techniques in urban areas due to the barriers of cost and knowledge to maintain them [18]. In urban gardens, the issues principally revolve around determining when and how much water needs to be applied to crops. For example, precision irrigation scheduling has been widely applied for efficient irrigation of large-scale monoculture crops. At the domestic and community garden scale, however, plants tend to be mixed and irrigation cannot be optimally scheduled to meet the needs of multiple crop water use patterns [19,20]. As a result, homeowners frequently apply more water than necessary, which leads to water loss as subsoil drainage [21]. Another major problem for urban growers is the labour invested in watering which may conflict with other activities [22]. In some regions, those interested in growing crops are not able to do so due to the presence of legacy contamination [23] or a shallow water table [24].

Capillary wick irrigation has been investigated for some years as a method to deliver water to plants in container gardens [25]. Existing approaches include wicking mats, self-watering pots, and sub-irrigated planters [24,26]. The more recent "wicking bed" (WB) [27] has been characterized by a saturated media-filled reservoir beneath an unsaturated soil (root zone). Water is delivered by capillary action to the root zone in response to the water requirement of the plant by allowing individual plants to uptake water according to their demand.

In principle, in a WB, each plant should get precisely the right amount of water at a particular time, making the system a possible solution to the urban irrigation issues identified above. Indeed, WBs are gaining popularity among the gardening community because they are relatively simple and scalable, and informal recommendations abound as to their construction details. A review of grey literature, including garden magazines and websites, reveals claims that WBs can save up to 90% of water compared to traditional surface irrigation [28]. Despite increasing enthusiasm for, and claims about, WB performance, there is no comprehensive study to investigate the performance of these systems in terms of crop growth, water use efficiency (WUE) or labour needs. Moreover, while recent research [24] on a sub-irrigated planter design (akin to, though not labelled, a WB) indicated that the system may have higher yield and lower maintenance than other container gardening approaches, it did not explicitly investigate the water use or labour efficiency of the irrigation system, and recommended further research into these areas.

A substantial knowledge gap exists relating to the engineering design of WBs, specifically the depth of both soil and saturated reservoir. We were unable to find any published research supporting design recommendations for these depths. It has long been understood that the capillary rise of water in soil reaches a maximum height, with moisture content decreasing with distance above the saturated water level [29,30]. It follows that in a WB reservoir, especially when the media used is coarse (e.g., gravel), there is a critical depth beyond which wicking into the soil will not occur, rendering the additional depth of storage useless. Meanwhile, insufficient soil may lead to evaporation from the surface, and excess soil may limit the rate of water delivery to the root zone. Thus, it is reasonable to expect an optimum soil bed height for a given plant rooting habit, which would balance water delivery and crop yield against minimal soil evaporation.

The objective of this study was therefore to evaluate, for the first time, the performance of WB systems relative to equivalent best-practice conventional surface irrigation techniques, with a focus on WUE, fruit yield, fruit quality, dry matter production and irrigation labour (indicated by the number of irrigation events required). The study also compared the effect of different reservoir and soil bed depths to better inform the design of WBs for maximum WUE. The outcomes of this research are of high significance in the context of water consumption and efficiency in urban agriculture, and should be of specific interest to decision-makers, practitioners and proponents of WBs as a water-efficient irrigation technology.

2. MATERIALS AND METHODS

2.1. Experimental Conditions

The experiment was conducted in a glasshouse at the University of South Australia, Mawson Lakes campus in Adelaide, Australia (34.81°S, 138.61°E). The glasshouse provided a protected environment in which to compare WB and surface irrigation efficiency without the influence of rainfall. The growing period was during spring and summer 2014–2015. Soil temperature, indoor air temperature and relative humidity were recorded at hourly intervals during the crop growing period. Soil temperature was measured in a single replicate of each treatment by installing TR 10 temperature data loggers at two depths—10 and 15 cm from the soil surface. Temperature was assumed not to vary significantly across the plots within the greenhouse (area 5 m × 3 m). Indoor air temperature (±1 °C) and relative humidity (±3%) were recorded using a USB temperature/humidity data logger (LCD—QP6014, Jaycar electronics, Sydney, Australia).

One of the most common varieties of tomato (*Solanum lycopersicum*), cv. "Mighty Red", was obtained from a local plant nursery in Virginia, South Australia. The experimental soil was purchased from a local supplier, and was a type especially developed for gardening, so was taken as representative of urban garden soils. The soil was a poorly graded sandy soil, according to the unified soil classification system (USCS) (90% sand and <5% fines). It had a field capacity of 30%, porosity of 49%, dry bulk density of 1.33 g/cm^3, organic matter of 4.5%, and pH of 7.7.

2.2. Experimental Design and Treatments

The main features of a WB are the reservoir, separation material, vertical pipe and overflow pipe. All beds were prepared using opaque plastic containers with an internal diameter of 56 cm and 100 cm initial height, which were cut down to provide the required soil plus reservoir depth. Different pots were built to compare the WUE of standard surface irrigation, conventional wicking beds, and wicking beds with a soil column, each with two soil bed heights (300, 600 mm) and two WB reservoir depths (150, 300 mm). A coarse-grained material (quartzite gravel) was used to fill the reservoir. The use of gravel adds strength to the WB to support the soil weight, conforms to the size and shape of the container, and creates a medium that is both porous and permeable so that it both holds a substantial volume of water and is fast to fill. A soil column within the reservoir and underneath the main soil body (Figure 1) to improve wicking consisted of a perforated polyvinyl chloride (PVC) pipe of 30 cm diameter, sealed at the base, and wrapped with a geotextile fabric. To measure evaporation solely from the soil surface, two pots with different soil bed heights (300 and 600 mm) were used without plants. Accordingly, there were eight treatments in total and each treatment was in triplicate. Treatments were labelled as treatment number (T1 to T8), followed by the irrigation method including depth of reservoir and soil bed in mm (Figure 1).

A transparent, flexible tube was fixed to the side of each wicking treatment as a manometer to view the water level in the reservoir. A vertical tube was placed in the reservoir through which water was added to the WB. A PVC access tube sealed at the base was installed into each pot to facilitate access by the soil moisture probe as shown in Figure 1 (this is not a part of typical WB). Aggregate (10 mm Quartzite) was then placed as the reservoir material and a geotextile fabric was laid over the aggregate. Soil was placed into pots in 10 cm layers, each layer being lightly compacted. Granulated slow-release fertilizer (20% N, 6.1% P, 11% K as *w/w*), was added in equal amounts to all pots (30 g/plant), and mixed into the soil layer from 10 to 20 cm below surface. The same plastic containers were used for the surface irrigated pots but with several holes drilled in the base to drain any surplus water. All pots were arranged inside the glasshouse with treatments in a randomized complete block design. The reservoirs were filled with water three days before the seedlings were transplanted. One tomato seedling, 4-weeks old, was transplanted into the centre of each bed.

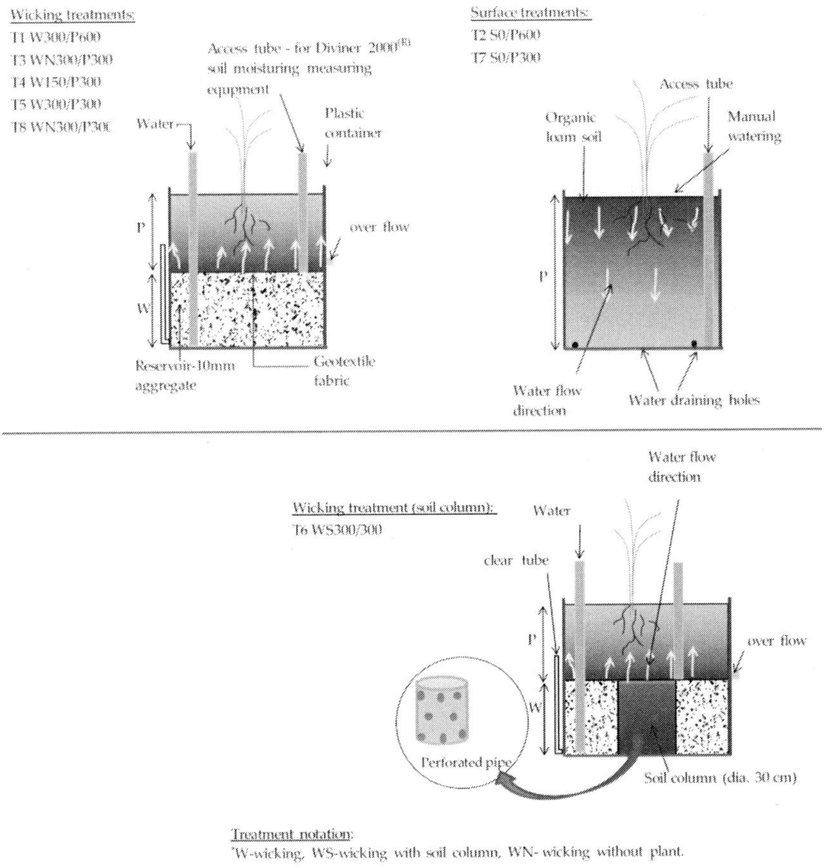

Figure 1. Treatment types and identification labels. The numbers following a letter for each irrigation method treatment indicate the depth of soil (P) or reservoir (W) in mm. S indicates surface irrigation.

2.3. Application of Irrigation

Pots were manually irrigated with small irrigation volumes immediately after each seedling was transplanted, and thereafter every fourth day during the crop establishment period to support early root growth (10.25 L for shallow beds and 26.25 L for deep beds in total). This period lasted about 14 days after transplanting (DAT) for wicking beds with a 300 mm soil bed depth and 21 DAT for 600 mm deep beds. The volume of water applied to each pot was manually recorded. Following the crop establishment period, irrigation scheduling and volume were determined based on the soil-water balance method outlined below.

The volumetric soil moisture content (VMC, *v*/*v*%) in each pot was monitored every day at 10 cm intervals using a portable Diviner 2000® soil moisture monitoring probe (Sentek Sensor Technol., Stepney, SA, Australia) [31]. The Diviner 2000® probe was calibrated gravimetrically for the experimental soil during the progress of the experiment. The summation of volumetric moisture content (VMC) in each layer was calculated and a decision on whether to irrigate the crops was made based when the total VMC in each treatment had decreased to 75% of the moisture content field capacity (*FC*) (Equation (1)):

If
$$If\ Total\ VMC = \sum_{i=1}^{n} (\theta_i) > \sum_{i}^{n} [0.75(\theta_{FC})],\ then\ no\ watering \quad (1)$$

where θ = volumetric moisture content (%), θ_i = volumetric soil moisture content (%) in layer *i*, *n* = number of soil layers, and *FC* = volumetric moisture content at field capacity.

The surface irrigation volume was determined according to Equation (2), which aimed to reach the *FC* in the upper layers and 75% of *FC* in the bottom layer, as:

$$Irrigation\ water\ (mm^3) = \sum_{i=1}^{n-1} (\theta_{FC} - \theta_i) \cdot A + \sum_{i=n} (0.75\theta_{FC} - \theta_n) \cdot A \quad (2)$$

where θ_i = soil moisture content in layer *i*, *n* = number of soil layers, *FC* = volumetric moisture content at field capacity, and *A* = pot area (mm^2).

The soil moisture data were collected until the end of the cropping period. A summation of daily moisture content depletion was used to determine the ET in all treatments. Evapotranspiration over the total growing period (ET_c) was calculated by summing up the ET values for each data recording period using Equation (3) below.

$$ET = I + \sum_{i=1}^{n} \frac{(\theta_{i,\,t-1} - \theta_{i,\,t})}{100} . (D_i) \quad (3)$$

where ET = evapotranspiration (mm), I = irrigation volume (surface treatment) or water added to the reservoir (wicking treatment) (mm), *n* = number of soil layers in each pot, $\theta_{i,t-1}$ = volumetric moisture content (%) in the ith layer on the previous day, $\theta_{i,t}$ = volumetric moisture content (%) in the ith layer on the current day, and D_i = depth of the ith layer (mm).

2.4. Plant Performance and Water Use Efficiency

The total crop growing period was divided into four stages according to experimental observations for each treatment. These stages were the seedling stage (Stage I, from transplant to first flowering), flowering stage (Stage II, from first flowering to first fruit set), fruit development stage (Stage III, from first fruit set to the first harvest) and fruit maturity and harvesting stage (Stage IV, from first harvest to end of the experiment). The average length of each stage was 14.7 ± 4.8, 18.2 ± 0.8, 39.9 ± 3.3 and 67.2 ± 2.8 days, respectively. First fruit was harvested about 72.8 ± 2.8 DAT. All plants were trained to one stem by pruning all side shoots and by topping plants at the end of the crop, taken as the point when each plant had formed 5 fruit bearing clusters [32]. Tomato fruits were harvested when more than 90% of the fruit were red in colour. The experiment was carried out for a period of 140 DAT.

The size of the tomato fruit was measured using two parameters: fruit weight and diameter. The fruit were weighed using a digital balance with an accuracy of ±0.01 g, and individual fruit length and crosswise diameter were recorded using a Vernier calliper. All harvested fruit were graded into various marketable and non-marketable grades according to Ontario commercial standard [33]. Marketable tomatoes had three grades based on diameter: small (40 to 50 mm), large (55 to 75 mm), and extra large (>75 mm). Unmarketable fruit were graded as culls which included the fruit that did not fit into the marketable grades above, and damaged or diseased fruit (blossom end rot or other visible marks or injury). Each fruit was cut into thin slices (~5 mm) and dried in a dehydrator at 55 °C to obtain the dry fruit weight.

Plant growth and performance were evaluated by measuring plant height and stem diameter and counting the number of leaves at 14 day intervals. The plant height was recorded from the soil surface to the apex, while the stem diameter was measured 5 cm above the soil surface. At the end of the experiment, the stem and foliage weight of each plant were weighed separately, and then were dried in a forced air oven at 55 °C for subsequent dry weight determination. The total biomass accumulation was calculated as the dry weight of each vegetative part plus the dry mass of fruit.

Water use efficiency was calculated as the ratio of total marketable fruit yield to either the amount of water used by the plants (WUE) or that was applied through irrigation (iWUE). WUE and iWUE were calculated using Equations (4) and (5) below [34]:

$$WUE = Y/ET_c \quad (4)$$

$$iWUE = Y/I \quad (5)$$

where Y = total marketable fruit yield (g), I = irrigation water volume (L), and ET_c = total crop evapotranspiration (L). ET_cvalues were obtained by multiplying Equation (3) above with the area of the pot (A).

2.5. Root Sampling and Analysis

The root distribution pattern of each treatment was investigated at the end of the experiment. Root samples were collected at the end of the cropping period from one representative plant per treatment. Soil cores were extracted using a 6 cm diameter soil auger at three different depths (0–10, 10–20, and 20–30 cm) for 30 cm soil beds and five depths (0 to 10, 10 to 20, 20 to 30, 30 to 45, and 45 to 60 cm) for the 60 cm soil beds. Cores were removed at eight different surface locations, along two perpendicular transects through the middle of the circular pot (Figure 2). Roots were separated from the soil by manually rinsing and sieving using the method of Smit et al. [35]. The root mass contained in each 10 cm soil core was measured with a precision of 10^{-5} g and used to calculate the root mass density (RMD) (g of root per L of soil). The average RMD in each layer was compared along the soil depth. RMD contour maps were plotted using the online "EZplot®" Excel contour plotting tool [36] to observe the root distribution pattern within the soil bed.

2.6. Statistical Analysis

Statistical analyses were performed using the one-way analysis of variance (One-Way ANOVA) procedure, computed using SPSS version 22 (IBM Corp., New York, NY, USA). The procedure was used to compare treatments with respect to total plant biomass accumulation, fruit yield, ET_c and WUE. Dixon's Q-test was used to decide whether suspected outliers between replicates could be legitimately rejected or not, with a confidence level of 95%. The test of homogeneity of variances was carried out to evaluate the confidence level of accepting the null hypothesis (i.e., variances are equal) and whether the homogeneity of variance assumption had been met. When the F value was statistically significant, Post-hoc multiple comparisons were performed using Fisher's Least Significant Difference (LSD) test at $P \leq 0.05$ to determine if treatment means were significantly different.

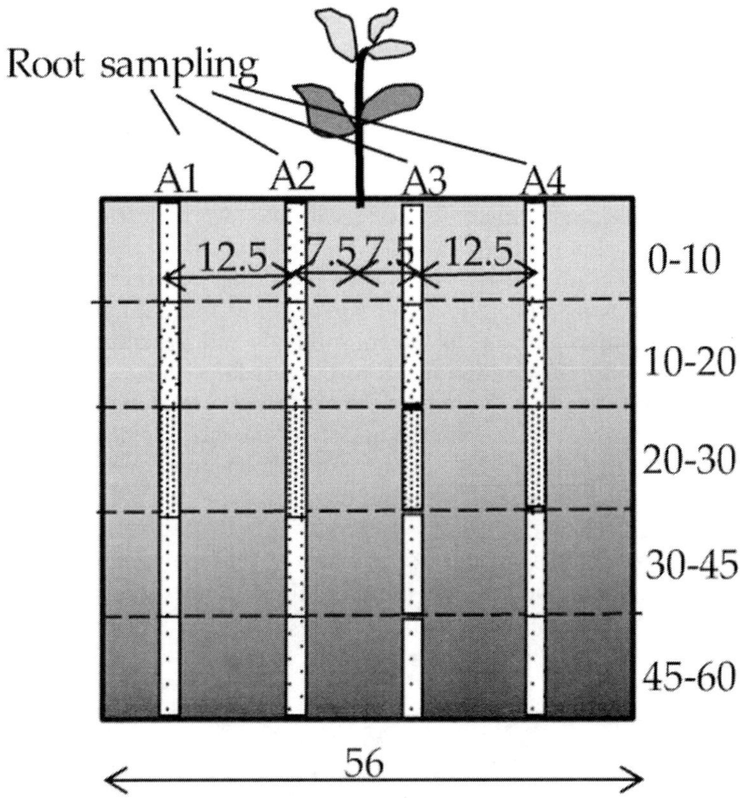

Figure 2. Root sampling diagram (all dimensions in cm).

3. RESULTS

3.1. Relative Humidity, Air and Soil Temperature

Soil temperature is an important parameter contributing to healthy plant growth. Figure 3 shows the variation in relative humidity (RH), indoor air temperature and soil temperature at 50 and 150 mm depths for 5 consecutive days during the fruit development stage (Stage III). The soil temperatures shown are only for treatments T2 S0/P600 and T5 W300/P300. The soil temperature of each soil depth followed the same trend as indoor air temperature but with a lag of about 1 to 6 h.

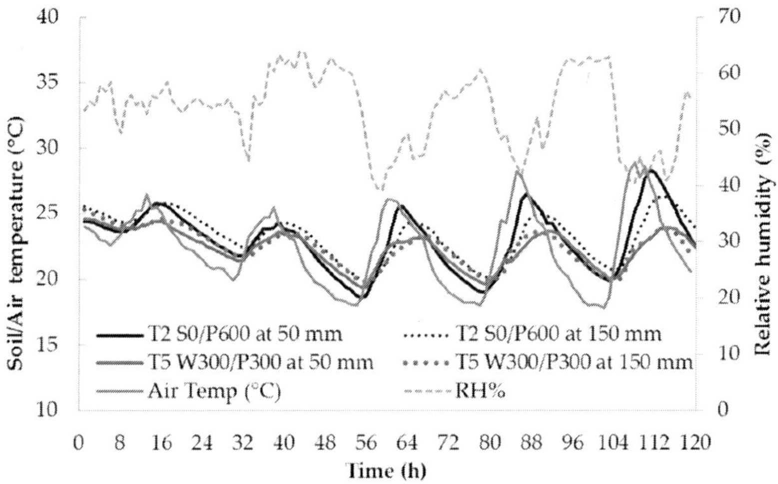

Figure 3. Daily relative humidity (%), indoor air temperature (°C) and soil temperature at 50 mm and 150 mm depth for five consecutive days in the fruit development stage (Stage III). See Figure 1 for treatment descriptions.

The difference in soil temperature between the two treatments was large when the air temperature was high—the wicking treatment T5 had a consistently lower maximum temperature than the surface treatment T2. This is most likely due to the greater thermal mass in the volume of water contained in the reservoir and saturated soil layers of T5. The difference decreased at lower temperatures. In both treatments, the highest soil temperature was typically recorded at 1500 h and the lowest was around 0800 h for the shallow soil bed at a depth of 50 mm. At the 150 mm depth, temperature maxima and minima lagged behind the shallower temperatures by about 2–3 h (Figure 4). The RH varied in the opposite direction to the air temperature, i.e., RH was high at low temperature, and vice versa (Figure 3).

3.2. Effect of Irrigation Treatment on Plant Growth, Yield and Fruit Quality

At 70 DAT, there was no significant difference among treatments for crop height, stem diameter and number of leaves. The crop height ranged from 137.7 ± 16.6 to 146.7 ± 14.6 cm. At the completion of the study period, the total and marketable fruit yields were higher for the WB treatments T4, T5, and T6 than for the surface-irrigated treatment T7, and marketable yield of T6 was greater than T1 and T7, with all other treatments at intermediate values. The highest number of marketable fruits was recorded from the WB treatment with no soil column (T5 W300/P300, 28.3 ± 1.5 fruit) and the lowest was with a surface treatment (T7 S0/P300, 22 ± 3.1 fruit).

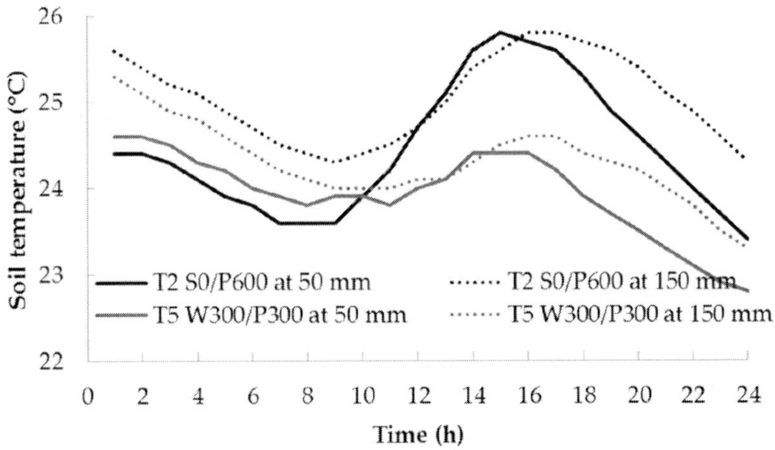

Figure 4. Diurnal variation of soil temperature 48 days after transplanting for surface (T2) and wicking (T5) treatments. See Figure 1 for treatment descriptions.

The diameter of red, ripe fruit ranged from 22 to 83 mm, and fruit weight ranged from 18 to 249 g across treatments. All WB treatments with the 300 mm soil bed produced more large-sized fruit than the surface treatment with the same soil depth (Figure 5a). Plants in the deep soil pots (600 mm) with surface or wick irrigation did not differ from the other treatments. More small fruit were produced by both surface irrigated treatments than the treatment with wick irrigation and a soil column (T6 WS300/P300). While yields of small and extra-large fruit were considerably lower than yields of large fruit, WB treatments T4, T5, and T6 produced greater marketable yields of large fruit than surface irrigated treatment T7 (Figure 5b). The wicking treatment with 150 mm reservoir depth (T4 W150/P300) produced more extra large fruit and greater marketable yield than wicking treatments T1 W300/P600 and T5 W300/P300 as well as the surface irrigated T7 S0/P300 treatment.

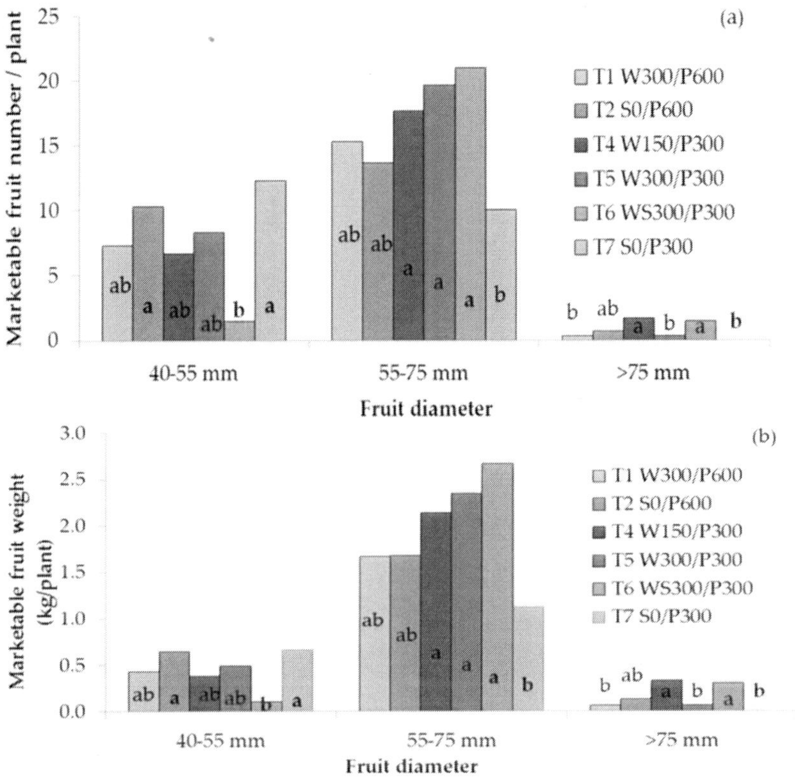

Figure 5. Number (**a**) and yield (**b**) of small, large and extra-large marketable tomato fruit as influenced by irrigation treatments. Columns within a group with different lower case letters are significantly different ($P \leq 0.05$, LSD). W = wicking bed; S = surface irrigation; WS = wicking bed with soil column; WN = wicking bed without plant; P = soil bed. The numbers following a letter for each treatment indicate the depth of reservoir (W) and soil (P), respectively, in mm.

The percentage of unmarketable yield of each treatment was calculated as a proportion of the total harvested yield using the data in Table 1. It varied widely across the treatments, with no discernible pattern. The lowest values were 0.60% ± 0.55% and 1.1% ± 1.9% unmarketable fruit in WBs with a 150 mm reservoir (T4 W150/P300) and wicking with a soil column (T6 WS300/P300), respectively. The wicking treatment with a 300 mm reservoir had 5.2% ± 3.2% rejected fruits, while the deep soil bed (both wicking and surface) had 7.6% ± 6.5 and 9.8% ± 3.8% rejected, respectively. The greatest value was 16.8% ± 7.2% unmarketable fruit for the shallow soil surface treatments (T7 S0/P300).

Table 1. ANOVA summary for total yield, marketable yield, total water, crop evapotranspiration (ET_c), irrigation water use efficiency (iWUE) and water use efficiency (WUE) as affected by each irrigation treatment at the completion of the study.

Treatment	Yield (kg/Plant)		Total Water (L)	Crop Water Use (L/kg)	ET_c (L)	iWUE (g/L)	WUE (g/L)
	Total	Marketable					
T1 W300/P600 [z]	2.36 ab [y]	2.18 b	123.1 c	0.06 ab	123.9 bc	17.8 bc	17.6 bc
T2 S0/P600	2.72 ab	2.46 abc	158.1 a	0.06 ab	160.6 a	15.5 c	15.3 c
T3 WN300/P600	* [x]	*	32.2 d	*	18.5 d	*	*
T4 W150/P300	2.86 a	2.85 ab	121.7 c	0.04 b	118.8 c	23.5 a	24.0 a
T5 W300/P300	3.06 a	2.89 ab	125.2 c	0.04 b	123.0 bc	23.2 a	23.6 a
T6 WS300/P300	3.08 a	3.08 a	143.0 ab	0.05 b	136.2 bc	21.5 ab	22.6 ab
T7 S0/P300	2.11 b	1.78 c	134.8 bc	0.08 a	137.4 b	13.2 c	13.0 c
T8 WN300/300	*	*	30.2 d	*	23.6 d	*	*

[z] W = wicking bed; S = surface irrigation; WS = wicking bed with soil column; WN = wicking bed without plant; P = soil bed. The numbers following a letter for each treatment indicate the depth of reservoir (W) and soil (P), respectively, in mm; [y] Means within columns followed by different letters are significantly different for irrigation treatments ($P \leq 0.05$, LSD); [x] An asterisk (*) indicates no data, or calculation was not possible as they relate to a treatment with no plant.

3.3. Effect of Irrigation Treatment on Water Use Efficiency

Neither WUE or iWUE were significantly different between the WB and surface treatments with 600 mm soil bed depths, T1 and T2, respectively (Table 1). However, there was a significant difference within the 300 mm soil bed depths where both WUE and iWUE were higher in WBs than the surface treatment, T4, T5, and T6 versus T7. The maximum ET_c value was by T2 S0/P600, a significantly higher value than all other treatments. The number of irrigation events over the growing period included water added at the surface (to all treatments) during the crop establishment period. Surface treatments needed 40 to 50 irrigation events (watering once every 2 to 3 days), compared with less than 26 for WB treatments (Figure 6). Table 2compares the plausible water saving by WBs with respect to the corresponding surface treatment. A yield increase of 62 to 73% was obtained for shallow soil WBs, with water use varying from a 6% loss to a 9% savings relative to surface irrigated plots. In contrast, the deeper WBs had a higher water saving (22%), but with an 11% loss of yield; hence, the lack of significant difference in WUE for these treatments.

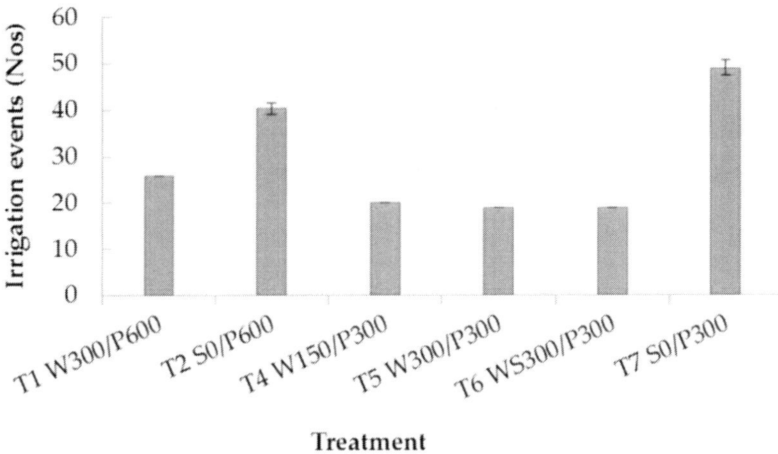

Figure 6. Histogram of average total number of irrigation events throughout the experimental period. Vertical bars indicate standard error. W = wicking bed; S = surface irrigation; WS = wicking bed with soil column; WN = wicking bed without plant; P = soil bed. The numbers following a letter for each treatment indicate the depth of reservoir (W) and soil (P), respectively, in mm. Error bars indicate standard deviation.

Table 2. Comparison of irrigation water reduction and tomato yield increase between wicking and surface treatments.

Treatment	Irrigation Water (L)	Marketable Yield (kg)	Water Reduction (S-W)/S %	Yield Increase (W-S)/S %
T1 W300/P600 [z]	123.1 [y]	2.18	22.1	−11.2
T2 S0/P600	158.1	2.46	-	-
T4 W150/P300	121.7	2.85	9.7	59.6
T5 W300/P300	125.2	2.89	7.1	62.5
T6 WS300/P300	143.0	3.08	−6.1	72.8
T7 S0/P300	134.8	1.78	-	-

[z] Treatment types are labelled with treatment number followed by the irrigation method, reservoir depth and the soil bed height. W = wicking bed; S = surface irrigation; WS = wicking bed with soil column; WN = wicking bed without plant; P = soil bed. The numbers following a letter for each treatment indicate the depth of reservoir (W) and soil (P), respectively, in mm; [y] Statistical differences for Irrigation Water and Marketable Yield are shown in Table 1. Percentage Water Reduction and Yield Increase for W versus S were calculated separately for each soil bed height.

3.4. Effect of Irrigation Treatments on Fruit and Shoot Dry Matter

The dry weight of fruit, stems, and foliage are presented in Table 3. There was no significant difference in dry stem weight between the treatments. The dry foliage weight of surface treatments (both 300 and 600 mm soil bed depths) was significantly higher than that of WB treatments.

However, overall dry biomass accumulation (stem, foliage and fruit) of T2 S0/P600 was significantly greater than T1 W300/P600 and T5W 300/P300. Generally, the higher the water consumption, the higher the biomass and foliage weight. Supporting this, results showed that the total dry biomass weight was linearly related to the amount of water applied in all other treatments ($R^2 > 0.99, P < 0.05$), excluding treatments T1 W300/P600 and T5 W300/P300. The lower foliage or biomass dry weight (in wicking treatments T1W300/600 and T5 W300/P300) may be due to less water availability and uptake by the plant in the former case, and lateral root growth and elongation observed in subsurface layers in the latter case (Figure 5 and Figure 6, discussed later).

Table 3. Above-ground dry biomass as affected by irrigation treatment at the completion of the study.

Treatment	Above-ground Average Dry Biomass Weight (g/Plant)			
	Fruits	Stem	Foliage	Total
T1 W300/P600	154.7 b	50.3 a	69.3 c	274.4 b
T2 S0/P600	203.6 ab	56.0 a	93.1 ab	352.7 a
T4 W150/P300	182.6 ab	58.8 a	75.9 bc	317.2 ab
T5 W300/P300	188.8 ab	45.2 a	61.0 c	294.9 b
T6 WS300/P300	204.6 a	55.3 a	77.6 bc	337.5 ab
T7 S0/P300	177.8 ab	54.9 a	99.2 a	331.9 ab

W = wicking bed; S = surface irrigation; WS = wicking bed with soil column; WN = wicking bed without plant; P = soil bed. The numbers following a letter for each treatment indicate the depth of reservoir (W) and soil (P), respectively, in mm; Means within columns followed by different letters are significantly different for irrigation treatments ($P \leq 0.05$, LSD).

The distribution of soil-water content in different layers during the fruit maturity and development stage (Stage IV, 74 to 134 days) is presented in Figure 7. In surface treatments (T2 S0/P600 and T7 S0/P300), a frequent fluctuation of moisture content was observed at the top (0 to 10 cm) layer, as would be expected due to the frequent irrigation events. In WB treatments, the soil water content was more stable and was driest at the top (0 to 10 cm). However, there was no clear pattern among the different WB treatments.

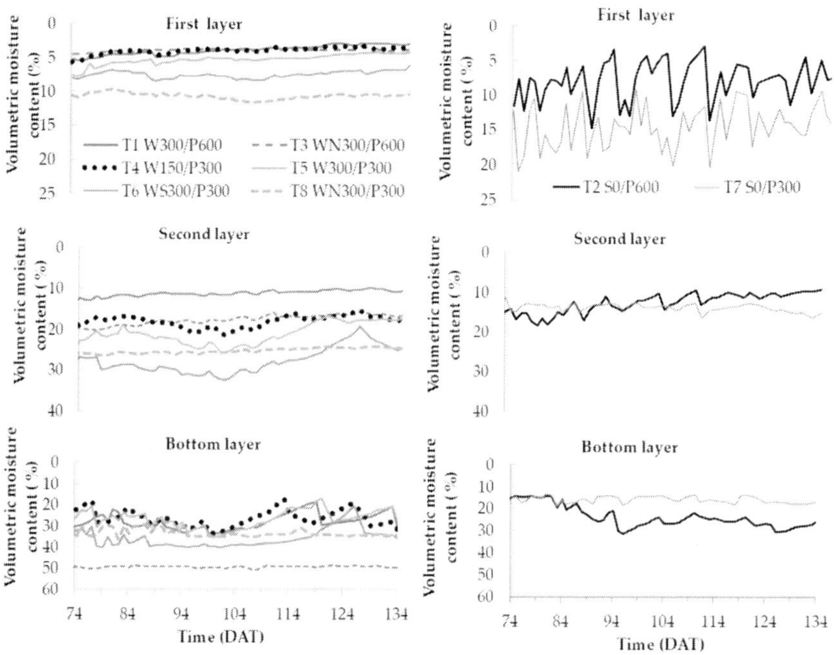

Figure 7. Average soil water content distribution (as % volumetric moisture content) at different soil depths during fruit maturity and harvest stage for wicking bed (**left**) and surface (**right**) treatments. W = wicking bed; S = surface irrigation; WS = wicking bed with soil column; WN = wicking bed without plant; P = soil bed. The numbers following a letter for each treatment indicate the depth of the reservoir (W) and soil (P), respectively, in mm. First layer—0 to 10 cm, second layer—10 to 20 cm, bottom layer—20 to 30 (300 mm bed) and 45 to 60 cm (600 mm bed).

In contrast to the surface treatments, the moisture content increased substantially with depth in the wicking treatments, as may be reasonably expected given that the source of water was at the bottom of the soil. In the second layer (10 to 20 cm), the difference in moisture content among different WB treatments was greater than in the first layer. A high moisture content was recorded for treatment T6 WS300/P300 followed by T5 W300/P300, T4 W150/P300 and T1 W300/P600 (for treatments with plants). The moisture content gradually decreased over time in this layer of the wicking treatments until the reservoir was refilled. For both surface treatments (deep and shallow), the moisture content in this layer corresponded to the lower end of the range of moisture content in the WBs. In WBs, the soil moisture content at the bottom layer was close to saturation, increasing when water was added and drying in the interim.

A vertical distribution of moisture content at the end of the experiment is shown in Figure 8. All treatments had a similar distribution pattern (moisture content increasing with depth). WBs of reservoir depths 150 and 300 mm and the WB with soil column had the same distribution pattern, with the soil column having a marginally (about 2%) higher moisture content. A higher moisture content at the surface level was recorded in the shallow wicking treatment without a plant, possibly due to the lack of interception and uptake of capillary water by roots, allowing more water to reach the surface.

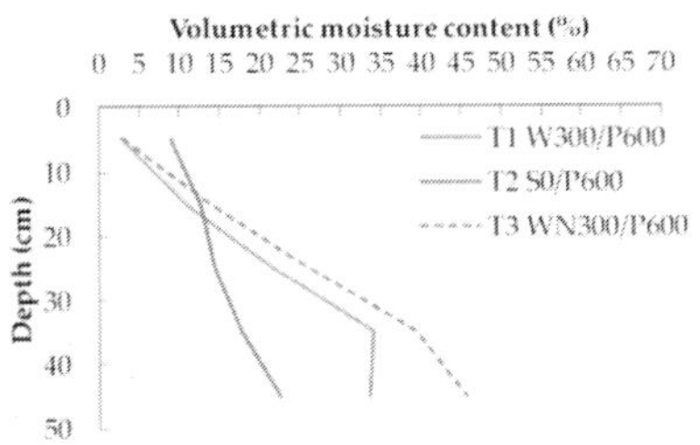

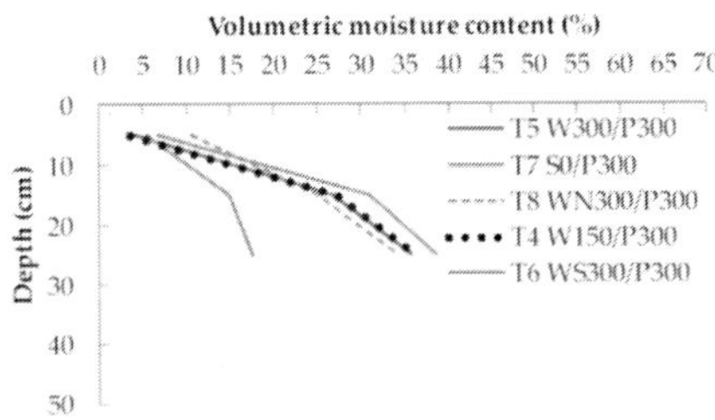

Figure 8. Average soil-water content distribution (as % volumetric moisture content) at the end of experiment for 600 mm soil bed depths (**left**) and 300 mm soil bed depths (**right**) treatments. W = wicking bed; S = surface irrigation; WS = wicking bed with soil column; WN = wicking bed without plant; P = soil bed. The numbers following a letter for each treatment indicate the depth of the reservoir (W) and soil (P), respectively, in mm.

3.5. Root Mass Density

The root mass density (RMD) for each treatment revealed a possible difference in rooting patterns among irrigation treatments (Figure 9). A large concentration of roots was generally found in the top layer (0 to 10 cm), except in treatment T6 WS300/P300. A higher RMD was observed near the surface for surface irrigated compared with WB treatments, with roots decreasing approximately exponentially with the soil depth. A contour plot of RMD across the bed showed that the RMD was preferentially concentrated near the plant and within the top 0 to 20 cm layer (Figure 10). Interestingly, no discernible differences or trends could be observed in RMD between the surface and wicking treatments, except that the wicking bed with soil column exhibited a more uniform, and generally lower, RMD compared to all other treatments. Although these results are therefore inconclusive, previous studies have indicated that a lower RMD was produced when there was more water available [37] near the water table, which may have been the case for the WB with soil column which had a higher water demand.

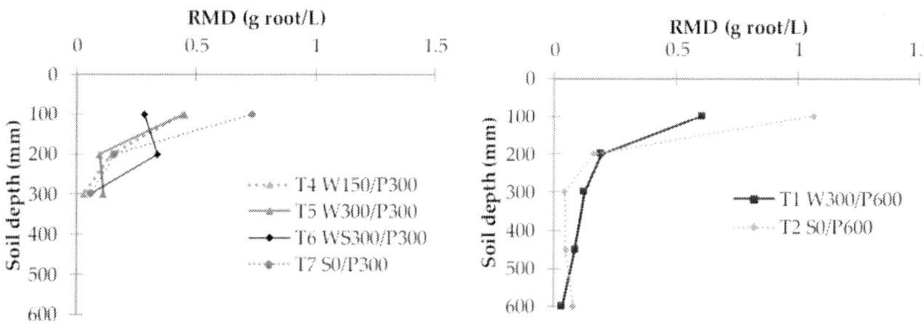

Figure 9. Root mass density (RMD) distribution for different treatments. W = wicking bed; S = surface irrigation; WS = wicking bed with soil column; WN = wicking bed without plant; P = soil bed. The numbers following a letter for each treatment indicate the depth of reservoir (W) and soil (P), respectively, in mm.

4. DISCUSSION

The results showed that there was a significant difference between the wicking and surface treatments for total biomass, yield and water use efficiency. The total water use was significantly different between surface and wicking treatments in the two deep soil bed depths (600 mm); however, the difference was not significant between the treatments for the 300 mm soil bed depth or between the two reservoir depths. An average irrigation amount of approximately 122 L/plant was the optimum amount of water for maximizing the WUE for tomatoes in these study conditions. This optimum WUE occurred in the 150 mm reservoir wicking bed (T4 W150/P300). The average amount of water needed per plant was about 0.9 L/plant/day in this WB treatment, depending on the growth stage and the season. In comparison, Snyder [38] recommended about 50 mL/day for newly transplanted crops and 2.7 L/plant/day at maturity on sunny days for glasshouse-grown tomatoes. Water requirements for drip irrigated tomatoes grown in greenhouse conditions in the tropics were reported to range from 0.3 to 0.4 L/plant/day, a figure derived from a pot area of 706.50 cm^2, similar to this study [39]. For context, the water requirement for an average field-grown tomato cultivar is 100 to 150 L/plant (a 400 to 600 mm irrigation rate) for a 90 to 120 day growth period (FAO, 2015).

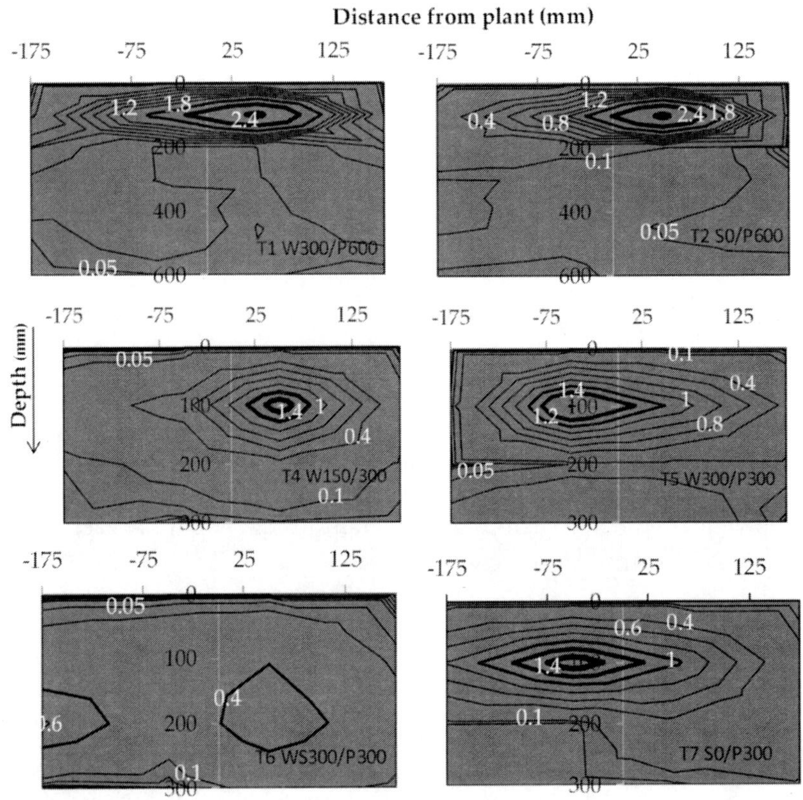

Figure 10. Tomato root mass density (RMD) distribution (mg/cm^3) as affected by soil depth and distance from the plant at the end of the experiment. Bolder contours indicated higher RMD. W = wicking bed; S = surface irrigation; WS = wicking bed with soil column; WN = wicking bed without plant; P = soil bed. The numbers following a letter for each treatment indicate the depth of reservoir (W) and soil (P), respectively, in mm.

Despite the application of larger irrigation volumes in surface treatments, yield and WUE were significantly greater in wicking treatments. The highest marketable yield of 3.1 kg/plant was obtained in the soil column WB (T6 WS300/300), while the lowest (1.8 kg/plant) occurred in the shallow soil (300 mm) surface irrigated treatment (T7 S0/P300). To place the yields in this experiment in some context, the FAO (2015) stated that a good commercial yield under irrigation was 1.1 to 1.6 kg/plant for field-grown tomatoes, while well-managed greenhouse-grown tomatoes in a soil-free culture should yield 3.3 to 6 kg/plant [40,41].

The deep soil bed wicking treatment (T1 W300/P600) had a lower yield than its corresponding surface treatment, and was also lower than the shallower wicking treatments. The reason for the low yield and WUE in T1 W300/P600 was initially considered to be insufficient supply of water to the root zone, due to the greater height and the resultant difficulty in maintaining water supply to the roots by capillary rise (noting that the amount of capillary water present in soil decreased with distance above the water table). In support of this theory, soil evaporation observed in the unplanted deep wicking treatment (T3 WN300/P600) was low when compared with the 300 mm WB T8 WN300/P300 treatment (Table 1), suggesting that less water was able to reach the surface by capillary rise. However, the ET_c in the deep soil WBs was not significantly different from the ET_c in the shallower wicking treatments, suggesting that a similar amount of water must have been delivered to the plant, despite the greater soil depth. Although there was about a 22% water reduction in T1 W300/P600 relative to the surface treatment (T2 S0/P600) which had an equivalent soil depth (Table 2), the wicking treatment delivered lower yields and thus failed to produce a significant improvement in WUE. It should be noted that the percentage of unmarketable fruits was lower in WBs compared to surface treatments for the shallow soil plots, but not for the deep soil treatments. On this basis, it is inferred that the use of wicking irrigation to grow tomatoes with a 600 mm soil depth was not effective. This depth of wicking bed may, however, be suitable for deeper rooted plants. It should be noted that the main reason for discarding fruit was blossom end rot. Previous studies have mentioned that insufficient calcium, or either too much or too little water, can cause blossom end rot in tomatoes [38,42]. It is therefore possible that the soil chemistry, the irrigation method, and/or the irrigation volume may have affected the reliability of fruit production in the experiment, and inferences drawn from these results (especially for the shallow 300 mm soil treatments) should be moderated accordingly.

The wicking bed with a soil column had a higher ET_c compared to other shallow wicking treatments. However, to compensate, it also had the highest yield, plant height and vegetative mass. Studies have shown that water moves upward slowly and over long distances in fine textured soils, but not so for coarse grained soils [43]. This may provide a plausible explanation for the improved performance as the soil column in the reservoir potentially carried more capillary water to the soil surface by maintaining a connection of saturated soil, thus allowing plants to consume more water than other treatments. This may be the reason for higher yield, plant vegetative mass, and water consumption (Table 1 and Table 2). The yield was highest in T6 WS300/P300 but, perhaps due to the increased vegetative mass, the WUE was not as high as the other wicking treatments. However, there was no significant difference in the crop water use with respect to yield among shallow WBs. As such, a wicking bed with a soil column would appear to be a better option for areas where there is little concern about water availability (Table 1). Nevertheless, we may expect a better result by optimizing this system design further. For example, changes could be made to the size of the soil column, the frequency of irrigation events or by the application of mulch to the soil surface.

There was no significant difference in WUE between the 150 mm and 300 mm reservoir depths. This finding has implications for design of cost-effective wicking beds, i.e., there was little gain from a deeper reservoir. Based on having a smaller reservoir depth and achieving an equal maximum WUE, treatment T4 W150/P300 may be considered the most economically efficient wicking bed system tested in this study. Overall, shallow wicking treatments were the more favourable treatments in terms of WUE and yield. This may be related to the relatively shallow (medium) rooting depth of tomatoes.

The labour efficiency of each treatment was evaluated by counting number of irrigation events over the growing period of each replicate and averaged. Surface treatments needed frequent watering at approximately 3 day intervals. In contrast, the wicking treatment pots lasted up to 4 weeks without watering during the developing stage (after the initial two weeks of surface watering to establish the plant), and from 1 to 2 weeks during the maturity stage. These results indicated that WBs reduced manual watering requirements. Most importantly, the WBs eliminated the problem of deciding when and how much water to irrigate. Efficient surface irrigation depended on precise soil moisture measurements and water balance calculations which may not be available to non-commercial growers. In practice, of course, the extent of the labour-savings will depend on the size of the pot, soil bed height and reservoir volume, and the type of crop or root behaviour of the plant selected.

The low soil moisture content in the top layer of all replicates was likely due to increased water evaporation from the soil surface along with transpiration, as the root density was highest at the upper soil layer (Figure 9 and Figure 10). Moreover, the moisture content in the upper layer was higher in the treatment with no plants than that of corresponding plant treatments (Figure 7). This suggests that transpiration was responsible for a significant amount of moisture depletion in the soil, and that the combined effect of evaporation and transpiration was responsible for lowering the moisture content in the soil.

A regular pattern of soil moisture content distribution was observed in wicking treatments without plants (T3 WN300/P600 and T8

WN300/P300) (Figure 7). The total water consumption of these pots may be assumed to be due to soil water evaporation. Soil moisture content of the first and second layer of treatment T3 WN300/P600 was always lower than that of treatment T8 WN300/P300. This was likely due to a decrease in matric potential (i.e., becoming more negative) when the soil bed depth increased (600 mm as opposed to 300 mm). Interestingly, this also led to reduced surface water evaporation (Table 1), as the evaporating water was likely not be replaced as quickly through the taller soil beds. As such, the bottom layer of treatment T3 WN300/P600 held a maximum moisture content, which was similar to the saturated moisture content of the soil.

The surface layer accumulates heat more during the day due to solar radiation than the underlying layer, resulting in a temperature gradient in the soil [44]. The soil temperature was always greater in T2 S0/600 than in the T5 W300/300 at high temperatures. The frequent application of water in the surface treatment increased the soil moisture content in the upper layer compared to the wicking treatment. Counterintuitively, the higher moisture content in the top layer of surface treatments may be the reason for the higher temperature, due to an increase in the thermal conductivity and diffusivity of wet soil [44,45,46]. Roots survive and grow where adequate water, temperature and oxygen are present [47]. The observed maximum RMD in the second layer of treatment T6 WS300/P300 may have been due to sufficient moisture content and other suitable growing conditions (Figure 9). The moisture content was very high (approaching saturation) at the bottom layer of wicking treatments, and as such, high moisture content may have lowered the amount of oxygen trapped in the soil and restricted root growth. As such, the unfavourable environment may have lead the root system to develop horizontally rather than vertically [48].

It is understood that container-grown plants have different root growth patterns compared to field-grown crops. For instance, where a container restricts roots, this may result in fewer primary roots and an increase in lateral roots [49,50]. The large diameter of the pot may have produced a high concentration of lateral rather than vertical roots, in which case a different root pattern may occur if smaller diameter pots were used. It is also noteworthy that there is a potential for salt accumulation at the surface of wicking beds, a phenomenon which has been previously noted in sub-irrigated systems. Follow-up research is being undertaken to determine if this occurs in WBs.

5. CONCLUSIONS

Wicking beds (WBs) have been widely promoted and adopted as an efficient irrigation system for urban agriculture. However, there has been little published research to support popular claims about the effectiveness of WBs. This study rigorously tested the performance of WBs relative to best-practice, precision surface irrigation systems in terms of total water use, marketable yield, fruit quality, water use efficiency (WUE) and irrigation frequency. The suitability of WBs for growing tomatoes was compared using two soil bed depths (300 vs. 600 mm), two reservoir depths (150 vs. 300 mm), and introduced a soil column in the reservoir to improve wicking. The WB with a 150 mm reservoir and 300 mm soil bed was the most effective of all treatments tested, as it was cost-effective (requiring minimal material) with a high WUE.

Overall, the results of this study indicated that WBs matched (or exceeded) WUE and yield achieved with best-practice surface irrigation, and offered a potentially substantial labour saving for gardeners. Moreover, given that surface irrigation in urban agriculture is likely to fall well short of the precision irrigation method used for comparison in this study, the relative WUE improvement of WBs in practice is probably greater. As WBs can be made out of recycled (and recyclable) material, they may be a low-tech and low-cost system. This study therefore provides a much-needed scientific basis for the widespread adoption of WBs in urban agriculture.

There are opportunities to further improve the system to provide more benefit to cities, not only for saving water in gardens, but ultimately to help facilitate the other benefits of urban food production. For example, they may also provide an interesting glasshouse model for smaller-scale commercial producers. There is a possibility to connect multiple WB pots in series and parallel for glasshouse settings. The reservoir can be filled manually or automatically, making it even more practical. However, sustainable irrigation practices will always be site specific, which depends on local climate, soil, topography, and the quality of water. One limitation of this trial is that this is a single site and a single season. Further studies are intended to improve our understanding of the effectiveness of the WB irrigation system for different crop types and environmental settings, as well as soil salinity and sodicity with the use of recycled wastewater.

ACKNOWLEDGMENTS

The authors would like to acknowledge that this research was conducted with the assistance of an Australian Government Endeavour Awards Scholarship. Funding was also provided by the Goyder Institute for Water Research, a partnership between the South Australian Government through the Department of Environment, Water and Natural Resources, CSIRO, Flinders University, the University of Adelaide, and the University of South Australia.

REFERENCES

1. Kadenyeka, M.V.; Omutimba, D.; Harriet, N. Urban agriculture livlihoods and household food security: A case of Eldoret, Kenya. *J. Agric. Biol. Sci.* **2013**, *8*, 90–96.
2. Cameron, R.W.F.; Blanuša, T.; Taylor, J.E.; Salisbury, A.; Halstead, A.J.; Henricot, B.; Thompson, K. The domestic garden—Its contribution to urban green infrastructure. *Urban For. Urban Green.* **2012**, *11*, 129–137.
3. Grewal, S.S.; Grewal, P.S. Can cities become self-reliant in food? *Cities* **2012**, *29*, 1–11.
4. Gibbons, S.; Mourato, S.; Resende, G.M. The amenity value of English nature: A hedonic price approach. *Environ. Resour. Econ.* **2014**, *57*, 175–196.
5. The Australia Institute. *Grow Your Own: The Potential Value and Impacts of Residential and Community Food Gardening*; The Australia Institute: Canberra, Australia, 2014.
6. Goldstein, M.; Bellis, J.; Morse, S.; Myers, A.; Ura, E. *Urban Agriculture: A Sixteen City Survey of Urban Agriculture Practices across the Country*; Turner Environmental Law Clinic: Atlanta, GA, USA, 2011.
7. Lovell, S.T. Multifunctional urban agriculture for sustainable land use planning in the United States. *Sustainability***2010**, *2*, 2499–2522.
8. Forget, G. Balancing the need for pesticides with the risk to human health. In *Impact of Pesticide Use on Health in Developing Countries*; Forget, G., Goodman, T., de Villiers, A., Eds.; International Development Resesrch Centre: Ottawa, ON, Canada, 1990.
9. Pimentel, D.; Culliney, T.W.; Bashore, T. Public Health Risks Associated with Pesticides and Natural Toxins in Foods. Available online: http://ipmworld. umn.edu/ pimentel-public-health (accessed on 25 June 2016).
10. Breuste, J.H.; Artmann, M. Allotment gardens contribute to urban ecosystem service: Case study Salzburg, Austria. *J. Urban Plan. Dev.* **2014**.
11. Taylor, J.R.; Lovell, S.T. Urban home gardens in the global north: A mixed methods study of ethnic and migrant home gardens in Chicago, IL. *Renew. Agric. Food Syst.* **2015**, *30*, 22–32.
12. Poulsen, M.N.; McNab, P.R.; Clayton, M.L.; Neff, R.A. A systematic review of urban agriculture and food security impacts in low-income countries. *Food Policy* **2015**, *55*, 131–146.
13. Prime Minister's Science, Engineering and Innovation Council (PMSEIC). *Recycling Water for Our Cities*; Prime Minister's Science, Engineering and Innovation Council (PMSEIC): Canbera, Australia, 2003; p. 44.

14. Dima, S.J.; Ogunmokun, A.A.; Nantanga, T. *The Status of Urban and Peri-Urban Agriculture in Windhoek and Oshakati*; Food and Agriculture Organisation (FAO): Windhoek, Nambia, 2002.
15. Ward, J.D.; Ward, P.J.; Saint, C.P.; Mantzioris, E. The urban agriculture revolution: Implications for water use in cities. *J. Aust. Water Assoc.* **2014**, *41*, 69–74.
16. Levidow, L.; Zaccaria, D.; Maia, R.; Vivas, E.; Todorovic, M.; Scardigno, A. Improving water-efficient irrigation: Prospects and difficulties of innovative practices. *Agric. Water Manag.* **2014**, *146*, 84–94.
17. Liu, F. Irrigation strategies for sustainable environmental and influence on human health. In *Encyclopedia of Environmental Health*; Nriagu, J.O., Ed.; Elsevier: Burlington, NY, USA, 2011; pp. 297–303.
18. Russo, T.; Alfredo, K.; Fisher, J. Sustainable water management in urban, agricultural, and natural systems. *Water***2014**, *6*, 3934–3956.
19. Jones, K.S.; Costello, L.R. *Irrigating the Home Landscape*; University of California Cooperative: Half Moon Bay, CA, USA, 2003.
20. Peters, T. *Drip Irrigation for the Yard and Garden*; Washington State University: Pullman, WA, USA, 2011; p. 7.
21. Hayden, L.; Cadenasso, M.L.; Haver, D.; Oki, L.R. Residential landscape aesthetics and water conservation best management practices: Homeowner perceptions and preferences. *Landsc. Urban Plan.* **2015**, *144*, 1–9.
22. Langellotto, G.A. What Are the Economic Costs and Benefits of Home Vegetable gardens? Available online: http://www.joe.org/joe/2014april/rb5.php (accessed on 2 June 2016).
23. Kessler, R. Urban gardening: Managing the risks of contaminated soil. *Environ. Health Perspect.* **2013**.
24. Sullivan, C.; Hallaran, T.; Sogorka, G.; Weinkle, K. An evaluation of conventional and subirrigated planters for urban agriculture: Supporting evidence. *Renew. Agric. Food Syst.* **2015**, *30*, 55–63.
25. Kirkham, M.; Gabriels, D. Water and nutrient uptake of wick-grown plants. *Hortic. Res.* **1979**, *19*, 3–13.
26. Schuch, U.K.; Kelly, J.J. Capillary mats for irrigating plants in the retail nursery—And saving water. *Southwest Hortic.***2006**, *23*, 24–25.
27. Austin, C. Wicking Bed—A New Technology for Adapting to Climate Change. Available online: http://www.waterright.com.au/ (accessed on 20 October 2015).
28. ModBOX © Modbox Raised Garden Beds. Available online: http://www.modbox.com.au/wicking-garden-beds/(accessed on 10 June 2016).
29. Lane, K.S.; Washburn, D.E. Capillarity tests by capillarimeter and by soil filled tubes. In *Highway Research Board Proceedings*; Highway Research Board: Washington, DC, USA, 1947; Volume 26, pp. 460–473.
30. Li, X.; Zhang, L.M.; Fredlund, D.G. Wetting front advancing column test for measuring unsaturated hydraulic conductivity. *Can. Geotech. J.* **2009**, *46*, 1431–1445.
31. Sentek Pty Ltd. *User Guide: Diviner 2000*, version 1.2; Sentek Pty Ltd.: Stepney, South Australia, Australia, 2003; p. 66.
32. Peet, M.; Welles, G. Greenhouse tomato production. In *Tomatoes (Crop production science in Horticulture)*; Heuvelink, E., Ed.; CAB International: Oxfordshire, UK, 2005; pp. 257–304.
33. Saha, U.K.; Papadopoulos, A.P.; Hao, X.; Khosla, S. Irrigation strategies for greenhouse tomato production on rockwool. *HortScience* **2008**, *43*, 484–493.
34. Abuarab, M.; Mostafa, E.; Ibrahim, M. Effect of air injection under subsurface drip irrigation on yield and water use efficiency of corn in a sandy clay loam soil. *J. Adv. Res.* **2013**, *4*, 493–499.
35. Smit, A.L.; Bengough, A.G.; Engels, C.; Noordwijk, M.; Pellerin, S.; Geijn, S.C. *Root Methods: A Handbook*; Springer: Berlin/Heidelberg, Germany, 2000; pp. 177–182.
36. Office Expander EZplot for Excel. Available online: http://www.officeexpander.com/ (accessed on 12 July 2015).

37. Sorenson, S.K.; Dileanis, P.D.; Branson, F.A. *Soil Water and Vegetation Responses to Precipitation and Changes in Depth to Ground Water in Owens Valley, California*; United States Geological Survey Water-Supply (USGPO): Washington, DC, USA, 1991.
38. Snyder, R.G. *Greenhouse Tomato Handbook*; Mississippi State University: Starkville, MS, USA, 2010.
39. Salokhe, V.; Babel, M.; Tantau, H. Water requirement of drip irrigated tomatoes grown in greenhouse in tropical environment. *Agric. Water Manag.* **2005**, *71*, 225–242.
40. Maboko, M.; Du Plooy, C.; Bertling, I. Comparative performance of tomato cultivars cultivated in two hydroponic production systems. *S. Afr. J. Plant Soil* **2011**, *28*, 97–102.
41. Rodriguez, J.C.; Cantliffe, D.J.; Shaw, N. Performance of greenhouse tomato cultivars grown in soilless culture in North Central Florida. *Proc. Fla. State Hortic. Soc.* **2001**, *114*, 303–306.
42. Whiting, D.; O'Meara, C.; Wilson, C. *Water Conservation in the Vegetable Garden*; Colorado State University: Fort Collins, CO, USA, 2015.
43. Kuo, J. *Practical Design Calculations for Groundwater and Soil Remediation*, 2nd ed.; Taylor and Francis: Hoboken, NJ, USA, 2014.
44. Chacko, T.P.; Renuka, G. Temperature mapping, thermal diffusivity and subsoil heat flux at Kariavattom of Kerala. *J. Earth Syst. Sci.* **2002**, *111*, 79–85.
45. Ismail, S.M.; Ozawa, K.; Khondaker, N.A. Influence of single and multiple water application timings on yield and water use efficiency in tomato (var. First Power). *Agric. Water Manag.* **2008**, *95*, 116–122.
46. Roxy, M.S.; Sumithranand, V.B.; Renuka, G. Estimation of soil moisture and its effect on soil thermal characteristics at Astronomical Observatory, Thiruvananthapuram, South Kerala. *J. Earth Syst. Sci.* **2014**, *123*, 1793–1807.
47. Phillips, L. Read about Root Development. Available online: http://gibneyce.com/1-read-about-root-development.html (accessed on 15 July 2015).
48. Stokes, A. *The Supporting Roots of Trees and Woody Plants: Form, Function and Physiology*; Springer: Dordrecht, The Netherlands, 2000.
49. Nesmith, D.S.; Duval, J.R. The effect of container size. *HortTechnology* **1998**, *8*, 495–498.
50. Peterson, T.A.; Reinsel, M.D.; Krizek, D.T. Tomato (Lycopersicon esculentum mill, cv. "Better Bush") plant response to root restriction. *J. Exp. Bot.* **1991**, *42*, 1233–1240.

CHAPTER 5

Soil Types Effect on Grape and Wine Composition in Helan Mountain Area of Ningxia

Rui Wang[1,2], Quan Sun[2], Qingrui Chang[1]

[1] *College of Resource and Environment, Northwest A&F University, Yangling, Shaanxi, 712100, P.R. China*

[2] *Grape and Wine Engineering Research Centre of Education Ministry, Ningxia University, Yinchuan, Ningxia, 750021, P.R. China*

ABSTRACT

Different soil types can significantly affect the composition of wine grapes and the final wine product. In this study, the effects of soil types on the composition of *Cabernet Sauvignon*grapes and wine produced in the Helan Mountains were evaluated. Three different representative soil types—aeolian, sierozem and irrigation silting soil were studied. The compositions of grapes and wines were measured, and in addition, the weights of 100-berry samples were determined. The grapes that grown on the aeolian and sierozem soils matured sooner than those grown on the irrigation silting soil. The highest sugar content, total soluble solids content, sugar to acid ratio and anthocyanin content were found in the grapes that grown on the aeolian soil. The wine produced from this soil had improved chroma and tone and higher-quality phenols. The grapes grown on the sierozem soil had the highest total phenol and tannin contents, which affected the wine composition. The grapes grown on the irrigation silting soil had higher acidities, but the remaining indices were lower. In addition, the grapes grown on the aeolian soil resulted in wines with better chroma and aroma. The sierozem soil was beneficial for the formation of wine tannins and phenols and significantly affected the wine composition. The quality of the grapes from the irrigation silting soil was relatively low, resulting in lower-quality wine.

Funding: This work was supported by the National Natural Science Foundation of China (31160417) and National Key Technology R&D Program in the 12th Five year Plan of China (2013BAD09B02).

INTRODUCTION

Soil provides the foundation for grape vine growth, providing the necessary water and nutrition [1]. The physical and chemical characteristics of soils, such as soil type, soil structure, soil depth, fertility, temperature and soil moisture, directly affect vine root growth and nutrient absorption [2]. Although wine grapes can adapt to many soil types, grape and wine composition are significantly affected by soil types, which influences the taste of the final product. Wine grapes grown on highly permeable soils and under the same environmental conditions with large diurnal temperature differences have faster photosynthetic rates, higher sugar concentrations, and improved chroma and palate [3,4]. Nutrients are more easily absorbed in slightly alkaline to neutral pH soils, which improves vegetative vine growth and fruit quality [5]. Soil fertility, particularly soil potassium and calcium concentrations, and soil microorganisms can influence the sugar and tannin contents of wine grapes [6,7]. Soils that are rich in P and Ca promote the accumulation of sugar and the formation of aromatics and anthocyanin in grapes [8,9].

The effects of regional characteristics such as soil type, climate and topography on grape phenolic compounds have been widely investigated [10,11]. All aspects of the soil-plant-atmosphere interaction are known to control grape berry composition [12]. Therefore, different soil types can lead to discrepancies in wine composition, even under the same climate conductions [5,12,13–15].

The composition of wine mainly depends on the composition of the grapes, followed by the wine-making technique [16]. The sugar, acid, tannins, anthocyanin and aromatics contents of the grapes and their interactions play key roles in the composition of wine [17]. Phenolic compounds, which mainly consist of anthocyanin, flavonoids, phenolic acids and stilbenes, play an important role in the composition of red grapes and wines [18]. These compounds contribute to the color and palatability of red wines [19,20]. Consequently, grape variety and fermentative and aging conditions affect the composition and content of phenolic compounds in wines. Regarding single-variety wine, the composition and content of phenolic compounds largely depend on the vineyard [21]. In Andalusia (southern Spain), there are more than 22 main wine-producing sub-regions, and each sub-region produces representative wines due to its local soil characteristics [22]. Several studies have investigated the effects of climate, canopy microclimate, soil types, and water status on the accumulation of phenolic compounds in grapes, revealing that most of these factors can affect the accumulation of phenolic compounds [2,23–25]. However, most of these studies focused on the effects of particular factors on phenol accumulation and were generally not concerned with the impacts of soil types. According to Li et al., understanding the impacts of soil types on the phenolic compounds in grapes and wines is necessary for identifying moderate grapevine cultivars and for developing an effective viticulture management program [21]. It is difficult to study the effects of soil types on grape and wine composition because they involve many factors. Currently, the pH, mineral content, and fertility of vineyard soils can be artificially modified to improve the quality of grapes. However, the soil types of a vineyard cannot be easily changed. Therefore, it is important to study the grape and wine composition from different soil types [8,15,21,26,27].

Different soils types have different soil physical structures and mineral nutrient contents. Because aeolian soil has high sand contents, it has highly permeable, low nutrient and plant available water contents (PAWC). Irrigation silting soil has equivalent percentages of clay, silt and sand, which makes it less permeable, with medium levels of PAWC. The structure of sierozem soil is loose and highly permeable, which results in moderate PAWC and fertility with low organic matter and mineral nutrient contents and high Ca concentrations [3,15]. Generally, the most favorable soil types for grape production in this arid region are gravelly and sandy loam soils, and grapes grown in gravelly soils result in better wine quality [3,28].

This study aimed to analyze the differences in wine grapes and wine composition among three soil types in the Helan Mountains. The objective of this research was to evaluate the relationships between soil types and wine grape composition and to provide practical information for adjusting vineyard growing practices and optimizing wine flavor.

MATERIALS AND METHODS

Ethics statement

The study was approved by the Grape and Wine Engineering Research Centre of Education Ministry and was conducted at their experimental station in Ningxia (NW China).

The study area

The vineyards are located on the Helan Mountain alluvial plain in Ningxia (NW China) (37°43′-39°23′ N, 105°45′-106°47′ E) at an altitude of approximately 1035 m with a cool, semi-arid climate and large diurnal temperature differences. The annual accumulated temperature varies from 3298–3351°C, with abundant sunshine of annual average sunshine 2800~3000 hours and low annual rainfall of 158–200 mm. Early frosts are common on the slopes of the Helan Mountains, and winter temperatures are very low (<-10). Therefore, the grapevines must be covered with soil to prevent injury. Any harvesting delay will reduce the available time for pruning and soil covering. In 2013, three wine-growing areas with distinct soil types were selected: an aeolian soil area (Yuquanying town, 38°23′37.38″N, 106°9′25.14″E), a sierozem area (Qingtongxia city, 38°27′0.28″N, 106°1′31.54″E) and an irrigation silting soil area (Luhua town, 38°34′13.73″N, 106°10′25.06″E). The three regions showing a triangular distribution and direct distance of them are less than 20 km, the altitude, light, rainfall and other climatic conditions are similar. All test points are our school experimental base, in order to understand the difference in quality wines under different soil types, all cultivation and management are the same, unified pruning, fertilization, irrigation and disease prevention. The basic soil physical and chemical parameters from these areas are shown in Table 1. The *Cabernet Sauvignon*grape variety was planted in 2006 in rows orientated north to south and with an individual vine density of 0.5 × 3.0 m.

Table 1. Basic physical and chemical parameters of three soil types.

Treatment	Organic matter g kg^{-1}	Available N mg kg^{-1}	Available P mg kg^{-1}	Available K mg kg^{-1}	Exchangeable Ca mg kg^{-1}	pH	Percentage of each grain size %		
							Clay <0.002 mm	Silt 0.002–0.02 mm	Sand 0.02–2 mm
Aeolian soil	2.74	22.35	18.64	68.44	402.27	8.43	3.45	6.11	90.44
Sierozem	4.46	48.25	32.79	97.27	438.91	8.41	12.65	38.62	48.73
Irrigation silting soil	10.57	76.73	88.92	198.31	415.67	8.40	28.97	40.37	30.66

Composition analysis of grapes and wines

In 2013, sixty plants were selected from each soil type in the vineyards as the study subjects. At the commencement of veraison (~mid-August), samples were collected weekly over a 8-week period until final harvest, which usually occurred during the first week of October. During the final harvest of each replicate, a random sub-sample of 100 grapes was weighed to determine the average grape weight. The samples were frozen and stored for later analyses, including tannin, anthocyanin, and iron-reactive phenolic content analyses. ^{0}Brix was determined using a Brix refractometer (PR32 Atago Co. Ltd., Japan). Reducing sugars were determined using titration with Fehling reagent, and titratable acidity was determined using standardized 0.1 N NaOH (end-point pH 8.2). The pH of the juice was recorded using a Metmorph 702SM automatic neutralizer (Titrino, Herisau, Switzerland) [3,8,15]. Whole grapes from the frozen samples were homogenized and analyzed for their anthocyanin, tannin, and iron-reactive phenolic concentrations, which were determined using spectrophotometry after extraction with ethanol [29,30]. The wine chroma was also measured using spectrophotometry, alcohol and dry extract contents were determined using the wine generic analytical method [21]. In this study, the details of winemaking process refer to Fig. 1.

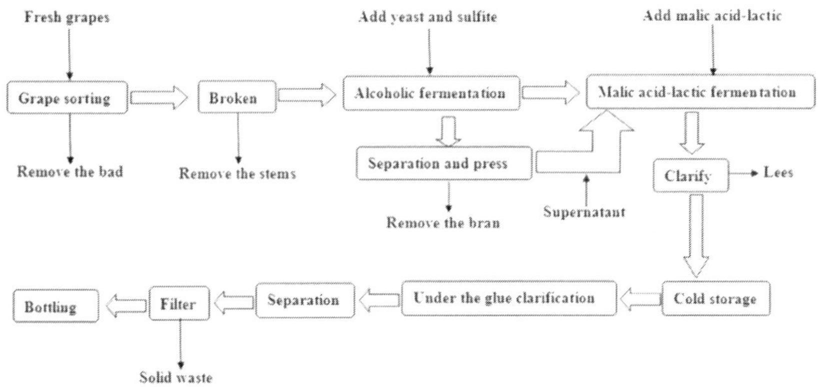

Figure 1. Wine making process schematic.

Statistical methods

Significant differences among soil types were random determined by methods of one-way ANOVA and two-way analysis of variance (ANOVA) followed by the Student-Newman-Keuls test using the SAS 8.1 software (SAS Institute Inc., Cary, NC, USA) followed by the least significant difference (LSD) test for multiple comparisons among groups. Differences with *p*values of less than 5% (p<0.05) were considered statistically significant.

RESULTS

100-berry weight

During the maturation period, the 100-berry weight varied significantly among the different soil types (Fig. 2a). It increased during the early maturation period and stabilized in mid-September. The grapes grown on the aeolian and sierozem soils had loose clusters with relatively lower 100-berry weights relative to the grapes grown on the irrigation silting soils (Table 2). Grapes grown on sierozem soil had the lowest 100-berry weight which was significantly lower than the grapes grown on the aeolian soil. The grapes from the irrigation silting soil vineyards had tight clusters and significantly greater 100-berry weights than the aeolian and sierozem soil sites.

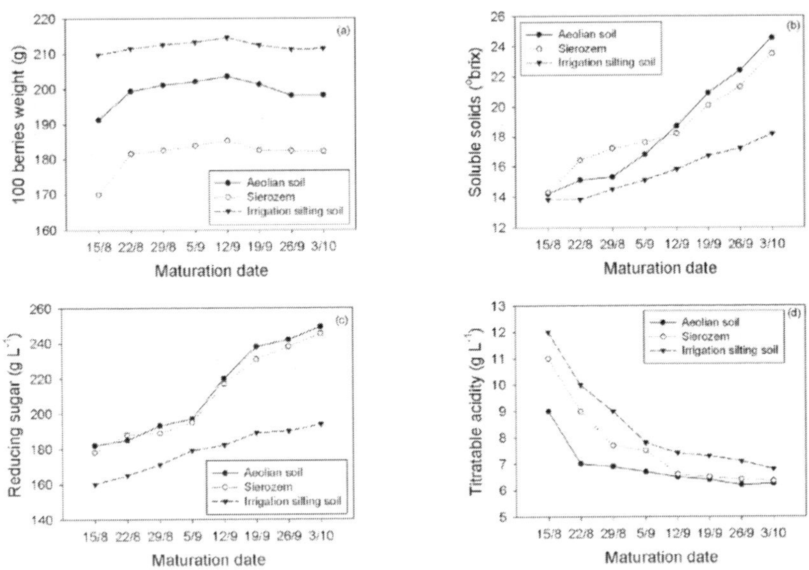

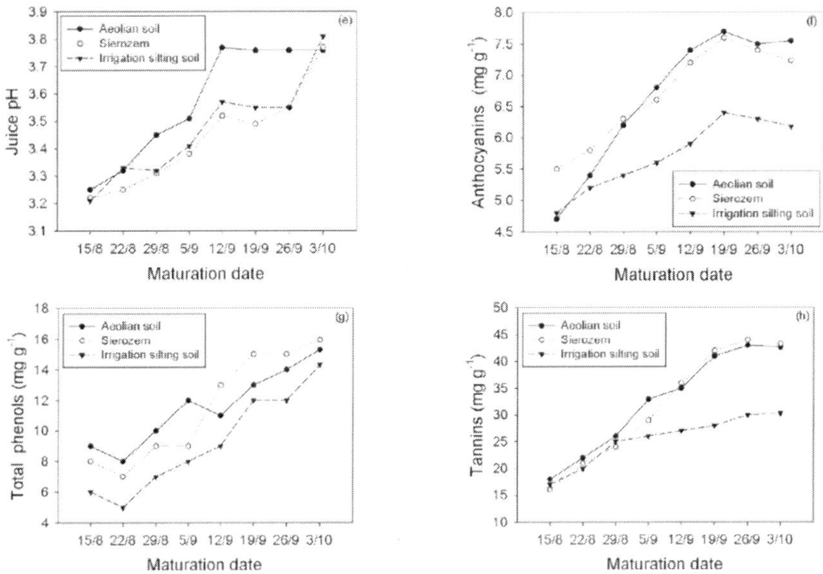

Fig 2. Changes in grape composition.

Table 2. Analytical composition of the grapes at harvest period on three soil types.

Treatment	Weight of 100 berries g	Soluble solids brix	Reducing sugar g L^{-1}	Titratable acidity g L^{-1}	Juice pH	Anthocyanins mg g^{-1}	Total phenols mg g^{-1}	Tannins mg g^{-1}
Aeolian soil	198.23±5.32b	24.55±0.45a	249.35±6.32a	6.25±0.01b	3.76 ±0.02a	7.55±0.01a	15.30±0.02b	42.65 ±0.05a
Sierozem	182.10±2.26c	23.50±0.26a	245.40±4.17a	6.35±0.02b	3.77 ±0.01a	7.23±0.01b	15.95±0.02a	43.25 ±0.04a
Irrigation silting soil	211.45±3.57a	18.18±1.31b	193.75±2.29b	6.80±0.01a	3.81 ±0.01a	6.18±0.02c	14.33±0.03c	30.35 ±0.07b

Note: Multiple comparisons were conducted among different levels of the same factor in one column ($p < 0.05$).

Total soluble solids and reducing sugar content

Generally, the total soluble solids increased during the maturation period of the grapes (Fig. 2b). During early maturation, the total soluble solids concentrations in the grapes grown on the aeolian soils were significantly lower than those grown on the sierozem soils. However, by mid-maturation, the situation was reversed. Over the entire maturation period, the total soluble solids concentrations from the grapes grown on the irrigation silting soil increased steadily (Fig. 2b), which was significantly lower than the total soluble solids concentrations of the grapes grown on the other soils (Table 2). The grapes harvested on October 3rd from the aeolian soil had the highest soluble solids content, which was not significantly different from that of sierozem soils.

During the first 3 weeks of maturation, the grape reducing sugar content increased steadily in all soil types. However, the accumulation rates increased during the subsequent 2 weeks for the grapes that were grown on the aeolian and sierozem soils (Fig. 2c). The reducing sugar accumulation rates of the grapes grown on the irrigation silting soil steadily increased (Fig. 2c). The reducing sugar contents in the final harvest grapes from the aeolian and sierozem soils were not significantly different. However, these reducing sugar contents were significantly greater than those of the grapes grown on the irrigation silting soils.

Titratable acidity and pH

The titratable acidity increased steadily during maturation and decreased rapidly during the first 3 weeks (Fig. 2d). The titratable acidities of the grapes grown on the irrigation silting soil, sierozem, and aeolian soils decreased respectively, during the first week. At final harvest, the titratable acidities of the grapes grown on the aeolian and sierozem soils were similar to each other, and they were significantly lower than the values of the grapes grown on the irrigation silting soils.

During the first 5 weeks of maturation, the pH of the grape juice generally increased by 0.10 pH units each week (Fig. 2e). Although the pH of the grapes grown on the aeolian soils stabilized at 3.76 during the remainder of the harvest period, the grapes grown on the other soils stabilized at a pH that was 0.30 pH units lower for 2 weeks before reaching an equivalent level (Table 2) in the final week.

Sugar to acid ratio

Sugar to acid ratios in the grapes from the aeolian and sierozem soils were generally higher than those in the grapes from the irrigation silting soil. Until September 5th, the values were similar, but they tended to diverge in the final 4 weeks (Table 3). At harvest, the sugar to acid ratio of the wine grapes from the aeolian soil reached 40.16, followed by 38.89 for the grapes from the sierozem soil and 28.38 for the grapes from the irrigation silting soil (Table 3). The sugar to acid ratio is a determinate of grape maturity and indicated that grape maturity occurred in the following order, from earliest to latest: aeolian, sierozem and irrigation silting soil.

Table 3. Changes in the sugar to acid ratio during the maturation period.

Treatment	Date							
	15/8	22/8	29/8	5/9	12/9	19/9	26/9	3/10
Aeolian soil	20.22±1.34a	26.43±0.24a	27.97±2.22a	29.40±0.98a	33.85±0.56a	37.19±0.37a	39.03±1.02a	40.16±0.57a
Sierozem	16.18±0.26b	20.89±0.06b	24.55±0.64b	26.00±1.37b	32.88±0.29a	35.54±0.22a	37.19±1.11a	38.89±0.52a
Irrigation silting soil	13.33±0.57c	16.50±1.33c	19.00±0.92c	22.95±1.11c	24.59±1.18b	25.89±1.15b	26.76±0.97b	28.38±1.01b

Note: Multiple comparisons were conducted among different levels of the same factor in one column ($p < 0.05$).

Analysis of anthocyanin

The anthocyanin contents in the grape during the first 5 weeks of maturation increased before decreasing over the final 2 weeks (Fig. 2f) in all treatments. Throughout the sampling period, the anthocyanin contents of the grapes grown on the aeolian and sierozem soils remained significantly greater than those grown on the irrigation silting soil. At harvest, the anthocyanin content decreased slightly in all treatments. The highest anthocyanin content appeared in the wine grapes from the aeolian soil, which was significantly higher than the contents obtained from those grown on the sierozem soil. The lowest anthocyanin content was observed in the grapes grown on the irrigation silting soil and was significantly lower than the content observed in the grapes grown on the sierozem soil (Fig. 2f).

Total phenol and tannin contents

During the early stages of the mature period, the total phenol contents in the wine grapes from the sierozem soil were lower than those grown on the aeolian soil (Fig. 2g). After September 12th, berry maturation was accelerated. Thus, the total phenol contents of the wine grapes grown on the three soil types increased rapidly. The grapes from the sierozem soil achieved the fastest rate, surpassing the rates of those grown on the aeolian soil. At harvest, the total phenol contents of the grapes from the sierozem soil reached 15.95 mg g^{-1}, which was significantly greater than that of the grapes grown on the aeolian soil. The total phenol contents of the grapes from the irrigation silting soil were the lowest among the treatments and were significantly lower than in the grapes grown on the aeolian soil.

Before August 29th, the tannin contents of the wine grapes from the three soil types were similar. However, after August 29th, the tannin contents of the wine grapes from the sierozem and aeolian soils rapidly increased, reaching a stable value by September 26th. However, the tannin contents of the wine grapes from the irrigation silting soil continued to increase gradually (Fig. 2h). At harvest, the tannin contents of the wine grapes from the sierozem soil reached 43.25 mg g^{-1}, followed by 42.65 mg g^{-1} from the aeolian soil. However, the difference between the sierozem and aeolian soils was not significant. The tannin content of the grapes from the irrigation silting soil was only 30.35 mg g^{-1}, which was significantly lower than that of the grapes from the other two soil types.

Effects of soil types on wine composition

According to Table 4, the alcohol contents of the wines made from the grapes grown on the aeolian and sierozem soils were significantly greater than from the grapes grown on the irrigation silting soil. No significant differences occurred regarding the remaining sugar contents, dry extract concentrations, and pH among the wines produced from the grapes grown on the different soil types. The total acid content of wine is influenced by maturity level and berry density and was greater in the wines made from the grapes grown on the irrigation silting soil relative to the sierozem and aeolian soils. The wine made from grapes grown on the aeolian soil had the highest anthocyanin content, followed by the wines made from the grapes grown on the sierozem and irrigation silting soils. The anthocyanin content was significantly lower in the wines from the irrigation silting soils relative to the other two soil types.

Table 4. Wine composition analysis that produced on three soil types.

Treatment	Alcohol % vol	Residual sugar g L^{-1}	Total acid g L^{-1}	Dry extract g L^{-1}	Chroma	Tone	pH	Anthocyanins mg L^{-1}	Total phenols mg L^{-1}	Tannins mg L^{-1}
Aeolian soil	12.29 ±0.02a	2.53±0.02a	6.17 ±0.11b	23.35 ±0.01a	8.23 ±0.04a	0.73 ±0.01a	3.72 ±0.02a	245.50±1.21a	1912.50 ±45.36b	2034.75 ±25.46b
Sierozem	12.48 ±0.06a	2.52±0.01a	6.24 ±0.06b	23.30 ±0.02a	8.18 ±0.02a	0.75 ±0.01a	3.72 ±0.02a	243.55±1.00a	2088.00 ±38.42a	2172.50 ±30.29a
Irrigation silting soil	11.40 ±0.01b	2.51±0.01a	6.66 ±0.09a	23.25 ±0.01a	6.06 ±0.63b	0.55 ±0.03b	3.71 ±0.01a	189.43±1.35b	1654.50 ±41.37c	1773.75 ±56.87c

Note: Multiple comparisons were conducted among different levels of the same factor in one column ($p<0.05$).

Soil types played an important role in the tannin and total phenol contents of the grapes. The wine made from grapes from the sierozem soil had the highest tannin and total phenol contents, followed by the aeolian soil and the irrigation silting soil, the latter of which was significantly lower than the other two soil types.

Similar to the phenolic compound results, the chromas of the wines made from the grapes grown on the aeolian and sierozem soils were not significantly different. The wines from the aeolian and sierozem soils both had chromas of approximately 8.20, which was significantly greater than that of the wine from the irrigation silting soil (Table 4). The wines that originated from the aeolian and sierozem soils had nearly the same tone, which were significantly higher than that of the wine originating from the silting soil (Table 4).

DISCUSSION

Soil composition as a determining factor for the sensory profile of a wine, the types and texture will influence the root systems and the soil water-holding capacity and mineral composition [31,32]. In addition, it can influence the sensory attributes of wine [13,33]. Light loam soils with coarse and fine sands are beneficial for improving grape quality. The photosynthetic rate and fruit sugar and anthocyanin contents of grapes grown on stony sandy loam soils are high. However, the tannin contents of grapes grown on aeolian soils are greater, stickier soil particles are less beneficial for grape composition [3,22,34]. Here, the moderate air and water permeability of the aeolian and sierozem soils resulted in the accumulation of high anthocyanin concentrations and a small berry size. The wine grapes from the irrigation silting soil had tight clusters and relatively high 100-berry weights, which were significantly greater than those of the grapes from the aeolian and sierozem soils.

Sugar, organic acid, phenolic compounds, anthocyanin and aroma substances are all important for wine composition, with sugar being the most important substance. Grapes with higher sugar contents can produce full-bodied wines [31,35]. Thus, our results indicated that the sugar contents of the berries increased rapidly during the maturation period, whereas the acidity decreased before remaining stable (Fig. 2b). The sugar contents and sugar acid ratios of the grapes from the aeolian and sierozem soils were greater than those from the irrigation silting soil. However, the total acid content in the former soils was lower than that of the irrigation silting soil. Berry acidity mainly decreased due to berry respiration, in which the organic acids were transformed into sugars [8,33,36]. Berry acidity not only influences wine flavor but also the fermentation process. The sugar to acid ratio was greater in the wine grapes from the aeolian and sierozem soils relative to the irrigation silting soil.

Some differences in the various phenolic compounds were observed in the regional wines, which suggested that the accumulation of phenolic compounds in grape berries is strongly affected by soil type [10,37–39]. The anthocyanin content, the total phenol content and the tannin contents in the grapes from the aeolian and sierozem soils were greater than in the grapes from the irrigation silting soil. In addition, the aeolian soil provided favorable conditions for the formation of anthocyanin and aroma substances, and the sierozem soil provided favorable conditions for the formation of tannin and phenolic compounds in the grapes, which corresponded with previous results [3,22].

Soil texture had the most wide-reaching influence on the wines from the various sites. However, consistency among vintages was not observed [40,41]. Zones with soils with high clay textures appeared to produce wines with more earthy and citrus aromas, whereas zones with sandy soils produce wines with floral and melon aromas and flavors [33]. It is difficult to discern the impacts of soil texture on wine composition. In clay zones, heavier berries are produced with slightly delayed fruit maturity, lower berry Brix values, higher acidic contents and lower pH values [33]. These previous findings are consistent with our study: the total acid content of the wine grapes from the irrigation silting soil were significantly higher than those of the wine grapes from the sierozem and aeolian soils, and the pH values of the wine grapes from the aeolian soil were greater than those of the wine grapes from sierozem and irrigation silting soils.

At the same grape-ripening level, soil may affect several wine characteristics. Wines from poorer soils with higher coarse fractions exhibit higher total phenolic contents and color intensity but lower resveratrol concentrations [2,41]. The degree of oxidation and composition can be determined based on the wine chroma and tone. The wine made from the grapes grown on the sierozem soil had the highest tannin and total phenol contents, and the wine made from the grapes grown on the irrigation silting soil had the lowest levels. The difference between the chromas of the wines made of the grapes from the aeolian and sierozem soils was not significant. The chroma values of the wines resulting from the aeolian and sierozem soils were significantly greater than the chroma resulting from the irrigation silting soil (Table 4). Furthermore, the wines made from the grapes from the aeolian and sierozem soils had similar tone values, which were significantly higher than that from the wine made from the irrigation silting soil grapes (Table 4). Thus, the wine made from the grapes grown on the sierozem soil had better chroma. If the anthocyanin and tannin contents are high, the tone of the wine will be rich; otherwise, the tone of the wine will be weak.

Different soil types resulted in significant differences in the total sugar and total acid contents of the grapes and, to some extent, the anthocyanin, tannin, phenolic and aroma contents in the berry skin. The grapes from the aeolian soil have small berry size and loose fruit clusters. The skin color was nicer and the skin fruit ratio was higher for the grapes from the sierozem soil. Therefore, the sugar and anthocyanin contents and the color density were greater, which resulted in better tasting and looking wine with a high tannin content. However, the composition of the wine made from the grapes grown on the irrigation silting soil was comparatively lower.

CONCLUSIONS

The grapes grown on the sierozem soil that matured early had high sugar and anthocyanin contents. Consequently, these grapes produced wine with better chroma. Furthermore, the grapes grown on the sierozem soil that matured during a moderate period had medium tannin contents and low acidity. In general, the sierozem soil was favorable for the formation of tannins and phenolic compounds, which influenced the mouthfeel and composition of the wine. The grapes grown on the irrigation silting soil had high acidity and lowest contents of other investigated substances.

ACKNOWLEDGMENTS

This work was supported by the National Natural Science Foundation of China (31160417, 31460552) and by the National Key Technology R&D Program in the 12th Five-year Plan of China (2013BAD09B02). We thank our colleagues for their comments regarding this paper and the journal's editors and anonymous reviewers for their critical reviews and comments regarding this manuscript.

REFERENCES

1. Li JM, Jiang WG, Yu Y, Liang DM, Liang HZ (2013) Effects of different soil texture on the quality of wine-making grape and grape wine. Liquor-Making Science and Technology 7: 37–41. (in Chinese.)
2. Ruth de Andrés-de Prado R, Yuste RM, Sort X, Andrés LC, Torres M, et al. (2007) Effect of soil type on wines produced from *Vitis vinifera* L. *cv.* Grenache in commercial vineyards. Journal of agricultural and food chemistry 55: 779–786. pmid:17263474
3. Li WC, Sun P, Wang ZP (2012) Effects of different soil condition on physiology and fruit quality of wine grapes. Journal of Fruit Science 29: 837–842. (in Chinese.)
4. Xu SW, Liu SQ, Yang ZX, Du G.Q, Chang JH, et al. (2009) Evaluation of Grape Quality and Relationship Between Grape Quality and Soil Texture. Soils 41: 790–795. (in Chinese.)
5. Wang XQ, Chen XB, Zhan JC, Huang WD (2006) Effects of ecological factors on quality of wine grape and wine. Food Science 27: 791–797. (in Chinese.)
6. Gómez-Míguez MJ, Gómez-Míguez M, Vicario IM, Heredia FJ (2007) Assessment of colour and aroma in white wines vinifications: Effects of grape maturity and soil type. Journal of food engineering 79: 758–764.
7. Kontkanen D, Reynolds AG, Cliff MA, King M (2005) Canadian terroir: sensory characterization of Bordeaux-style red wine varieties in the Niagara Peninsula. Food research international 38: 417–425.
8. Li H (2008) Viticulture. Beijing, China Agriculture Press. (in Chinese.)

9. Jiang B, Zhang ZW (2012) Comparison on phenolic compounds and antioxidant properties of *cabernet sauvignon* and merlot wines from four wine grape-growing regions in China. Molecules 17: 8804–8821. pmid:22832882
10. Gambelli L, Santaroni GP (2004) Polyphenols content in some Italian red wines of different geographical origins. Journal of Food Composition and Analysis 17: 613–618.
11. Rastija V, Srečnik G (2009) Polyphenolic composition of Croatian wines with different geographical origins. Food Chemistry 115: 54–60.
12. Tramontini S, Leeuwen C, Domec JC, Destrac-Irvine A, Basteau C, et al. (2013) Impact of soil texture and water availability on the hydraulic control of plant and grape-berry development.Plant and soil 368: 215–230.
13. Willwerth JJ, Reynolds AG, Lesschaeve I (2010) Terroir factors: Their impact in the vineyard and on the sensory profiles of Riesling wines. Le Progrès agricoleet viticole 127: 159–168. pmid:15552406
14. Mendez-Costabel MP, Wilkinson KL, Bastian SE, McCarthy M, Ford CM, et al. (2013) Seasonal and Regional Variation of Green Aroma Compounds in Commercial Vineyards of Vitis vinifera L. Merlot in California. American Journal of Enology and Viticulture 64: 430–436.
15. Wang LN, Zhang ZW (2011) Research on quality of wine grape of main cultivated in Ningxia in 2009. Northern Horticulture 3: 4–8. (in Chinese.)
16. Cortell JM, Sivertsen HK, Kennedy JA, Heymann H (2008) Influence of vine vigor on Pinot noir fruit composition, wine chemical analysis, and wine sensory attributes. American journal of enology and viticulture 59: 1–10.
17. Douglas D, Cliff MA, Reynolds AG (2001) Canadian terroir: characterization of Riesling wines from the Niagara Peninsula. Food Research International 34: 559–563.
18. Soleas GJ, Grass L, Josephy PD, Goldberg DM, Diamandis EP (2002) A comparison of the anticarcinogenic properties of four red wine polyphenols. Clinical biochemistry 35: 119–124. pmid:11983346
19. Gawel R (1998) Red wine astringency: a review. Australian Journal of Grape and Wine Research 4: 74–95.
20. Lesschaeve I, Noble AC (2005) Polyphenols: factors influencing their sensory properties and their effects on food and beverage preferences. The American journal of clinical nutrition 81: 330S–335S. pmid:15640499
21. Li Z, Pan Q, Jin Z, Mu L, Duan C (2011) Comparison on phenolic compounds in Vitis vinifera cv. *Cabernet Sauvignon* wines from five wine-growing regions in China. Food chemistry 125: 77–83.
22. Fernández-Marín MI, Guerrero RF, García-Parrilla MC, Puertas B, Ramírez P, et al. (2013) Terroir and variety: Two key factors for obtaining stilbene-enriched grapes. Journal of Food Composition and Analysis 31: 191–198.
23. Buttrose MS, Hale CR, Kliewer WM (1971) Effect of temperature on the composition of '*Cabernet Sauvignon*' berries. American Journal of Enology and Viticulture 22: 71–75.
24. Kennedy JA, Matthews MA, Waterhouse AL (2002) Effect of maturity and vine water status on grape skin and wine flavonoids. American Journal of Enology and Viticulture 53: 268–274.
25. Yokotsuka K, Nagao A, Nakazawa K, Sato M (1999) Changes in anthocyanins in berry skins of Merlot and *Cabernet Sauvignon* grapes grown in two soils modified with limestone or oyster shell versus a native soil over two years. American journal of enology and viticulture 50: 1–12.
26. Conradie WJ, Carey VA, Bonnardot V, Saayman D, Van Schoor LH (2002) Effect of different environmental factors on the performance of Sauvignon blanc grapevines in the Stellenbosch/Durbanville districts of South Africa. I. Geology, soil, climate, phenology and grape composition. South African journal for enology and viticulture 23: 78–91.

27. Bramley RGV, Ouzman J, Boss PK (2011) Variation in vine vigour, grape yield and vineyard soils and topography as indicators of variation in the chemical composition of grapes, wine and wine sensory attributes. Australian Journal of Grape and Wine Research 17: 217–229.
28. Zou JF, Peng ZX, Du HJ, Duan CQ, Reeves MJ, et al. (2012) Elemental Patterns of Wines, Grapes, and Vineyard Soils from Chinese Wine-Producing Regions and Their Association. American Journal of Enology and Viticulture 63: 232–240.
29. Harbertson JF, Picciotto EA, Adams DO (2003) Measurement of polymeric pigments in grape berry extract sand wines using a protein precipitation assay combined with bisulfite bleaching. American Journal of Enology and Viticulture 54: 301–306.
30. Orak HH (2007) Total antioxidant activities, phenolics, anthocyanins, polyphenoloxidase activities of selected red grape cultivars and their correlations. Scientia Horticulturae 111: 235–241.
31. Jackson DI, Lombard PB (1993) Environmental and management practices affecting grape composition and wine quality-a review. American Journal of Enology and Viticulture 44: 409–430.
32. Van Leeuwen C (2010) Terroir: The effect of the physical environment on vine growth, grape ripening and wine sensory attributes. In Managing Wine Quality. Vol. 1. Viticulture and Wine Quality. Reynolds A. (ed.) pp: 273–315.
33. Reynolds AG, Taylor G, Savigny C (2013) Defining Niagara Terroir by Chemical and Sensory Analysis of Chardonnay Wines from Various Soil Textures and Vine Sizes. American Journal of Enology and Viticulture 64: 180–194.
34. Jiang B, Xi Z, Luo M, Zhang Z (2013) Comparison on aroma compounds in *Cabernet Sauvignon* and Merlot wines from four wine grape-growing regions in China. Food Research International 51: 482–489.
35. .Seguin G (1986) 'Terroirs' and pedology of wine growing. Experientia 42: 861–873.
36. Roullier GC, Boutegrabet L, Gougeon RD, Schmitt KP (2014) A grape and wine chemodiversity comparison of different appellations in Burgundy: Vintage vs terroir effects. Food chemistry 152: 100–107. pmid:24444912
37. Goldberg DM, Karumanchiri A, Tsang E, Soleas G.J (1998) Catechin and epicatechin concentrations of red wines: regional and cultivar-related differences. American Journal of Enology and Viticulture 49: 23–34.
38. Gonzalez-San Jose ML, Santa MG, Diez C (1990) Anthocyanins as parameters for differentiating wines by grape variety, wine-growing region, and wine-making methods. Journal of food composition and analysis 3: 54–66.
39. McDonald MS, Hughes M, Burns J, Lean ME, Matthews D, et al. (1998) Survey of the free and conjugated myricetin and quercetin content of red wines of different geographical origins. Journal of agricultural and food chemistry 46: 368–375. pmid:10554248
40. Rapisarda P, Tomaino A, Lo Cascio R, Bonina F, De Pasquale A, et al. (1999) Antioxidant effectiveness as influenced by phenolic content of fresh orange juices. Journal of Agricultural and Food Chemistry 47: 4718–4723. pmid:10552879
41. Wang HG, Li JM, Jiang WG, Gao M, Liang HZ (2013) Effects of different soil textures on wine quality of Cabernet Gernischt. Sino-Overseas Grapevine Wine 4: 24–27. (in Chinese.)

CHAPTER 6

Relative Salt Tolerance of Seven Strawberry Cultivars

Youping Sun [1], Genhua Niu [1,*], Russ Wallace [2], Joseph Masabni [3] and Mengmeng Gu [4]

[1] *Texas A&M AgriLife Research, El Paso Research and Extension Center, 1380 A&M Circle, El Paso, TX 79927, USA*

[2] *Texas A&M AgriLife Extension, Lubbock Research and Extension Center, 1102 East FM 1294, Lubbock, TX 79403, USA*

[3] *Texas A&M AgriLife Extension, Overton Research and Extension Center, 1710 FM 3053 N, Overton, TX 75684, USA*

[4] *Department of Horticultural Sciences, Texas A&M AgriLife Extension, College Station, 2134 TAMU, TX 77843, USA*

ABSTRACT

Strawberry (*Fragaria × ananassa*) cultivars ("Albion", "Benicia", "Camarosa", "Camino Real", "Chandler", "Radiance", and "San Andreas") were evaluated for salt tolerance in a greenhouse environment. Plants were irrigated with a nutrient solution with an electrical conductivity (EC) of 1.1 $dS{\cdot}m^{-1}$ (control) or a nutrient solution with the addition of salts (salt solution) with ECs of 2.2, 3.3, or 4.4 $dS{\cdot}m^{-1}$ for four months. Salinity reduced plant growth and fruit yield of strawberry; however, the magnitude of reduction varied with cultivar. For example, at an EC of 4.4 $dS{\cdot}m^{-1}$, "Benicia" and "Chandler" had 39% and 44% less shoot dry weight (DW) respectively, compared with control plants. At ECs of 3.3 and 4.4 $dS{\cdot}m^{-1}$, "Camino Real" had equal shoot DW, which was about 50% lower than that of the control. The fruit yield of "Benicia" and "Camino Real" at 4.4 $dS{\cdot}m^{-1}$ was reduced by 56%, while the other salt treatments did not affect their shoot DW or fruit yield. To distinguish differences among the cultivars with respect to their tolerance to salinity, cluster analysis was performed based on growth parameters and visual quality. The results indicated that "Albion", "Camarosa", and "San Andreas" were more salt tolerant, while “Camino Real”, “Benicia”, “Chandler”, and “Radiance” were less salt tolerant.

Keywords: chloride toxicity; *Fragaria* × *ananassa*; gas exchange; salinity

1. INTRODUCTION

Strawberry (*Fragaria* × *ananassa*) is an economically important crop that covered an estimated 58,560 acres in the U.S. in 2013 [1]. The commercial production in California and Florida accounted for approximately 82% of total strawberry acreage [1]. Due to its economic importance and the demand for locally-grown berries, growers in other states are starting to produce more strawberries. In Texas, strawberry is still a minor crop with less than 150 acres representing 0.02 percent of national production [2,3]. With the large size of Texas, there is great potential for strawberry production to expand into traditionally non-producing regions. One of the key production constraints for strawberry is high salinity levels, which are often found in soils in arid and semi-arid regions and irrigation water. In recent years, strawberry growers in California have also faced decreased irrigation water quality and increased soil salinity, possibly due to the deterioration of coastal groundwater and low rainfall. Selecting salt-tolerant strawberry cultivars may be an effective approach for preventing yield and quality reductions.

Strawberry is categorized as one of the most salt-sensitive crops with varying degrees of tolerance among cultivars. Salinity causes leaf edge burn, necrosis, nutrient imbalance or specific ion toxicity, reduction in fruit quality and yield, and potential plant death if salinity stress persists or increases. In a two-year field study by Saied *et al.* [4], fruit yield was reduced up to 27% and 64% in the strawberry cultivars “Korona” and “Elsanta”, respectively, when the plants were exposed to NaCl salinity. Fruit quality, characterized as taste, aroma, and texture by a consumer-type panel, decreased by more than 24% in “Elsanta”, but differences in “Korona” were not significant. The reduction in shoot growth between these two strawberry cultivars was also different, up to 90% in “Elsanta” and 40% in “Korona”. Orsini *et al.* [5] compared the strawberry varieties “Elsanta” and “Elsinore” grown in the presence of 0, 10, 20, and 40 mM NaCl (electrical conductivities (ECs) of 0.45 to 3.9 $dS \cdot m^{-1}$). The shoot dry weight and leaf area of both cultivars decreased linearly as the EC of the irrigation solution increased. However, the reduction in growth was smaller in “Elsanta” (19%) than in “Elsinore” (90%). Many other studies have shown differences in salt tolerance among strawberry cultivars: “Korona” was more tolerant than “Elsanta” [6], “Toro” was more tolerant than “Douglas” [7], and “Yalova-104”, “Yalova-15”, “Yalova-416”, and “Arnavutkoy” were more tolerant than "Douglas", "Dorit", and "Aliso" [8]. Turhan and Eris [9] found that "Camarosa" was more tolerant than "Chandler" to NaCl at 8.5, 17.0, or 34.0 mM (equivalent to ECs of 0.8, 1.6, or 3.1 $dS \cdot m^{-1}$). In a field study, Ferreira *et al.* [10] reported that 'Albion' was relatively tolerant among five cultivars based on growth, yield, and the calculated salinity level (EC50) that would reduce fruit yield per hectare by 50%.

The above studies have indicated variation in salt tolerance among strawberry cultivars and the importance of cultivar selection when soil or water salinity is too high. The objective of this study was to determine the relative salt tolerance of seven commercial strawberry cultivars by irrigating plants with a nutrient solution or saline solution at selected levels of salinity. Gas exchange and leaf sodium (Na), potassium (K), calcium (Ca), and chloride (Cl) accumulation were also determined.

2. EXPERIMENTAL SECTION

2.1. Plant Materials

On 29 October 2013, plugs of seven strawberry cultivars ("Albion", "Benicia", "Camarosa", "Camino Real", "Chandler", "Radiance", and "San Andreas") were obtained from the Goodson Farm and Nursery (Damascus, AR, USA). Plants (~5 leaves, ~15 cm wide) were transplanted into 3.8-L (15.8-cm diameter) black plastic containers filled with LM-40 high porosity growing mix (Canadian sphagnum peat moss 60%, horticultural perlite 40%, limestone, dolomite, wetting agent, micro & macronutrients; Lambert Peat Moss Inc., QC, Canada). All dead leaves, runners, flowers and/or fruits were trimmed off at transplanting. Plants were grown in a greenhouse with temperature maintained at 26.5 ± 6.1 °C (mean ± standard deviation) during the day and 19.5 ± 6.3 °C at night. The daily light integral (photosynthetically active radiation) was 1.1 ± 3.7 $mol \cdot m^{-2} \cdot d^{-1}$ and relative humidity was 30.3 ± 16.8%. Plants were watered with a nutrient solution until salt treatments were initiated. The nutrient solution with an EC of 1.1 ± 0.1 $dS \cdot m^{-1}$ was prepared by adding 1 $g \cdot L^{-1}$ of 15 N-2.2 P-12.5 K (Scotts Peters 15-5-15; Marysville, OH, USA) to reverse osmosis (RO) water.

2.2. Treatments

On 25 November 2013, treatments were initiated by irrigating plants with 1 L of nutrient solution (control) or salt solutions to maintain 10% to 20% leaching fraction. Plants were then irrigated once a week with the nutrient or salt solutions for a total of six times. On 10 January 2014, plants were flushed with 1 L RO water due to high salt accumulation in the root zone (Figure 1). From 17 January to 31 January, plants were watered with 1 L of nutrient solution. Plants were then irrigated once a week with the nutrient or salt solutions for another three times (nine times in total). Thereafter, the nutrient solution was applied until the end of the experiment on 11 March 2014. The nutrient solution at an EC of 1.1 ± 0.1 $dS \cdot m^{-1}$ (control, EC1.1) was prepared as described above. The salt solution was prepared by adding sodium chloride (NaCl) and calcium chloride ($CaCl_2$) at a 2:1 ratio (molar ratio) to the nutrient solution. Although the composition of salts in salt-affected soil or poor irrigation water varies with location and source of water, NaCl is always dominant among other salts such as $CaCl_2$. All solutions were prepared in 100-L tanks with confirmed ECs of 2.3 ± 0.3 $dS \cdot m^{-1}$ (EC2.2), 3.3 ± 0.5 $dS \cdot m^{-1}$(EC3.3), and 4.5 ± 0.5 $dS \cdot m^{-1}$ (EC4.4). The leachate of the substrate EC was determined periodically using the pour-through method according to Wright [11].

2.3. Growth, Yield, and Visual Quality

Two perpendicular widths (cm), leaf count, and number of crowns of strawberry plants were recorded before initiating treatment (25 November 2013) and at the end of the experiment (11 March 2014). Mature berries were harvested starting on 7 January 2014, and their number and fresh weight (g) were recorded. The number of mature berries and fresh weight from each harvest were combined for total yield. To determine if salt treatments influenced the sugar content of strawberry fruit, Brix values of four berries were collected on 7 Mar. for plants in the control and EC4.4 treatments only, using a RF15 Brix refractometer (Extech Instruments Corporation, Nashua, NH, USA). Brix value is a measure of total soluble solids (TSS) in juice, including sugars such as sucrose, fructose, and glucose [12].

At the end of the experiment, foliar salt damage (leaf edge burn, necrosis, and discoloration) was rated using a visual score of each plant from 0 to 5, where 0 = dead; 1 = over 90% foliar damage; 2 = moderate (50% to 90%) foliar damage; 3 = slight (less than 50%) foliar damage; 4 = good quality with minimal foliar damage; 5 = excellent without any foliar damage. All immature fruits were harvested, and their number and fresh weight (g) were recorded.

Upon termination, shoots were severed at the substrate surface. Leaf area (cm^2) was determined using an LI-3100C area meter (LI-COR® Biosciences, Lincoln, NE, USA). Shoot dry weight (DW) was determined after shoots were oven-dried at 65 °C.

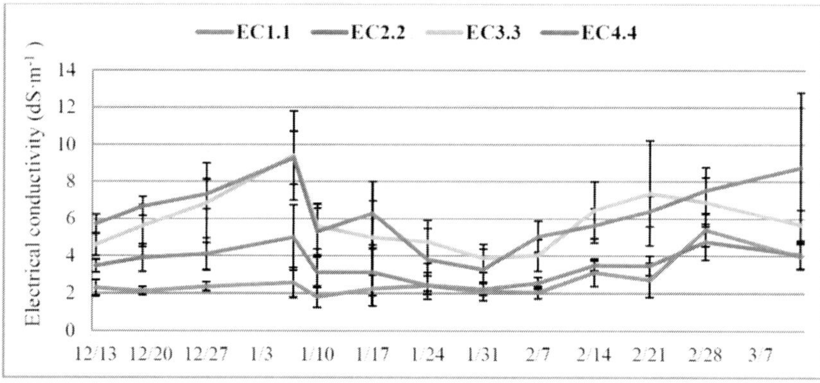

Figure 1. Leachate electrical conductivity (EC) from 13 Dec. 2013 to 7 Mar. 2014. EC1.1, EC2.2, EC3.3, and EC4.4 represent treatment solutions with ECs of 1.1 $dS·m^{-1}$, 2.2 $dS·m^{-1}$, 3.3 $dS·m^{-1}$, and 4.4 $dS·m^{-1}$, respectively. Error bars represent the standard error (SE) of five leachate samples. During the experimental period, treatment solutions were applied nine times. Treatments were initiated on 25 November 2013, and plants were then irrigated weekly with the nutrient or salt solutions for a total of six times. On 10 January, plants were flushed with reverse osmosis (RO) water due to high salt accumulation in the root zone. From 17 Jan. to 31 Jan., all plants were watered with the nutrient solution. From 7 February to 21 February, all plants were irrigated weekly with the nutrient or salt solutions for another three times. On 28 February and 7 March, plants were irrigated with the nutrient solution.

2.4. Gas Exchange and Leaf Greenness

Leaf transpiration rate (E), stomatal conductance (g_s), and net photosynthesis (P_n) of four plants per treatment by cultivar were measured after the 6th salt treatment and one week before the harvest started (eight weeks after the 1st measurement) using a CIRAS-2 portable photosynthesis system (PP Systems, Amesbury, MA, USA) with an automatic universal PLC6 broad leaf cuvette. Fully expanded, healthy leaves were chosen for measurements. The environmental conditions in the cuvette were controlled at a leaf temperature = 25 °C, photosynthetic photon flux (*PPF*) = 1000 $\mu mol·m^{-2}·s^{-1}$, and CO_2concentration = 375 $\mu mol·mol^{-1}$. Data were recorded when the environmental conditions and gas exchange parameters in the cuvette became stable. These measurements were taken on sunny days between 1000 HR and 1400 HR, and the plants were well-watered to avoid water stress.

Leaf greenness (or relative chlorophyll content) of all plants was measured using a hand-held SPAD chlorophyll meter (Minolta Camera Co., Osaka, Japan) one week before the harvest started. Three healthy, fully-expanded leaves were chosen from each plant.

2.5. Mineral Analysis

Four plants per treatment by cultivar were selected for mineral analyses. All dried leaves were ground to pass a 40-mesh screen with a stainless Wiley mill (Thomas Scientific, Swedesboro, NJ, USA). About 300 mg of plant leaf tissues were digested using the Environmental Protection Agency method 3051 with 1 mL nitric acid and 4 mL H_2O_2 using a microwave acceleration reaction system (CEM Corporation; Matthews, NC, USA) for determining alkaline earth metals (Na, K, Ca). The plant tissues were extracted with 2% acetic acid (EM Science, Gibbstown, HJ) for determining anions (Cl) using methods described in Gavlak *et al.* [13]. Na, K, and Ca in the digested samples were analyzed by inductively coupled plasma-optical emission spectrometry (Perkin-Elmer Optima 4300 DV, Shelton, CT, USA). Chloride was determined using a M926 chloride analyzer (Cole Parmer Instrument Company, Vernon Hills, IL, USA).

2.6. Experimental Design and Statistical Analysis

The experiment followed a split-plot design with salinity as the main plot and cultivar as subplot. Four or five plants were used as replications per salinity level by cultivar. All data were analyzed by two-way ANOVA using PROC GLM. Means separation among cultivars and treatments was conducted using Tukey's honest significant difference (HSD) test. Relative shoot dry weight for each plant in the salt treatments was calculated as:

$$\text{Relative shoot dry weight (\%)} = \frac{\text{Shoot dry weight in salt treatment}}{\text{Averaged shoot dry weight in control}} \times 100\%$$

Similarly, relative percent of perpendicular width, leaf count, leaf area, number of crowns, and cumulative number and fresh weight of mature and immature fruits was calculated. These relative values were used as salt tolerance indexes for hierarchical cluster analysis [14]. A dendrogram of the seven strawberry cultivars was obtained based on the Ward linkage method and squared Euclidian distance of the means of the salt tolerance indexes for nine multivariate parameters including all relative growth data. All statistical analyses were performed using JMP (Version 12, SAS Institute Inc., Cary, NC, USA).

3. RESULTS AND DISCUSSION

3.1. Leachate EC

The leachate EC data are shown in Figure 1. From 13 December 2013 to 7 January 2014, leachate EC increased from 2.3 to 2.6 $dS \cdot m^{-1}$ for plants irrigated with the nutrient solution (control, EC1.1) (Figure 1). When plants were watered with the saline solution at 2.2, 3.3, or 4.4 $dS \cdot m^{-1}$, the leachate EC increased from 3.8 to 5.0 $dS \cdot m^{-1}$, 4.6 to 9.4 $dS \cdot m^{-1}$, or 5.7 to 9.3 $dS \cdot m^{-1}$, respectively. From 7 February to 28 February, the leachate EC increased from 2.0 to 5.4 $dS \cdot m^{-1}$ for EC1.1, 2.6 to 4.8 $dS \cdot m^{-1}$ for EC2.2, 4.0 to 6.9 $dS \cdot m^{-1}$ for EC3.3, and 5.0 to 7.5 $dS \cdot m^{-1}$ for EC4.4, respectively. The final leachate ECs recorded on 12 March were 4.0, 4.1, 5.7, and 8.8 $dS \cdot m^{-1}$ for EC1.1, EC2.2, EC3.3, and EC4.4, respectively. Salinity in the root zone of a container varies with type of substrate, irrigation frequency (water use of the plants and evaporation), leaching fraction, and salinity of the irrigation water [15]. For most substrates containing materials such as peat, salt accumulation is inevitable. To prevent excessive salt accumulation, the salinity of leachate should be monitored periodically and leaching should be performed as needed.

3.2. Plant Growth and Visual Quality

Salt treatment reduced visual quality with variations among cultivars (Table 1). The cultivars "Albion" and "San Andreas" had similar visual quality scores (0 = dead; 5 = excellent) across treatments, and their visual scores were higher than "Camino Real" and "Radiance". For "Benicia", no differences in visual score were found among treatments, although all scores were below 4.0 even in the control. The lowest visual score (1.8) was observed for "Radiance" at EC4.4. "Benicia" at EC4.4, "Camino Real" at EC3.3 and EC4.4, and "Chandler" at EC4.4 also had visual quality scores lower than 3.0. Foliar salt damages including leaf edge burn, necrosis, and/or discoloration have been observed in other strawberry cultivars [7,8].

Elevated salinity reduces the growth of strawberry plants [4,6,8,9,16]. In our experiment, plant width, leaf count, leaf area, and shoot dry weight were significantly different between the salt treatments and among cultivars, but no interactions occurred (Table 1). Salt treatment did not affect the perpendicular width in most cultivars with the exception of "Camino Real" and "Chandler". Compared with the control plants, EC4.4 reduced the perpendicular width of "Camino Real" and "Chandler" by 52% and 67%, respectively. All strawberry cultivars except "Camarosa" and "Camino Real" had similar numbers of leaves across treatments. "Camarosa" plants at EC3.3 and EC4.4 had 52% and 42% fewer leaves, respectively, than the control, while "Camino Real" plants had 65% and 64% fewer leaves, respectively. Salt treatment did not significantly reduce leaf area in individual cultivars except "Camino Real", although leaf area was reduced at higher salinity levels. "Albion", "Camarosa", "Radiance", and "San Andreas" produced similar shoot biomass in all treatments, but salt treatment at EC4.4 reduced shoot biomass of "Camino Real" and "Chandler" by 49% and 44%, respectively. In addition, when averaged across all cultivars, EC3.3 and EC4.4 decreased shoot biomass by 25% and 38%, respectively. Overall, salt treatment had no effect on the number of crowns ($p = 0.27$); however, "Chandler" and "San Andreas" developed more crowns than other cultivars ($p < 0.0001$) (data not shown).

Table 1. Visual score, plant perpendicular width, number (No.) of leaves, leaf area, and shoot dry weight (DW) of strawberry cultivars irrigated with the nutrient solution (electrical conductivity (EC) =1.1 $dS{\cdot}m^{-1}$) or salt solutions (EC = 2.2, 3.3, or 4.4 $dS{\cdot}m^{-1}$) in the greenhouse.

Variety	Treatment ($dS{\cdot}m^{-1}$)	Visual Score	Width (cm)	No. of Leaves	Leaf Area (cm^2)	Shoot DW (g)
Albion	1.1	4.1 a [z]	10.8 a	9 a	663 a	11.9 a
	2.2	3.9 a	11.2 a	6 a	565 a	9.7 a
	3.3	4.0 a	11.5 a	5 a	526 a	8.6 a
	4.4	3.5 a	8.3 a	6 a	444 a	6.9 a
	Mean	3.9 A [y]	10.5 AB	7 C	555 BC	9.4 BC
Benicia	1.1	3.6 a	8.6 a	8 a	420 a	6.8 ab
	2.2	3.3 a	8.2 a	9 a	404 a	7.5 a
	3.3	3.5 a	9.0 a	11 a	357 a	7.6 a
	4.4	2.9 a	3.0 a	4 a	230 a	4.1 b
	Mean	3.3 ABC	7.0 B	8 BC	349 D	6.4 CD
Camarosa	1.1	4.2 a	12.4 a	17 a	756 a	11.9 a
	2.2	4.2 a	12.8 a	12 ab	704 a	10.2 a
	3.3	3.6 ab	11.7 a	8 b	646 a	9.3 a
	4.4	3.0 b	10.0 a	10 b	608 a	9.1 a
	Mean	3.8 AB	11.8 A	12 AB	684 AB	10.2 AB
Camino Real	1.1	3.7 a	12.9 a	18 a	779 a	14.8 a
	2.2	3.8 a	11.1 ab	11 ab	673 ab	11.4 ab
	3.3	2.6 b	12.2a	6 b	408 b	7.5 b
	4.4	2.7 b	6.2 b	6 b	443 b	7.2 b
	Mean	3.2 BC	10.5 AB	11 ABC	580 BC	10.3 AB
Chandler	1.1	4.0 a	11.4 a	16 a	854 a	15.2 a
	2.2	4.1 a	11.9 a	15 a	861 a	16.0 a
	3.3	3.7 a	9.1 ab	17 a	759 a	11.3 ab
	4.4	2.6 b	3.8 b	10 a	558 a	8.5 b
	Mean	3.6 ABC	9.2 AB	15 A	831 A	12.8 A

Radiance	1.1	3.3 ab	7.5 a	6 a	433 a	6.1 a
	2.2	3.9 a	11.2 a	10 a	507 a	6.9 a
	3.3	3.1 ab	8.0 a	7 a	381 a	5.1 a
	4.4	1.8 b	7.1 a	5 a	369 a	5.3 a
	Mean	3.0 C	8.6 AB	7 BC	428 CD	5.9 D
San Andreas	1.1	3.9 a	11.6 a	16 a	814 a	12.3 a
	2.2	4.1 a	12.3 a	17 a	774 a	13.3 a
	3.3	3.9 a	12.2 a	13 a	645 a	10.3 a
	4.4	3.8 a	7.9 a	11 a	558 a	8.5 a
	Mean	3.9 A	11.0 A	14 A	691 AB	11.0 AB
Salt treatment	1.1	3.8 a [x]	10.9 a	13 a	706 a	11.5 a
	2.2	3.9 a	11.2 a	11 ab	653 ab	10.6 ab
	3.3	3.5 a	10.5 a	10 ab	538 bc	8.6 bc
	4.4	2.9 b	6.5 b	8 b	457 c	7.1 c
Cultivar		*** [w]	***	***	***	***
Treatment		***	***	***	***	***
Cultivar × Treatment		*	NS	NS	NS	NS

[z] For each cultivar, means with the same lowercase letters within a column are not significantly different between treatments according to Tukey's HSD test at $p < 0.05$; [y] means with same uppercase letters within a column are not significantly different according to Tukey's HSD test at $p < 0.05$; [x] means with the same lowercase letters within a column are not significantly different according to Tukey's HSD test at $p < 0.05$; [w] NS, *, **, ***: not significant, or significant at $p < 0.05$, 0.01, and 0.001, respectively.

Strawberry yield (number and fresh weight of mature fruits) and potential yield (number and fresh weight of immature fruit) were significantly different among salt treatments and among cultivars, but no interactions were evident (Table 2). All strawberry cultivars except "Benicia" and "Camino Real" had similar cumulative numbers of mature berries across treatment levels. The number of mature berries of "Benicia" and "Camino Real" at EC4.4 decreased by 56% and 38%, respectively, compared with the control. Salt treatment did not significantly reduce the yield of "Albion", "Camarosa", "Chandler", "Radiance", and "San Andreas", although there was a lower total mature fruit fresh weight at EC4.4. The fresh weight of mature fruit of "Benicia" and "Camino Real" at EC4.4 was reduced by 57% and 56%, respectively, compared with the control. Salt treatment had no effect on immature fruit number. The fresh weight of immature berries of "San Andreas" at EC4.4 was reduced by 58%. The other cultivars had similar fresh weights among treatments. These results further demonstrate that salinity negatively impacted the yield of immature fruit strawberries, but the level of yield reduction varied with cultivar. Similar results on strawberry yield reduction with increasing salinity levels were reported by others [4,6,8].

Fruit Brix values for plants at EC4.4 were similar to those in the control treatment ($p = 0.11$), but significantly different among cultivars ($p = 0.02$) (data not shown). In addition, there was no interaction effect between treatment and cultivar ($p = 0.95$). Our results indicated that salt treatment did not impact the sweetness of harvested berries. Ferreira *et al.* [10] also found similar results. However, Saied *et al.* [4] reported that Brix values for the strawberry cultivars "Korona" and "Elsanta" decreased significantly with salinity. The differences may be a result of the different salt concentrations used in these studies.

Hierarchical cluster analysis of the seven strawberry cultivars was conducted using multivariate parameters including all relative growth data [14]. The dendrogram showed three distinguishable clusters (Figure 2). "Albion", "Camarosa", and "San Andreas" were clustered together and were determined to be the most salt tolerant group. "Camino Real" was separated from all others and considered to have moderate salt

Table 2. Cumulative number of mature berries, cumulative fresh weight (FW) of mature fruit, number of immature fruit, and FW of immature fruit of strawberry cultivars irrigated with nutrient solution (electrical conductivity (EC) = 1.1 dS·m^{-1}) or salt solutions (EC = 2.2, 3.3, or 4.4 dS·m^{-1}) in the greenhouse.

Variety	Treatment (dS·m^{-1})	No. of Mature Fruits	FW of Mature Fruits (g)	No. of Immature Fruits	FW of Immature Fruits (g)
Albion	1.1	7 a ^{z}	103.5 a	5 a	14.5 ab
	2.2	7 a	105.2 a	6 a	23.9 a
	3.3	6 a	89.4 a	5 a	12.1 ab
	4.4	7 a	81.8 a	4 a	9.8 b
	Mean	7 CD ^{y}	95.2 A	5.1 C	14.9 ABC
Benicia	1.1	11 a	130.4 a	5 a	10.3 a
	2.2	7ab	82.7 ab	4 a	9.4 a
	3.3	8 ab	94.4 ab	5 a	14.8 a
	4.4	5 b	56.4 b	3 a	8.3 a
	Mean	8 CD	88.6 AB	4.1 C	10.5 C
Camarosa	1.1	11 a	122.8 a	7 a	10.6 a
	2.2	12 a	126.3 a	6 a	9.0 a
	3.3	12 a	103.1 a	5 a	7.0 a
	4.4	12 a	100.9 a	5 a	7.3 a
	Mean	12 A	114.5 A	5.7 C	8.6 C
Camino Real	1.1	8 a	138.3 a	6 a	17.6 a
	2.2	6 ab	90.7 ab	8 a	21.0 a
	3.3	7 ab	112.2 ab	4 a	10.5 a
	4.4	5 b	60.6 b	7 a	18.2 a
	Mean	7 CD	100.3 A	6.2 BC	16.7 ABC

Chandler	1.1	11 a	135.0 a	10 a	27.5 a
	2.2	9 a	127.2 a	12 a	28.7 a
	3.3	9 a	112.0 a	9 a	21.1 a
	4.4	7 a	69.5 a	9 a	16.0 a
	Mean	9 BC	112.3 A	9.8 A	23.1 A
Radiance	1.1	13 a	128.4 a	7 a	12.7 a
	2.2	11 a	124.7 a	7 a	17.6 a
	3.3	12 a	119.0 a	6 a	11.3 a
	4.4	9 a	90.4 a	6 a	10.5 a
	Mean	11 AB	117.8 A	6.3 BC	13.3 BC
San Andreas	1.1	6 a	66.8 a	10 a	31.7 a
	2.2	6 a	70.1 a	9 a	18.7 ab
	3.3	5 a	54.4 a	10 a	17.8 ab
	4.4	5 a	41.0 a	6 a	13.3 b
	Mean	5 D	57.6 B	8.6 AB	19.8 AB
Salt treatment	1.1	10 a [x]	118.8 a	7 a	17.8 a
	2.2	8 ab	103.5 a	7 a	17.3 a
	3.3	8 ab	97.3 a	6 a	13.8 a
	4.4	7 b	69.0 b	6 a	12.1 a
Cultivar		*** [w]	***	*	**
Treatment		***	***	***	***
Cultivar × Treatment		NS	NS	NS	NS

[z] For each cultivar, means with the same lowercase letters within a column are not significantly different among treatments according to Tukey's HSD multiple comparison at $p < 0.05$; [y] means with the same uppercase letters within a column are not significantly different according to Tukey's HSD test at $p < 0.05$; [x] means with the same lowercase letters within a column are not significantly different among treatments according to Tukey's HSD test at $p < 0.05$; [w] NS, *, **, ***: Differences between means for main effects and their interaction were not significant or significant at $p < 0.05$, 0.01, and 0.001, respectively.

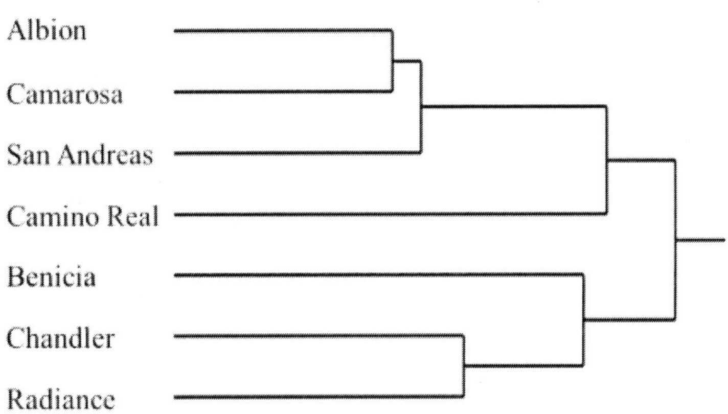

Figure 2. Hierarchical cluster analysis of seven strawberry cultivars using multivariate parameters including all relative growth data. The dendrogram is based on Ward linkage using the squared Euclidian distance of the means of multivariate parameters.

tolerance. "Benicia", "Chandler", and "Radiance" were classified as salt sensitive. These results agree with previous reports on the salt tolerance of strawberry cultivars. Gulen *et al.* [17] found that "Camarosa" was more tolerant than "Chandler" to NaCl treatment at 8.5, 17.0, or 34.0 mM. Ferreira *et al.*[10] reported that in a field study, "Albion" was relatively tolerant among five cultivars based on growth, yield, and the calculated EC50 (the salinity level that would reduce fruit yield per hectare by 50%).

3.3. Gas Exchange and Leaf Greenness

Gas exchange recorded after the strawberry cultivars were watered six times with the salt solutions showed that salt treatment had no effect on P_n (p = 0.47; data not shown). However, P_n differed among cultivars (p = 0.003). "Albion" had the highest P_n of 14.7 $\mu mol \cdot m^{-2} \cdot s^{-1}$, while "San Andreas" had the lowest P_n of 10.6 $\mu mol \cdot m^{-2} \cdot s^{-1}$. Gas exchange was measured again one week prior to the final harvest (eight weeks after the first measurement). Leaf, stomatal g_s, and P_n were different among treatments (Table 3). Salt treatment generally decreased the overall transpiration rate or stomatal conductance across all cultivars at EC4.4 or EC3.3, respectively, although individual cultivars did not show salt effects on P_n. As shown in Table 3, all three parameters, P_n, g_s, and E were reduced at EC3.3 and EC4.4 across all cultivars. Salt treatment did not affect the overall cultivar chlorophyll leaf content, but the SPAD meter readings varied among cultivars (Table 3). When averaged across all salt treatments, "San Andreas" and "Camino Real" had higher leaf chlorophyll content with SPAD readings of 49.2 and 48.3, respectively, while "Camarosa" and "Chandler" were lower at 45.6 and 41.9, respectively. By comparison, Turhan and Eris [18] reported that total chlorophyll content was unaffected in "Camarosa" and "Tioga" strawberries at the end of 10 weeks of NaCl treatments at 500, 1000, and 2000 $mg \cdot L^{-1}$.

3.4. Ion Analysis

Leaf Na and Ca concentrations were significantly different among salt treatments and cultivars, but there were no interactions (Table 4). Compared with the plants in the control treatment, leaf Na concentrations of "Camarosa", "Chandler", and "San Andreas" at EC4.4 increased by 323%, 423%, and 90%, respectively, while those of "Camino Real" and "Chandler" at EC3.3 increased by 199% and 156%, respectively. The other salt treatments did not affect the leaf Na concentration of "Camarosa", "Camino Real", "Chandler", and "San Andreas". The highest Na concentration (2.21 $mg \cdot g^{-1}$DW) was measured

Table 3. Leaf transpiration rate (E, mmol·m^{-2}·s^{-1}), stomatal conductance (g_s, mmol·m^{-2}·s^{-1}), net photosynthesis (P_n, μmol·m^{-2}·s^{-1}), and relative chlorophyll content (SPAD) of strawberry cultivars irrigated with a nutrient solution (electrical conductivity (EC) =1.1 dS·m^{-1}) and a salt solution (EC = 2.2, 3.3, or 4.4 dS·m^{-1}) in the greenhouse.

Variety	Treatment	E (mmol·m^{-2}·s^{-1})	g_s (mmol·m^{-2}·s^{-1})	P_n (μmol·m^{-2}·s^{-1})	SPAD
Albion	1.1	7.9 a [z]	422.6 a	16.2 a	46.3 a
	2.2	7.1 a	396.3 a	15.6 a	46.6 a
	3.3	6.3 a	306.4 a	13.2 a	45.5 a
	4.4	6.1 a	323.8 a	13.9 a	47.2 a
	Mean	6.9 A [y]	360.5 AB	14.7 A	46.4 BC
Benicia	1.1	7.8 a	478.0 a	16.0 a	43.1 a
	2.2	7.1 a	391.6 a	15.2 a	47.4 a
	3.3	7.3 a	424.8 a	15.6 a	46.6 a
	4.4	6.3 a	299.4 a	13.4 a	47.0 a
	Mean	7.1 A	392.6 A	14.9 A	46.2 BC
Camarosa	1.1	7.3 a	369.0 a	14.7 ab	45.1 a
	2.2	7.6 a	366.6 a	15.8 a	44.2 a
	3.3	6.4 a	269.8 a	11.4 b	46.2 a
	4.4	6.2 a	314.3 a	12.2 ab	47.2 a
	Mean	6.9 A	334.1 AB	13.7 A	45.6 C
Camino Real	1.1	7.9 a	369.6 a	14.0 ab	49.4 a
	2.2	7.3 a	420.5 a	17.0 a	49.4 a
	3.3	6.6 a	304.8 a	12.4 b	47.1 a
	4.4	7.4 a	393.8 a	15.8 ab	47.3 a
	Mean	7.3 A	373.2 AB	14.8 A	48.3 AB
Chandler	1.1	7.1 a	394.0 a	15.8 ab	43.2 a
	2.2	7.1 a	407.8 a	16.5 a	44.1 a
	3.3	6.1 a	306.8 a	11.3 b	40.5 a
	4.4	6.2 a	359.0 a	15.4 ab	40.0 a
	Mean	6.6 A	365.1 AB	14.6 A	41.9 D
Radiance	1.1	7.5 ab	331.8 ab	14.1 ab	42.6 a
	2.2	8.1 a	375.0 a	14.9 a	48.9 a
	3.3	5.7 b	272.8 ab	11.9 ab	43.6 a
	4.4	5.9 ab	234.3 b	9.4 b	45.6 a
	Mean	6.8 A	305.7 AB	12.7 A	45.8 BC
San Andreas	1.1	7.1 a	325.0 a	13.7 ab	48.5 ab
	2.2	7.3 a	356.6 a	15.2 a	52.2 a
	3.3	6.2 a	284.8 a	13.0 ab	50.0 ab
	4.4	6.0 a	245.2 a	11.1 b	47.2 b
	Mean	6.6 A	301.7 B	13.2 A	49.2 A
Salt treatment	1.1	7.5 a [x]	384.8 a	14.9 a	45.8 a
	2.2	7.4 a	385.8 a	15.7 a	47.4 a
	3.3	6.3 b	307.8 b	12.6 b	45.5 a
	4.4	6.3 b	310.7 b	13.1 b	46.1 a
Cultivar		NS [w]	***	*	***
Treatment		***	***	***	NS
Cultivar × Treatment		NS	NS	NS	*

[z] For each cultivar, means with the same lowercase letters within a column are not significantly different among treatments according to Tukey's HSD test at $p < 0.05$; [y] means with the same uppercase letters within a column are not significantly different according to Tukey's HSD test at $p < 0.05$; [x] means with the same lowercase letters within a column are not significantly different according to Tukey's HSD test at $p < 0.05$; [w] NS, *, **, ***: Differences between means for main effects and their interaction were not significant, or significant at $p < 0.05$, 0.01, and 0.001, respectively.

in "Radiance" at EC4.4. Turhan and Eris [9] reported that NaCl treatments at 8.5, 17.0, or 34.0 mM increased the Na concentration in the leaf tissue of "Camarosa" and "Chandler" with a sharp increase in "Chandler".

Leaf Ca concentrations in the cultivars "Albion", "Benicia", and "Camarosa" were not statistically different among treatments (Table 4). Compared with control plants, the leaf Ca concentration of "Chandler", "Radiance", and "San Andreas" at EC3.3 increased by 25%, 27%, and 30%, respectively, while that of "Camino Real", "Chandler", and "San Andreas" at EC4.4 increased by 31%, 39%, and 33%, respectively. The highest Ca concentrations (25.84 $mg \cdot g^{-1}$ DW) were measured in "Radiance" at EC3.3. In contrast, Keutgen and Pawelzik [6] reported that leaf Ca concentrations for "Korona" and "Elsanta" were not affected when they were exposed to NaCl at 40 and 80 $mmol \cdot L^{-1}$.

Leaf Cl concentrations were significantly affected by the salt treatments and cultivars and their interaction (Table 4). Applying salt treatments increased the leaf Cl concentration of all strawberry cultivars. The three cultivars with relatively low leaf Cl concentrations (below 15 $mg \cdot g^{-1}$) at EC4.4 were "San Andreas", "Albion", and "Camarosa". Coincidentally, these three cultivars had visual scores above 3.0 and were clustered together in the relatively tolerant group, while the other four cultivars had visual scores below 3.0 with leaf Cl concentrations at EC4.4 above 15.0 $mg \cdot g^{-1}$. The highest Cl concentration (19.77 $mg \cdot g^{-1}$ DW) was found in "Chandler" at EC4.4. Turhan and Eris [9] similarly reported that the Cl concentration in the leaves of "Camarosa" and "Chandler" progressively increased with increasing NaCl concentrations from 8.5 to 34.0 mM.

Na exclusion and tolerance of tissue to accumulated Cl are two mechanisms of plant adaptation to salinity [19]. Strawberry is considered an Na excluder [4] and has an extremely low chlorine requirement [8]. In our study, all cultivars had relatively low leaf Na concentrations, generally lower than 1.5 $mg \cdot g^{-1}$ dry mass. However, leaf Cl concentrations at EC4.4 were much higher, especially for "Benicia", "Camino Real", "Chandler", and "Radiance". Similar results were reported by Saied *et al.* [4] who conducted a two-year field study with "Elsanta" and "Korona" under salinity levels of 0.3, 2.6, and 5.1 $dS \cdot m^{-1}$. In their study, Na concentrations below 3 $mg \cdot g^{-1}$ were reported, while Cl concentrations increased up to 70 $mg \cdot g^{-1}$ in "Korona" and 80 $mg \cdot g^{-1}$ in "Elsanta" plants. They also reported that "Korona" retained most of its Cl in the roots and crowns, while the highest concentration of Cl was detected in "Elsanta" petioles. Strawberry plants are sensitive to high Cl levels [8] and a leaf Cl concentration higher than 0.5% is associated with leaf necrosis and yield reduction in many cultivars [20]. The foliar injury and growth reduction observed in "Benicia", "Camino Real", "Chandler", and/or "Radiance" were most likely caused by high leaf Cl accumulation, similar to that reported by Martinez Barroso and Alvarez [7] and Kepenek and Koyuncu [8].

Table 4. Leaf ion concentration of strawberry cultivars irrigated with the nutrient solution (electrical conductivity (EC) = 1.1 dS·m^{-1}) or salt solution (EC = 2.2, 3.3, or 4.4 dS·m^{-1}) in the greenhouse.

Variety	Treatment	Ion Concentration (mg·g^{-1})			
		Na	Ca	Cl	K
Albion	1.1	0.56 a [z]	17.29 a	3.63 c	27.03 a
	2.2	0.82 a	22.47 a	8.58 b	23.46 ab
	3.3	0.66 a	22.08 a	10.58 ab	23.34 ab
	4.4	0.92 a	21.37 a	13.80 a	21.16 b
	Mean	0.75 A [y]	20.80 A	9.14 A	23.75 A
Benicia	1.1	0.73 a	18.52 a	4.73 d	23.79 a
	2.2	0.94 a	18.49 a	8.60 c	21.95 ab
	3.3	1.41 a	18.64 a	11.58 b	21.28 ab
	4.4	1.55 a	21.58 a	16.08 a	19.46 b
	Mean	1.16 A	19.31 AB	10.24 A	21.62 AB
Camarosa	1.1	0.43 b	16.09 a	3.10 c	22.36 a
	2.2	0.86 ab	18.39 a	9.20 b	21.49 ab
	3.3	1.06 ab	21.98 a	14.30 a	19.29 b
	4.4	1.35 a	21.86 a	13.43 a	18.64 b
	Mean	0.89 A	19.25 AB	9.78 A	20.44 B
Camino Real	1.1	0.61 b	17.08 b	4.00 c	26.36 a
	2.2	0.93 b	17.66 ab	9.85 b	23.95 ab
	3.3	1.83 a	22.30 ab	16.38 a	21.26 b
	4.4	1.13 b	22.42 a	17.55 a	20.54 b
	Mean	1.13 A	19.86 A	11.94 A	23.03 AB
Chandler	1.1	0.40 c	16.65 b	3.63 d	24.33 a
	2.2	0.62 bc	18.42 ab	9.48 c	21.91 b
	3.3	1.02 ab	20.74 a	13.00 b	20.88 b
	4.4	1.31 a	21.05 a	19.77 a	21.82 b
	Mean	0.80 A	19.10 AB	10.91 A	22.24 AB
Radiance	1.1	0.83 a	21.27 b	5.38 b	25.67 a
	2.2	1.11 a	19.53 b	9.70 b	24.53 a
	3.3	1.43 a	25.84 a	9.37 b	21.09 a
	4.4	2.21 a	23.96 ab	17.23 a	21.29 a
	Mean	1.27 A	22.20 A	10.00 A	23.41 A
San Andreas	1.1	0.58 b	13.85 b	2.98 c	25.80 a
	2.2	0.80 ab	15.31 ab	7.83 b	24.75 a
	3.3	0.82 ab	17.94 a	11.33 a	22.80 ab
	4.4	1.11 a	18.37 a	13.78 a	20.13 b
	Mean	0.83 A	16.37 B	8.98 A	23.37 A

Salt treatment	1.1	0.59 b [x]	17.25 b	3.92 d	25.05 a
	2.2	0.87 b	18.61 b	9.03 c	23.15 b
	3.3	1.18 a	21.16 a	12.47 b	21.42 c
	4.4	1.30 a	21.32 a	15.84 a	20.37 c
Cultivar		*** [w]	***	***	***
Treatment		***	***	***	***
Cultivar × Treatment		NS	NS	*	NS

[z] For each cultivar, means with the same lowercase letters within a column are not significantly different among treatments according to Tukey's HSD test at $p < 0.05$; [y] means with the same uppercase letters within a column are not significantly different according to Tukey's HSD test at $p < 0.05$; [x] means with the same lowercase letters within a column are not significantly different according to Tukey's HSD test at $p < 0.05$; [w] NS, *, **, ***: Differences between means for main effects and their interaction were not significant, or significant at $p < 0.05$, 0.01, and 0.001, respectively.

Potassium (K) plays an important role in turgor-pressure-driven solute transport in the xylem and water balance of plants [21]. Plants exposed to NaCl inevitably accumulate high amounts of Na, which subsequently interferes with K uptake, causing a reduction in K content [22]. In our experiment, leaf K concentrations decreased significantly with increasing EC levels in all cultivars except "Radiance" (Table 4). Compared with the control plants, the leaf K concentration at EC4.4 for all strawberry cultivars decreased by 10% to 26%, except for Radiance. However, K concentrations for "Camarosa", "Camino Real", and "Chandler" were reduced by 14% to 19% when irrigated with a salt solution of only EC 3.3 $dS \cdot m^{-1}$. Interestingly, a salt solution at an EC of only 2.2 $dS \cdot m^{-1}$ decreased the leaf K concentration of "Chandler" by 10%. Turhan and Eris [9] reported that the potassium content decreased in the aerial part of "Camarosa" plants with increasing NaCl levels from 8.5 to 34.0 mM, but with "Chandler", a solution of 8.5 mM NaCl increased the K content compared with the control. Keutgen and Pawelzik [6] reported that "Korona" strawberry also had a significant increase in the K content of the leaves. These results suggest that the efficiency of K uptake or the ability of strawberry adaptation to increasing levels of salinity is cultivar dependent.

The results of this study further indicated that variations in tolerance to salinity exists among cultivars. Growers who are considering growing strawberries in soils with relatively high salinity levels or who will be irrigating with saline water should consider selecting cultivars that have demonstrated salt tolerance. When produced under optimal conditions, strawberries have a high value of return; however, costs of production are also high and the increased salinity levels found in many regions like Texas could experience reduced yields, berry quality and, subsequently, grower profitability. Similar research is needed on additional selected cultivars that are being considered for strawberry production in Texas and other regions where high salt levels can potentially reduce crop performance.

4. CONCLUSIONS

The plant growth and fruit yield across all strawberry cultivars were reduced by the increased salinity of the irrigation water in this study, but the level of reduction varied with cultivar and the level of salinity. The salt solution at an EC of 4.4 $dS \cdot m^{-1}$ significantly reduced the shoot DW of "Camino Real" and "Chandler" as well as the fruit yield of "Benicia" and "Camino Real". The salt solution at an EC of 3.3 $dS \cdot m^{-1}$ significantly reduced the shoot DW of "Camino Real", but not the other cultivars. Three distinguishable strawberry groups were obtained using cluster analysis, which showed that "Albion", "Camarosa", and "San Andreas" were the most salt tolerant cultivars, while "Benicia", "Chandler", and "Radiance" were the least tolerant. Although "Camino Real" was classified as moderately salt tolerant by the cluster analysis, its visual scores were lower and leaf Cl concentrations were higher, and thus it probably should be grouped as less tolerant. "Chandler" is considered one of the industry standards for producers in Texas, but its susceptibility to salinity may reduce its widespread use as production acreage expands. "Camarosa" is similar to "Chandler" in overall production, and may be a better choice due to higher salt tolerance.

ACKNOWLEDGMENTS

This research was partially supported by a grant from the National Strawberry Sustainability Initiative with funding provided by the Walmart Foundation and administered by the University of Arkansas and U.S. Department of Agriculture National Institute of Food and Agriculture Hatch project TEX090450. The content is solely the responsibility of the authors and does not necessarily represent the official views of the funding agencies.

REFERENCES

1. USDA, National Agricultural Statistics Service. *Noncitrus Fruits and Nuts 2013 Summary*; U.S. Department of Agriculture: Washington, DC, USA, 2014.
2. Wallace, R.W.; Stein, L. Strawberry production in Texas. In *Production Guide for Texas-Grown Strawberries*; Wallace, R.W., Anciso, J., Eds.; Texas A&M AgriLife Extension: College Station, TX, USA; Available online: http://www.agrilifebookstore.org/ (accessed on 25 May 2015).
3. Wallace, R.W.; Webb, C.J. Strawberries grown under protective cultivation on the Texas High Plains. *J. Amer. Pomol. Soc.* **2013**, *67*, 7–10.
4. Saied, A.S.; Keutgen, A.J.; Noga, G. The influence of NaCl salinity on growth, yield and fruit quality of strawberry cvs. "Elsanta" and "Korona". *Sci. Hortic.* **2005**, *103*, 289–303.
5. Orsini, F.; Alnayef, M.; Bona, S.; Maggio, A.; Gianquinto, G. Low stomatal density and reduced transpiration facilitate strawberry adaptation to salinity. *Environ. Exp. Bot.* **2012**, *81*, 1–10.
6. Keutgen, A.J.; Pawelzik, E. Quality and nutritional value of strawberry fruit under long term salt stress. *Food Chem.***2009**, *107*, 1413–1420.
7. Martinez Barroso, M.C.; Alvarez, C.E. Toxicity symptoms and tolerance of strawberry to salinity in the irrigation water. *Sci. Hortic.* **1997**, *71*, 177–188.
8. Kepenek, K.; Koyuncu, F. Studies on the salt tolerance of some strawberry cultivars under glasshouse. *Acta Hortic.***2002**, *573*, 297–304.
9. Turhan, E.; Eris, A. Changes of growth, amino acids, and ionic composition in strawberry plants under salt stress conditions. *Commun. Soil Sci. Plant Anal.* **2009**, *40*, 3308–3322.
10. Ferreira, J.F.S.; Liu, X.; Suarez, D.L. Salinity tolerance of five commercial cultivars of strawberry (*Fragaria* × *ananassa* Duch.). In Proceedings of the 3rd International Salinity Forum, Riverside Convention Center, CA, USA, 16–18 June 2014; pp. 170–171.
11. Wright, R.D. The pour-through nutrient extraction procedure. *Hortscience* **1986**, *21*, 227–229.
12. Harrill, R. *Using a Refractometer to Test the Quality of Fruits & Vegetables*; Pineknoll Publishing: Keedysville, MD, USA, 1998.
13. Gavlak, R.G.; Horneck, D.A.; Miller, R.O. *Plant, Soil, and Water Reference Methods for the Western Region*; Western Regional Extension Publication (WREP): Corvallis, OR, USA, 1994; Volume 125.
14. Zeng, L.; Shannon, M.C.; Grieve, C.M. Evaluation of salt tolerance in rice genotypes by multiples agronomic parameters. *Euphytica* **2002**, *127*, 235–245.
15. Niu, G.; Cabrera, R. Growth and physiological responses of landscape plants to saline water irrigation: A review. *HortScience* **2010**, *45*, 1605–1609.
16. Kaya, C.; Higgs, D.; Saltali, K.; Gezerel, O. Response of strawberry growth at high salinity and alkalinity to supplementary potassium. *J. Plant Nutr.* **2002**, *25*, 1415–1427.
17. Gulen, H.; Turhan, E.; Eris, A. Changes in peroxidase activities and soluble proteins in strawberry varieties under salt stress. *Acta Physiolagiae Plant.* **2006**, *28*, 109–116.
18. Turhan, E.; Eris, A. Changes of micronutrients, dry weight, and chlorophyll contents in strawberry plants under salt stress conditions. *Commun. Soil Sci. Plant Anal.* **2005**, *36*, 1021–1028.
19. Munns, R.; Tester, M. Mechanisms of salinity tolerance. *Annu. Rev. Plant Biol.* **2008**, *59*, 651–681.
20. Ulrich, A.; Mostafa, M.A.E.; Allen, W.W. *Strawberry Deficiency Symptoms: A Visual and Plant Analysis Guide to Fertilization*; Agricultural Experiment Station,

Division of Agriculture and Natural Resources, University of California: Berkeley, CA, USA, 1980.

21. Marschner, H. *Mineral Nutrition of Higher Plants*, 2nd ed.; Academic Press: London, UK, 1995.
22. Hasegawa, P.M.; Bressan, R.A.; Zhu, J.K.; Bohnert, H.J. Plant cellular and molecular responses to salinity. *Annu. Rev. Plant Physiol. Plant Mol. Biol.* **2000**, *51*, 463–499.

CHAPTER 7

Liquid Organic Fertilizers for Sustainable Agriculture: Nutrient Uptake of Organic versus Mineral Fertilizers in Citrus Trees

BeleÂn MartõÂnez-AlcaÂntara, Mary-Rus MartõÂnez-Cuenca, Almudena Bermejo, Francisco Legaz, Ana Quiñones

Department of Citriculture and Vegetal Production, Instituto Valenciano de Investigaciones Agrarias, Moncada, Valencia, Spain

ABSTRACT

The main objective of this study was to compare the performance of two liquid organic fertilizers, an animal and a plant-based fertilizer, with mineral fertilization on citrus trees. The source of the fertilizer (mineral or organic) had significant effect in the nutritional status of the organic and conventionally managed mandarins. Nutrient uptake, vegetative growth, carbohydrate synthesis and soil characteristics were analyzed. Results showed that plants fertilized with animal based liquid fertilizers exhibited higher total biomass with a more profuse development of new developing organs (leaves and fibrous roots). Liquid organic fertilization resulted in an increased uptake of macro and micronutrients compared to mineral fertilized trees. Moreover, organic fertilization positively affected the carbohydrate content (fructose, glucose and sucrose) mainly in summer flush leaves. Liquid organic fertilization also resulted in an increase of soil organic matter content. Animal-based fertilizer, due to intrinsic composition, increased total tree biomass and carbohydrate leaves content, and led to lower soil nitrate concentration and higher P and Mg exchangeable in soil extract compared to vegetal-based fertilizer. Therefore, liquid organic fertilizers could be used as an alternative to traditional mineral fertilization in drip irrigated citrus trees.

INTRODUCTION

Management techniques of agricultural production are nowadays focused on a greater commitment to environmental sustainability. On this way, organic agriculture, accepted by the EU and the FAO as an alternative system to conventional agriculture, appears to be an environmentally friendly growing system [1] since mineral fertilizers abuse and misuse are responsible for health problems and environmental pollution [2]. It has been claimed that organic agriculture is the fastest growing agriculture in the world. In the decade from 2001 to 2011, the total worldwide organic agricultural hectares (ha) grew by 135% [3], which equates to an 8.9% per annum compound growth over the decade. Ecological surface in Spain was 1845 10^3 ha in 2011, representing an annual increase of 11.76% during the last decade, being positioned as the first country of the EU in number of ha in organic farming.

In organic systems, soil management involves the use of mowed or tilled cover crops, animal manures, composts and the application of organic fertilizers which increase soil-organic matter (SOM) whilst provide a steady release of nutrients to the crops as the organic matter breaks down. Exogenous organic matter applications are known to improve soil chemical and physical properties and biological functions, [4,5]. Ameliorated soil physical conditions enhance root growth facilitating nutrient uptake [6,7,8]. In this sense, organic manures have been demonstrated to improve the uptake efficiency of fertilizers [9]. Nutrient uptake efficiency may be defining as total element recovery in plant (mg) per fertilizer applied (mg). Moreover, the application of organic fertilizers has gained more importance not only because of its impact on soil quality but also because its role in carbon sequestration [10], mitigating atmospheric CO_2levels [11]. Organic fertilization has also been reported to have an influence on the phyto-nutritional quality of crops, enhancing the production of antioxidant metabolites in plants [12]. Yang et al. [13] obtained a remarkable increase in leaf yield, quality parameters, soluble sugar content, and thus in economic benefits, in *Stevia rebaudiana*, a perennial herb that it is an excellent source of sugars, under organic cultivation compared to traditional inorganic fertilization. In the medicinal plant *Labisia pumila*, organic fertilizer enhanced the production of total phenolics, flavonoids, ascorbic acid, saponin and gluthathione, when compared to inorganic fertilization [14]. However, liquid organic fertilizers to be used in fertigation with different origin are scarce and insufficiently tested.

Citrus (*Citrus* spp.) play an important role in organic farming system, being one of the most highly demanded products on the market for organic produce [15]. Nowadays, organic citrus makes about 2–7% of the global production [16] and 0.8% of the world citrus cultivated area [17], being thus a niche crop. In this context, the organic citrus sector has undergone a dynamic development in the last decade in the EU. The sector is concentrated around few Member States only: Italy, Greece, Spain and Cyprus. The biggest citrus areas are situated in Italy (more than 21900 ha) and Spain, where the sector amounted to around 6000 ha in 2011 and is under development.

In the past, citrus production was focused exclusively on maximizing the yield for commercial markets, and excessive fertilizer rates have been thus supplied to the crops with concomitant salt build up, phytotoxic effects on plant growth and ground water contamination [18,19,20]. In this context, efficient use of nitrogenous fertilizers has become a first-order concern in modern citrus production due to nitrate contamination of ground and surface waters [21,22]. Nowadays, fertilization studies are addressed to match and synchronize crop demand with nutrient supply. Interest in improving utilization of fertilizers, and specially N by citrus has been particularly widespread in Mediterranean areas, were citrus cultivation predominates [21,23,24], and most wells show nitrate concentrations clearly above the limit of the World Health Organization [25]. For nutrient uptake studies, the adoption of stable isotope techniques (^{15}N) enables tracing the movement of fertilizer-N in the plant-water-soil system. Under organic fertilization ^{15}N-labelled manure has allowed direct measurement and accurate estimation of N recovery not only in soil but also in crops [26]. But, due to the arduousness of the labelling procedure of organic residues and compost, very few assays on NUE of organic-derived fertilizers have been carried out, and all of them in horticultural crops [27,28,29,30].

The purpose of this study was to test the performance of two liquid organic fertilizers, a vegetal and an animal-based fertilizers, on citrus nutrient uptake, vegetative growth and soil characteristics, when compared to mineral fertilization under drip irrigation.

MATERIAL AND METHODS

Ethics statement

The experiments were conducted in the Department of Citriculture and Vegetal Production from Valencian Institute of Agrarian Research (Moncada, Spain). Dr. Ana Quiñones was response for experimental analysis in this manuscript and can be contacted in the future. The authors declare that this manuscript does not matter the any ethic issue and it does not involve endangered or protected species.

Experimental conditions, plant material and treatments

The study was carried out in 2010/2011 at the experimental station of Valencian Institute of Agricultural Research in Moncada (39° 33' N; 24° 24' W; Valencia). Twenty homogeneous 4-year-old "Nules Clementine" mandarin (*Citrus reticulata* Blanco) with 18 cm of canopy diameter grafted on Carrizo citrange (*Citrus sinensis* x *Poncirus trifoliata*) rootstock were grown individually in 50 L pots containing a loam soil characterized by sand 45.1%, silt 38.1%, clay 16.8%; pH 8.4 with 0.37% total organic carbon concentration. At the beginning of the assay, the average canopy diameter at breast height (1 m above the soil surface) was 70 ± 8 cm and the diameter measured at 4 cm from the graft trunk area was 3.8 ± 0.3 cm for the rootstock and 2.8 ± 0.2 cm for graft. The containers were arranged outdoors on benches under polycarbonate shelter to exclude rain.

Two liquid organic fertilizers were tested, a vegetal (VO) and an animal-based (AO) fertilizer (Table 1). Forage maize (*Zea mays* L.) grown under ^{15}N-labelled fertilizer supply was used as raw material for VO fertilizer production, and also as ^{15}N-labelled sheep feed to obtain ^{15}N-labelled manure. The labelled faeces fraction was used as raw material for the AO fertilizer. The VO fertilizer was obtained after an acidic and an enzyme-driven hydrolysis [31]. The AO fertilizer was obtained after acidic hydrolysis [31]. Liquid organic fertilizers (VO and AO) were compared to two mineral solutions containing 55 and 95% of N as ammonium sulphate and the remaining 45 and 5% as potassium nitrate, which were used as mineral controls for vegetal (VMC) and animal (AMC) fertilizer, respectively. The N fertilizer rate was 20 g N $year^{-1} \cdot tree^{-1}$according to tree canopy size [30], in order to fulfill plant N requirements.

Table 1. Total nitrogen and among fractions, ^{15}N excess and macro and micronutrient concentration of plant (VO) and animal (AO) based organic fertilizer ($mg \cdot L^{-1}$) §, †.

	N total ($mg\ L^{-1}$)	N-NH^{+}_{4} ($mg\ L^{-1}$)	N-NO^{-}_{3} ($mg\ L^{-1}$)	NH_4/NO_3	N-organic ($mg\ L^{-1}$)	^{15}N (atom % excess)
VO††	330.8±31.6	26.5±0.6†	21.6±0.03	1.23	282.6±6	2.62±0.07
AO‡	495.7±47.1	88.2±0.4	4.2±0.4	21.00	403.3±0.0	2.17±0.01
	C ($mg\ L^{-1}$)	C/N	P	K	Mg	Ca
VO	3504±43	10.6±1.43	47±1	923±14	60±2	386±9
AO	6046±801	12.2±2.3	365±26	921±65	257±19	545±28
	Fe	Zn	Mn	Cu	B	
VO	8.3±0.3	2.5±0.5	1.22±0.02	0.13±0.00	0.16±0.00	
AO	22.6±2.1	18.0±1.4	5.7±0.5	0.04±0.01	0.52±0.04	

§ Each value is a mean of three samples ± standard error.

† Volume applied: 60.5 and 40.3 L of vegetal and animal fertilizers, respectively.

†† Ammonium and nitrate nitrogen accounted for 55 and 45% of released total inorganic nitrogen, respectively.

‡ Ammonium and nitrate nitrogen accounted for 95 and 5% of released total inorganic nitrogen, respectively.

Other micro and macronutrients present in the vegetal and animal-based organic fertilizers were also supplied in similar amounts in their respective control. Nitrogen (N) and potassium (K) were supplied as potassium nitrate and ammonium sulphate. Phosphorus (P) fertilizer demand was applied as phosphoric acid and Magnesium (Mg) and Calcium (Ca) requirements as magnesium and calcium sulphate. The basic iron needs per tree were distributed throughout the growing cycle in a similar way to N in chelate form. Foliar spray treatments of zinc (Zn) and manganese (Mn) were applied as organic commercial fertilizer at 0.5% w/v (Zn: 6.6% w/w and Mn: 4.8% w/w) to correct deficiencies. In this way, organic fertilizers and their mineral controls differed in the nutrient form applied with the same concentration. Similar N and Ca amounts were added in all treatments (20 and 23 $g.plant^{-1}$). However, higher K rate and lower P and Mg were applied with vegetal fertilizers (56 vs. 37, 3 vs. 15 and 4 vs. 10, $g.plant^{-1}$ respectively). Plants nutritional requirements were supplied by treatments. Vegetal- and animal-based liquid organic fertilizers were ^{15}N labeled (2.62 and 2.17%^{15}N excess, respectively), while their respective controls were also labeled in the same extent. As a result, the experiment consisted of four treatments with five uniform trees per treatment, which were randomized across the experimental area. Mineral and organic fertilizers were supplied between March (spring growth resumption) and October. Plants were watered to field capacity every 2–3 days through 2 drip emitters per tree with deionized water, to avoid isotopic dilution of fertilizer-15N with water N. Mineral fertilizers were diluted into deionized water according to the following percentages [30], March (5%), April (10%), May (15%), June (22%), July (18%), August (15%), September (10%), October (5%). Similar quantities of organic fertilizers were manually added to each pot. The soil water potential was controlled daily using a ThetaProbe PR2 (Delta-T Devices, UK) and irrigation was scheduled when the matric potential at 30 cm depth attained -10 kPa [32,33].

With the aim of quantifying nutrient losses associated with abscised parts, tree litter (flowers, petals and fruitlets) was caught in nets from onset of flowering (1st April) until the end of fruit setting (4th July). Abscised organs were dried, weighed, milled and stored for subsequent nutrient and ^{15}N analysis.

Plant harvesting, sample preparation and vegetable analysis

At the end of the labelling period, during dormancy (December), trees were destructively harvested to determine nitrogen uptake efficiency (NUE) and nutrient plant uptake. Young (flowers/fruits, leaves and twigs of new shoots) and old organs (leaves and twigs of previous years, trunk and root system) were separated and sampled to quantify total dry biomass. All samples were washed in non ionic detergent solution followed by several rinses in deionized water, weighed, frozen into liquid nitrogen, freeze-dried and dry-weighed.

Vegetal samples were ground with a water-refrigerated mill, then sieved through a 0.3 mm mesh sieve and stored at -20°C for further analysis.

Macronutrients (P, K, Ca, Mg) and micronutrients (Fe, Zn, Mn) were measured in simultaneous inductively coupled plasma atomic emission spectrometry (ICAP-AES 6000, Thermo Scientific, Cambridge, United Kingdom) [34], after nitric-perchloric digestion. Dried plant material (0.5 g) was pre-digested overnight with 10 mL HNO_3 on a digestion block at 120°C. The samples were cooled down to room temperature and 2.0 mL of a 70% ultra-trace-metal-grade $HClO_4$ was added and re-digested at 220°C until white fumes were produced. Digest product was diluted to 25 mL with ultrapure water [35] and nutrient concentrations were subsequently measured.

Determinations of total N and C concentration and ^{15}N abundance were performed with an Elemental Analyzer (NC 2500 Thermo Finnigan) coupled to an Isotope Ratio Mass Spectrometer (Delta Plus, Thermo Finnigan). Results were expressed as percentage (macronutrient) or parts per million (micronutrients) of dry weight (DW).

Carbohydrates determination: soluble sugars in leaves

Leaf samples were washed, lyophilized, ground and stored at 4°C. Soluble sugars were extracted (100 mg DW) with 5 mL of ethanol 80% (v/v) at 60°C for 15–30 min, and then mixtures were centrifuged at 10000 rpm, for 30 min at 4°C. For recovery purposes, known amounts of fucose (Sigma Quimica, Madrid, Spain), a sugar absent in the extracts, were added to extracts as an internal standard. The supernatant was removed and the pellet was extracted twice (the extraction was repeated three times). The combined supernatants were collected, evaporated in vacuo at 45°C [36]. Residues were redissolved in 1 mL of water, and then filtered through 0.45 μm nylon filter and analyzed by HPLC by refraction index, using a column Tracer Carbohydr 250 mm x 4.5 mm, 5 μm (Teknokroma, Barcelona, Spain) and a mobile phase composed by acetonitrile:water (75:25) at a flow rate of 1 mL min^{-1}. An HPLC system equipped with a Waters 515 HPLC pump, a Waters 2414 refractive index detector and a 20 μL loop Rheodyne injector was used for sugar analysis. Empower 2 software (Waters, Spain) was used for data processing. Fructose, glucose and sucrose sugars were identified by comparing their retention time with a standard and quantified using an external calibration curve.

Soil sampling, sample preparation and analysis

On July, when half of the N was supplied, soil samples were extracted with a stainless steel cylindrical auger of 4 cm in diameter. At the end of the experiment, on December, soil contained in the pot of each harvested plant was thoroughly mixed after completely removing the fine roots, and then weighed before sampling. Each sample consisted of three subsamples, which were air-dried at room temperature, dry weighed, crushed through a 2 mm screen and stored for analysis.

The mineral nitrogen of the soil (NO_3-N and NH_4-N) was measured with flow-injection analysis (FIAstar 5000, Foss Tecator, Höganäs, Sweden) in KCl extracts (2 M), according to Raigon et al. [37]. In order to determine $^{15}N/^{14}N$ isotopic composition of both mineral fractions, soil extracts were steam-distilled (2200 Kjeltec, Auto Distillation Unit, Foss Tecator, Höganäs, Sweden); $^{15}NH_4$-N and $^{15}NO_3$-N were recovered in boric acid [38]. Aliquots were acidified with 0.32 N H_2SO_4 and reduced to dryness in an oven (P Selecta, Barcelona, Spain) at 65°C before analysis using the mass spectrometer mentioned above. After KCl extraction, the residual soil samples were washed with 50 mL distilled water to remove the remaining extractant solution, air-dried, milled, and analyzed for the organic N concentration and $^{15}N/^{14}N$ isotopic composition [39] using the Elemental Analyzer coupled to an Isotope Ratio Mass Spectrometer aforementioned.

CALCULATIONS

Based on data of dry weight (DW, g) and total nutrient concentration ([Nutrient], %, w/w) for each plant compartment, nutrient content was calculated:

$$Nutrient\,(g) = \frac{[Nitrogen] \times DW}{100}$$

The ^{15}N content per plant compartment was calculated as follows:

$$^{15}N_{plant\ compartment}\,(mg) = \frac{[Nitrogen] \times DW \times atom\ ^{15}N_{excess}}{10}$$

Where *atom* $^{15}N_{excess}$ was calculated by subtracting the natural abundance of ^{15}N from the atom % ^{15}N in each sample. The natural abundance of ^{15}N was considered to be the abundance of atmospheric N_2, 0.3663 atom %, according to the International Atomic Energy Agency [40].

The fraction N which is derived from the fertilizer (Ndff) was calculated according to Hardarson [41]:

$$N_{dff}\,(\%) = \frac{100 \times atom\ \%^{15}N_{excess\ plant\ compartment}}{atom\ \%^{15}N_{excess\ fertilizer}}$$

Total plant recovery of applied ^{15}N-fertilizer represents the proportion of applied ^{15}N that is taken up by the tree and embodies its fertilizer-nitrogen uptake efficiency (NUE). NUE was calculated by the formula:

$$N_{dff}\,(\%) = \frac{100 \times atom\ \%^{15}N_{excess\ plant\ compartment}}{atom\ \%^{15}N_{excess\ fertilizer}}$$

$$NUE(\%) = \frac{100 \times {}^{15}N_{whole\ plant}\,(mg)}{^{15}N_{fertilizer}\,(mg)}$$

Where:

$$^{15}N_{whole\ plant}\,(mg) = \sum {}^{15}N\,(mg)_{plant\ compartments}$$

The amount of ^{15}N recovered in each soil N fraction (Organic-^{15}N, NO_3-^{15}N, NH_4-^{15}N) was determined as follows:

$$^{15}N_{soil\ fraction}\,(mg) = \frac{N_{soil\ fraction}\left(\frac{mg\ N}{kg\ soil}\right) \times soil\ DW\ (kg) \times {}^{15}N_{excess}}{100}$$

Then, nitrogen retained in soil profile was calculated by the formula:

$$N_{dff\ soil} = \frac{100 \times {}^{15}N_{soil\ fraction}\,(mg)}{^{15}N_{fertilizer}\,(mg)}$$

Statistical Analysis

Data were subjected to ANOVA to test for significant differences between treatments. Before carrying out any statistical analysis, the normality of all the data was studied using the Kolmogorov-Smirnov test. In case the hypothesis of normality was discarded at the 95% confidence level, the data were transformed according to the logarithmic function. Otherwise, the data analyses were carried out with the variables measured in their natural scales. The variance of the transformed or non-transformed data was partitioned through a variance analysis (ANOVA, Statgraphics Centurion for Windows, Statistical Graphics Corp.) into one source of variability. The experiment consisted of two factors i) the source of the nutrients, organic versus mineral fertilizers, ii) the origin of organic fertilizer (vegetal- or animal-based). The significance of the comparisons made among treatments was analyzed using Fisher's least significance difference (LSD) test at $P < 0.05$.

RESULTS AND DISCUSSION

Plant Biomass

Different sources of nutrient (organic or mineral) had significant effect on total tree biomass, in spite of an equal N dosage per tree was applied in all treatments (Fig 1). Plants fertilized with organic fertilizers showed higher total biomass than mineral fertilized trees, as a result of a more profuse development of new organs and fibrous roots. Plant growth stimulation under organic management has been previously reported.

In citrus, enhancements of the physical growth characters (trunk and shoot diameter, shoot length, tree canopy and leaf area) have been recorded under organic fertilization based either on poultry manure [42] and compost plus humic acid [4] when compared to mineral NPK fertilizers. Also in peach trees, plant biomass was enhanced by supplying different organic fertilizers [6]. Biomass stimulation is the consequence of the hormone like effect that humic acids present in organic fertilizers play on the whole plant and especially on root growth [6,43]. In this sense, Hassan et al. [44] with mineral nitrogen plus humic acid as soil application observed higher leaves dry weight per plant than the other treatments in olive.

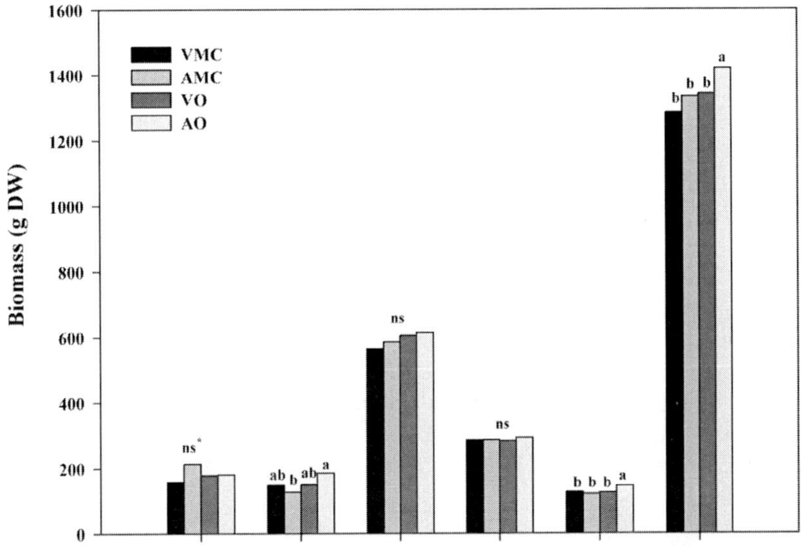

	Fruits	New Organs	Old Organs	Coarse R	Fibrous R	Total Tree
M vs. O[†]	ns[‡]	<0.05	ns	ns	<0.05	<0.05
V vs. A[†]	ns	<0.10	ns	ns	<0.05	<0.05

[*]Each value is an average of five plants. Old organs: old leaves+branches+trunk+abscised organs. Freedom degrees: between groups 1, within groups 8

[†]ANOVA is a partition of variance amongst the treatment and block factors. Factor 1: the source of the nutrients, Mineral (M) versus Organic (O) fertilizers. Factor 2: the fertilizer origin, Vegetal (V) or Animal (A) based organic fertilizer.

[‡]Within each group of organs, the treatments containing the same letter form a group of means within there are no statistically significant differences. The comparisons among treatments were made using Fisher's least significance difference (LSD) test at P ? 0.05 and not significant P>0.05 (ns).

Figure 1. Distribution of dry biomass (g DW plant^{-1}) among the main organs of Nules clementine trees harvested in December (at dormancy) receiving vegetal (VO) or animal-based (AO) organic fertilizer and their respective mineral controls (VMC and AMC, vegetal and animal mineral controls, respectively)*.

According to the origin, animal-based organic fertilizer led to a greater total biomass than vegetal-derived, due to the enhancement of new organs and fibrous roots development. However, Baldi et al. [45] found organic fertilization improved plant biomass, without differences between cow manure and compost obtained from municipal solid waste mixed with pruning material. Nevertheless, it is worth mentioning that in this experiment animal manure was pre-digested in order to obtain a liquid fertilizer, which would have increased nutrient availability compared to a direct application of cow manure.

Nitrogen uptake efficiency

Foliar nitrogen values were in the optimal range according to the standards for citrus nutritional status diagnosis established by Quiñones et al. [30] without significant differences between treatments (Fig 2a). No significant differences were also observed in nitrogen content in whole tree (Fig 2b). However, other authors found a effect of concentration and source of N added (NO_3^- or NH_4^+) on nutrient concentrations in leaves from citrus trees [46,47]. Nitrogen concentrations in leaves were highest when plants were provided with either NO_3^- or NH_4^+ as a unique source of N. However, lowest N concentration in leaves was found with a 75:25 NO_3^-:NH_4^+ratio.

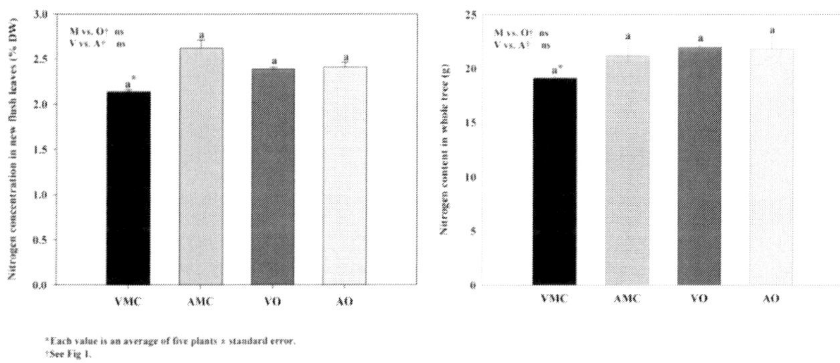

Figure 2. Nitrogen concentration (a) in new flush leaves (NFL) and total content in whole tree (b) of Nules clementine trees harvested in December (at dormancy) receiving vegetal (VO) or animal-derived (AO) organic fertilizer and their respective mineral controls (VMC and AMC, vegetal and animal mineral controls, respectively)*.

Nitrogen derived from fertilizer (Ndff), that is the relative contribution of fertilizer N to the total content of this element in plant organs, was lower in new flush leaves (Fig 3a) and on the average of the whole tree (Fig 3b) under organic fertilization when compared to mineral fertilization at the end of the cycle. This lower contribution of fertilizer-N to the total content of this element results from the diminished availability of N from fertilizer [48], since more than 80% of the total N supplied in the vegetal and animal-based liquid fertilizers was in organic form. The fact of the lower contribution of fertilizer-N in organic than in mineral supplied trees, together with the lack of differences in total N concentration among treatments, points to an enhanced remobilization of reserve N in organically fed trees. Similar result was found in nitrogen uptake efficiency (NUE) in whole tree (Table 2) with slightly greater values in plants in which mineral fertilizers were supplied. Bosshard et al. [49] found a similar pattern, 37, 10 and 47% of ^{15}N applied as urine, faeces and mineral fertilizer were recovered, respectively, in wheat.

Table 2. Nitrogen uptake efficiency (NUE) in whole tree and nitrogen recovered in different soil fractions of Nules clementine trees harvested in December (at dormancy) receiving vegetal (VO) or animal-derived (AO) liquid organic fertilizer and their respective mineral controls (VMC and AMC, vegetal and animal mineral controls, respectively) §.

	VMC	AMC	VO	AO	M vs. O†	V vs. A‡
NUE	33.87±1.5a††	34.47±0.50a	29.90±0.90b	29.59±0.21b	**	ns
Soil NH_4-N	0.33±0.04bc	0.26±0.04c	0.58±0.07a	0.49±0.08ab	**	ns
Soil NO_3-N	12.32±1.19a	8.37±1.16b	1.53±0.14c	1.17±0.17c	***	**
Soil Organic-N	14.08±0.69b	13.87±0.80b	32.05±2.41a	30.90±1.39a	***	ns

§Each value is an average of five soil samples plants ± standard error.
†ANOVA. Partition of variance amongst the treatment and block factors according to Fisher's least significance difference (LSD) test at $P \leq 0.05$ (*). $P \leq 0.01$ (**), $P \leq 0.001$ (***) and not significant $P>0.05$ (ns). Factor 1: the source of the nutrients, Mineral (M) versus Organic (O) fertilizers
‡Factor 2: the organic fertilizer origin with two levels, Vegetal (V) or Animal (A) based organic fertilizer.
††Within each row, different letters denote differences among means according to Fisher's least significance difference (LSD) test at $P \leq 0.05$ and not significant $P>0.05$ (ns).

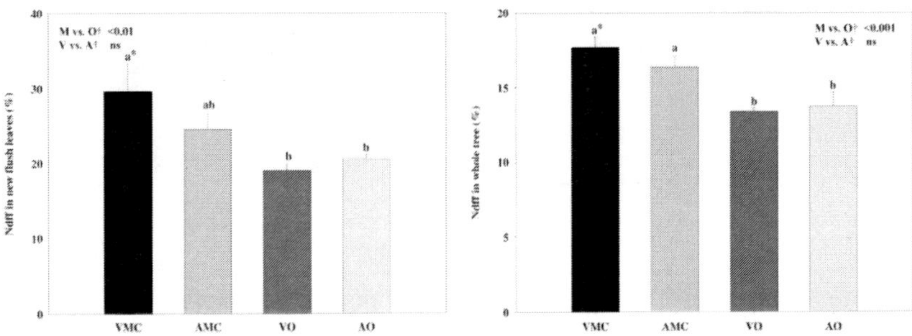

*Each value is an average of five plants ± standard error.
†See Fig 1.

Figure 3. Nitrogen derived from fertilizer (Ndff, %) in (a) new flush leaves (NFL) and (b) whole tree of Nules clementine trees harvested in December (at dormancy) receiving vegetal (VO) or animal-derived (AO) liquid organic fertilizer and their respective mineral controls (VMC and AMC, vegetal and animal mineral controls, respectively)*.

Neither Ndff (Fig 3a and 3b) in new flush leaves and in the whole tree nor NUE (Table 2) were affected by the origin (vegetal or animal based) of the liquid organic fertilizers tested.

Macro and micronutrient content and carbon fixation in plant organs

According to the analysis of spring leaves, the source of the fertilizer (mineral or organic) had a significant effect on the nutritional status of the organic and conventionally managed citrus trees. Phosphorous (P) content was higher in organic management, mainly due to both slightly higher concentration and greater biomass developed in plants fertilized with organic compost (Fig 4a and 4b). In micronutrients (Fig 5a and 5b), a significantly lower iron (Fe), manganese (Mn) concentration and zinc (Zn) and Mn content were found in plants fertilized with mineral fertilizers.

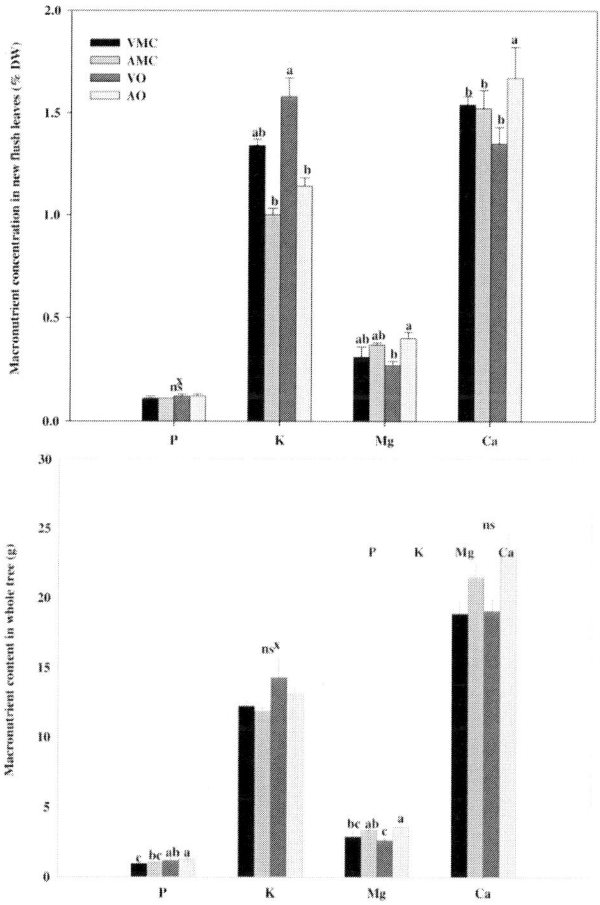

Concentration	P	K	Mg	Ca	Content	P	K	Mg	Ca
M vs. O†	<0.01	ns	ns	ns		ns	ns	ns	ns
V vs. A†	ns	<0.05	<0.01	<0.05		ns	<0.01	<0.05	<0.05

Figure 4. Macronutrient concentration (% DW) in new flush leaves (a) and total content (g) in whole tree (b) of Nules clementine trees harvested in December (at dormancy) receiving vegetal (VO) or animal-derived (AO) organic fertilizer and their respective mineral controls (VMC and AMC, vegetal and animal mineral controls, respectively)*. *, †See Fig 1.

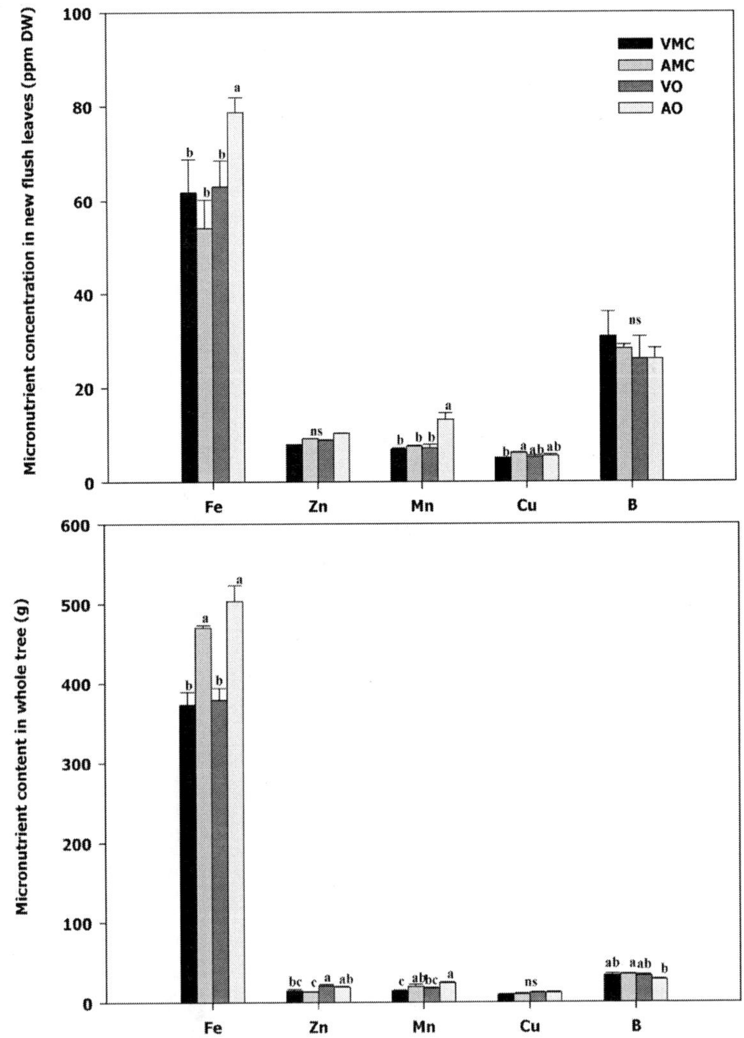

Concentration	Fe	Zn	Mn	Cu	B	Content	Fe	Zn	Mn	Cu	B
M vs. O†	<0.05	ns	<0.05	ns	ns		ns	<0.01	<0.05	ns	ns
V vs. A†	ns	ns	<0.05	ns	ns		<0.05	ns	<0.001	ns	ns

Figure 5. Micronutrient concentration (% DW) in new flush leaves (a) and total content (g) in whole tree (b) of Nules clementine trees harvested in December (at dormancy) receiving vegetal (VO) or animal-derived (AO) organic fertilizer and their respective mineral controls (VMC and AMC, vegetal and animal mineral controls, respectively)*. *, †See Fig 1.

Serna et al [47] found different behaviour with increasing proportions of NH_4^+ in the N supply, where leaf nutrients such as P and Fe increased, whereas Ca, K, Mn and Zn decreased. No consistent results with this respect have been found in the present study probably due to the fact that about 80% of total N supplied was in organic form.

These results are similar to that found by Barakat et al. [4] on Newhall and Canali et al. [15] on Tarocco and Navelina oranges on a 4-year average investigation, in citrus. These authors showed an improvement on plant nutritional status with organic fertilization compared to the chemical fertilizers through increasing their contents of phosphorus and potassium. Helail et al. [50] also showed an enhancement in leaf mineral content by application of organic manure of Washington Navel orange. Baldi et al. [51] observed no differences in Ca, Mg and Fe accumulation in relation to non-composted trees. However, Canali et al. [52] on Valencia late grafted on sour orange and Gasparatos et al. [53] on apple trees showed no significant difference between treatments for leaf N, K and P content whereas significant differences for Ca, Mg and micronutrients were observed. These different results may be due to the broad variety of organic fertilizers and differences in nutrient compositions depending on the origin of the fertilizer, moreover, differences in physical properties of the animal or vegetal subproducts result in a great variability in solubility and thus in elements availability for plant uptake [54]. As shown above, in comparison with mineral fertilizers, they provide lower levels of N, P and K; however, their addition can provide minerals to the soil (calcium, sulfur, iron, boron, and zinc), which will be continuously available to the growing plant. Unlike, all these elements are not supplied when unique mineral fertilizers are added [10].

Concerning to the origin of the organic fertilizers (vegetal vs. animal-based), significant differences were observed on K and Mg concentration according to the different mineral composition of both organic fertilizers. Vegetal-based fertilizer supplied higher amounts of K whereas in animal-based fertilizer Mg content was reinforced. In the case of Ca, the amount supplied was similar in both treatments. However, calcium uptake was higher (concentration and content) in plants receiving the animal-based fertilizer. This finding is attributable to a greater availability of calcium in faeces than in vegetal-based fertilizer. Canali et al. [51] found no differences in macronutrient concentration in Valencia late fertilized with citrus by-product compost (as vegetable fertilizer) and poultry manure (as animal manure) and Baldi et al. [45] also found similar N and P concentration in plant addressed with compost and cow manure, but with higher K leaf concentration in plant fertilized with compost. This response was probably due to the great K amount of K added with compost fertilization that inhibited Ca and Mg uptake. Moreover, K concentration was lower in compost fertilizer than in cow manure but with a master release of K. In this assay, as said above, manure was previously digested which increased their availability.

However, all values for macro and micronutrients were in the optimal range according to the international standard citrus nutritional status [30].

With respect to carbon fixation, total carbon in trees supplied with organic fertilizers was higher than in conventional fertilization, and greater when liquid animal manure was applied (Fig 6), these results logically paralleled those of tree biomass. Similar results were obtained by Baldi et al. [51] in nectarine.

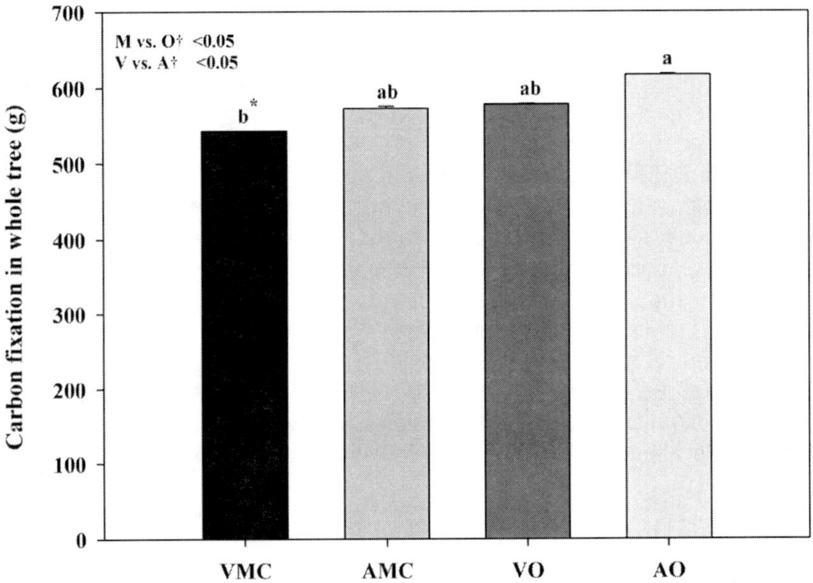

Figure 6. Carbon fixation in whole tree of Nules clementine trees receiving vegetal (VO) or animal-based (AO) organic fertilizer and their respective mineral controls harvested in December (at dormancy). Each value is an average of five plants ± standard error. †,‡See Fig 1.

Soluble foliar carbohydrates content

In order to determine the carbohydrates content according to the fertilizer applied, soluble sugars were analyzed in leaves from different flushes (spring and summer flush and old leaves). Under organic fertilization, soluble carbohydrates contents were significant greater in summer flush leaves and slightly higher in the other sproutings than conventionally fertilizer plants (Table 3).

Table 3. Soluble carbohydrate contents (g 100 g^{-1} DW) among the main flushes of Nules clementine trees harvested in December (at dormancy) receiving vegetal (VO) or animal-derived (AO) liquid organic fertilizer and their respective mineral controls (VMC and AMC, vegetal and animal mineral controls, respectively) §.

	Carbohydrate	VMC	AMC	VO	AO	M vs. O†	V vs. A‡
Spring flush leaves	Fructose	0.61±0.13a††	0.82±0.18b	0.64±0.21bc	1.10±0.27c	*	**
	Glucose	0.64±0.18b	0.44±0.104a	0.45±0.13a	0.70±0.13b	ns	ns
	Sucrose	1.77±0.47b	1.97±0.59b	1.84±0.21b	2.01± 0.58c	ns	ns
Summer flush leaves	Fructose	0.42±0.08a	0.58±0.18a	0.49±0.11ab	0.72± 0.14b	**	*
	Glucose	0.37±0.06a	0.4 ± 0.08a	0.44± 0.10a	0.84± 0.33a	**	**
	Sucrose	0.82±0.30a	1.11±0.43a	1.13±0.26a	1.62± 0.49b	*	*
Old leaves	Fructose	0.47±0.06a	0.40±0.03a	0.42±0.06a	0.55±0.08a	ns	ns
	Glucose	0.42±0.06a	0.39± 0.06a	0.36± 0.04a	0.45±0.08a	ns	ns
	Sucrose	0.60±0.05a	0.81± 0.05a	0.75± 0.03a	0.77± 0.20a	ns	ns

§Each value is an average of five plants ± standard error.

†, ‡, ††See Table 2.

Other authors have reported higher levels of primary metabolites and phenolic contents in horticultural cultivars, strawberries [55], marionberries [56], apples [57] and sweet pepper [58], among others, grown under organic conditions compared with inorganic fertilization. Generally, these results suggest that the usage of organic fertilizers can enhance the production of secondary and primary metabolites, and positively affect carbohydrate content. Zekri and Obreza [59] founded a positive influence of Magnesium fertilization on the synthesis of carbohydrates in leaves. Higher soluble carbohydrate content in leaves may be due to a greater Mg rate applied with animal derived liquid fertilizer (O). Moreover, source of the organic fertilizer also affected on the sugars content, mainly in summer flush leaves, with higher values when animal-based fertilizer was supplied. Summer leaves showed the most remarkable differences since according to timing of fertilizer application, these flush sprouts when approximately 50% of the total fertilizer rate has been supplied. No differences were found in old leaves that developed the prior year.

Soil characteristics and N recovered

The analysis of soil characteristics at the end of the growing cycle indicated higher values of soil organic matter (SOM) and N, P, Ca and Mg exchangeable in those soils corresponding to organic fertilized plants, compared to mineral fertilized (Table 4). No significant differences in pH, and electrical conductivity (EC), K, and Na exchangeable were found among mineral and organic fertilized soils, with similar values to other agricultural fields. Numerous authors have found K deficiencies under organic fertilization management [51,60,61]. But the lower values of these variables under organic management system seem to be due to an insufficient input of organic sources [51]. In this assay, K rate was similar in both managements (organic or mineral) being available for plant uptake.

Table 4. Effect of vegetal (VO) or animal-derived (AO) liquid organic fertilizers and their respective mineral controls (VMC and AMC, vegetal and animal mineral controls, respectively) on soil characteristics at the end of the growing cycle§.

Analysis	Initial	VMC	AMC	VO	AO	M vs. O†	V vs. A†
pH	8.4±0.01	8.2±0.0b	8.1±0.0b	8.5±0.0a	7.9±0.0	ns	ns
EC ($dS \cdot m^{-1}$) ‡	0.25±0.03	0.66±0.03b	0.81±0.02b	0.83±0.09b	0.82±0.01a	ns	ns
C (%)	0.37±0.02	0.36±0.02b	0.36±0.00b	0.50±0.02a	0.48±0.02a	***	ns
N (% dw) ††	0.04±0.00	0.04±0.00b	0.04±0.00b	0.05±0.01a	0.05±0.00a	**	ns
C:N	9.25±0.12	9.00±0.25b	9.00±0.14b	10.00±0.08a	9.6±0.18a	**	ns
P (mg kg^{-1} dw) ‡‡	13.5±0.7	27.5±1.5b	30.5±0.5b	31.7±0.9b	64.3±8.5a	*	*
K exchange*	0.27±0.02	0.95±0.05a	0.71±0.01b	1.03±0.04a	0.52±0.03c	ns	***
Mg exchange*	1.27±0.03	1.31±0.07c	1.77±0.01ab	1.71±0.03bc	2.18±0.18a	*	**
Ca exchange*	4.89±0.69	4.32±0.78a	5.56±0.10a	4.76±0.27a	6.04±0.35a	*	*
Na exchange*	0.23±0.01	0.34±0.01a	0.33±0.01a	0.46±0.11a	0.21±0.06a	ns	ns

§Each value is an average of five soil samples plants ± standard error.

†See Table 2

‡Electric conductivity in ext. 1:5 H_2O at 25°C.

††Total Nitrogen Kjeldahl.

‡‡Phosphorous Olsen.

*Exchangeable cation in meq·100 g^{-1}.

In general, most agricultural benefits from compost application to soil are derived from improved physical properties related to increase SOM rather than its value as a fertilizer [4,62] and is directly related to soil quality [63,64]. Initial results of the introduction of organic farming on soil quality of organically managed citrus orchards in the Mediterranean region were reported by Intrigliolo et al. [65]. They reported that organic management induced only slight differences in the main physical and chemical characteristics of conventionally managed soil. However, Canali et al. [52], in their study based on the comparison between organic and conventional citrus orchards, found significant differences in SOM content and soil aggregates stability, and no significant differences were found for clay, sand, pH and EC [13].

Nitrogen fractions in soil samplings on July and December revealed differences due to the source of the fertilizer applied (Fig 7). Organic fertilization led to greater concentrations on N in the ammonium and organic fractions in both sampling events. However, nitrate-N concentration in soils fertilized with organic liquid fertilizers was almost one third of that in mineral fertilized soils on July. Nevertheless, at the end of the cycle, once the nitrate supplied has been preferentially uptaken, differences in nitrate concentration were almost negligible because of the slow nitrification of the ammonium-N forms.

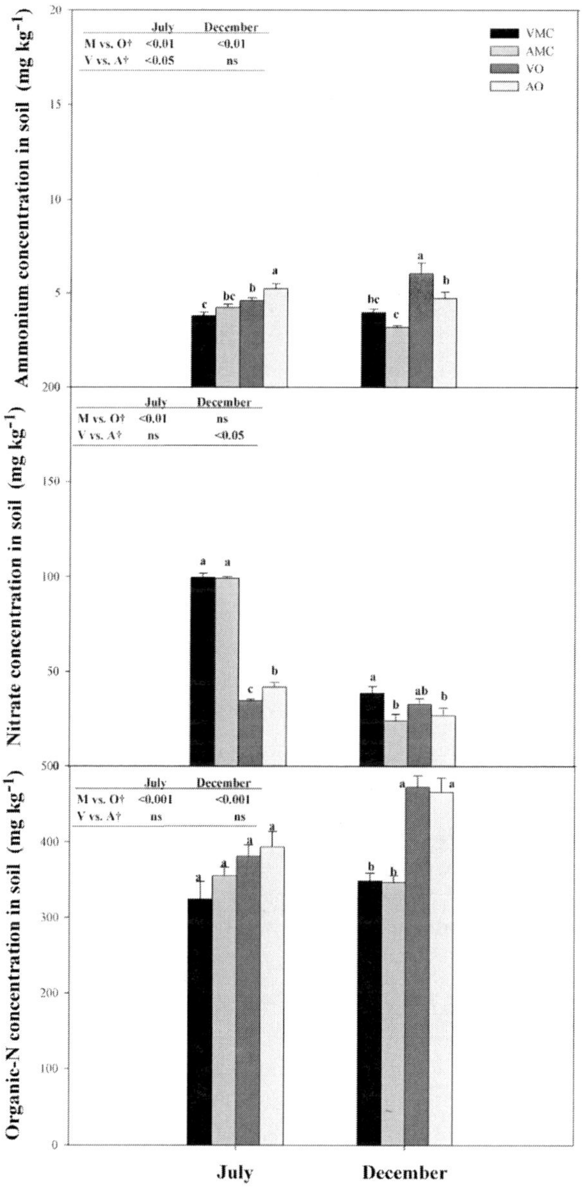

Figure 7. Nitrogen concentration in different N fractions in soils receiving vegetal (VO) or animal-based (AO) organic fertilizer and their respective mineral controls (VMC and AMC, vegetal and animal mineral controls, respectively)*. *, †, ‡See Fig 1.

The origin of liquid organic fertilizers also affected nitrogen partitioning among soil N pools. Animal-based fertilizer resulted in an increased concentration of N present in the ammonium fraction on July sampling event as a consequence of a reinforced supplied of ammonium-N form if compared to vegetal-based organic fertilizer. At the end of the cycle nitrate-N concentration was higher in plants supplied with vegetal-based fertilizer due to the nitrification of ammonium-N supplied.

Similarly, the percentage of fertilizer-N recovered in different soil-N fractions at the end of the assay (Table 2) showed that the amount retained as nitrate-N was significantly higher under mineral fertilization when compared to organic fertilization (10.35% and 19.15%, respectively). The reverse pattern was found in organic-N, with organic fertilized soils exhibiting the greater values (1.35% in mineral and 31.07% in organic fertilization). No differences were found on fertilizer-N compartmentation in soil N fractions attending to the origin (animal or vegetal) of the organic liquid fertilizer. Soil residual nitrate-N was higher in the plants supplied with the VMC than in those receiving AMC where N was predominantly (95%) in ammonium form. Nevertheless, no differences in NUE were found probably due to the fact that in VMC N was equally supplied under ammonium (55%) and nitrate (45%) forms. This lower residual NO_3-N accumulation under organic management is advantageous in comparison with conventional fertilization from an environmental point of view. Numerous authors linked main source of groundwater pollution with NO_3 leached from intensive agricultural areas [66,67]. So, the application of organic fertilizers can enhance nutrient uptake, mainly N, by reducing mineral leaching. Moreover, some nutrients in the water-soluble form required by plants are readily leached from mineral soil particles, whereas they are effectively held on the surface of humified organic matter [10]. The pre-digestion of organic residues in order to obtain water soluble fertilizers leads to an equivalent availability of nutrient elements to that in traditional mineral fertilizers.

It is worth mentioning that only a small fraction of the labeled fertilizer N was recovered as ammonium in all treatments, being higher in organic fertilized plants. This can be due to the process of fertilizer nitrate immobilization and later mineralization of labelled soil organic matter [24]. Davidson et al. [68] found a rapid turnover of a small NO_3-pool in intact soil cores due to a rapid phase of immobilization immediately following the addition of ^{15}N tracers to soils. The labelled ammonium-N found at the end of the trial accounted for less than 1% of applied-^{15}N for all treatments.

CONCLUSION

Liquid organic fertilizer obtained from maize residues (vegetal-based) and faeces sheep manure (animal-based) promoted biomass production and nutrient concentration in citrus plants. Organic fertilizer also resulted in an increase of soil organic matter. Moreover, organic fertilization positively affected the carbohydrate content. Plant fed with animal-based fertilizer, due to intrinsic composition, displayed a better biomass development and mineral nutrition.

The presented data support the idea that liquid organic fertilizers can be successfully used as a substitute of mineral fertilizers in citrus trees nutrient management under drip irrigation, since they enhance soil chemical fertility, prevent excessive nitrate-N concentration, promote plant growth and C fixation in the plant. Moreover, these fertilizers would allow not only to reduce the use of chemicals, but also to re-use crop residues and animal manure, conferring them an added value. Nevertheless, further studies should be addressed in order to evaluate these results in field conditions.

ACKNOWLEDGMENTS

We thank the nutrition team of the Valencian Institute for Agricultural Research: Mª Carmen Prieto, Josefa Giner, Teresa Garcia Estellés and Ramón Pardo for technical assistance.

REFERENCES

1. Polat E, Demiri H, Erler F. Yield and quality criteria in organically and conventionally grown tomatoes in Turkey. Sci Agricola 2010;67(4): 424–429.
2. Zahoor WA, Khanzada H, Bashir U., Aziz K, Zahir S, Faheem AK, et al. Role of Nitrogen Fertilizer in Crop Productivity and Environmental Pollution. Int J Agric For. 2014;4(3): 201–206
3. Willer H, Kilcher L In The World of Organic Agriculture: Statistics and Emerging Trends. Bonn: International Federation of Organic Agriculture Movements (IFOAM); Frick, Switzerland: Research Institute of Organic Agriculture (FiBL). 2011.
4. Barakat MR, Yehia TA, Sayed BM. Response of newhall naval orange to bio-organic fertilization under newly reclaimed area conditions I: Vegetative growth and nutritional status. J Hortic Sci Ornamental Plants. 2012;4(1): 18–25.
5. Diacono MA, Montemurro F. Review Effectiveness of Organic Wastes as Fertilizers and Amendments in Salt-Affected Soils. Agriculture 2015, 5, 221–230;
6. Baldi E, Toselli M, Eiseenstat DM, Marangoni B. Organic fertilization leads to increase peach root production and lifespan. Tree Physiol. 2010;30: 1373–1382. pmid:20921024

7. Baldi E, Toselli M, Marcolini G, Quartieri M, Cirillo C, Innocenti A, et al. Compost can successfully replace mineral fertilizers in the nutrient management of commercial peach orchard. Soil Use and Manage. 2010;26(3): 346–353, dio
8. Baldi E, Toselli M. Root growth and survivorship in cow manure and compost amended soils. Plan Soil Environ. 2013;59: 221–226.
9. Abd-Alla HM, Yan F, Schubert S. Effects of sewage sludge application on nodulation, nitrogen fixation and plant growth of faba bean, soybean and lupin. J Appl Bot 1999;73: 69–75.
10. Favoino E, Hogg D. The potential role of compost in reducing greenhouse gases. Waste Manage Res 2008;26(1): 61–69.
11. Stockmann U, Adams MA, Crawford JW, Field DJ, Henakaarchchi N, Jenkins M, et al. The knowns, known unknowns and unknowns of sequestration of soil organic carbon. Agric Ecosyst Environ. 2013;164: 80–99.
12. Dumas Y, Dadomo M, Di Lucca G, Grolier P. Effects of environmental factors and agricultural techniques on antioxidant content of tomatoes. J Sci Food Agric. 2003;83: 69–382.
13. Yang J, Liu X, Shi Y. Effect of different mixed fertilizer on yield, quality and economic benefits in Stevia rebaudiana bertoni. Adv J Food Sci Tech 2013;5(5): 588–591.
14. Ibrahim MH, Jaafar HZE, Karimi E, Ghasemzadeh A. Impact of organic and inorganic fertilizers application on the phytochemical and antioxidant activity of Kacip fatimah (Labisia pumila benth). Molecules 2013;18: 10973–10988. pmid:24013410
15. Canali S, Roccuzzo G, Tittarelli F, Ciccia C, Stagno F, Intrigliolo F. Organic Citrus: Soil fertility and plant nutrition management. In: Srivastava AK, editors. Advances in citrus nutrition. 2012. pp. 353–368.
16. Kilcher L. Organic citrus: challenges in production and trade. In: Cuaderno de Resúmenes I Conferencia Internacional de Citricultura Ecológica BIOCIITRICS, 2005. pp. 22–27.
17. FAOSTAT. FAO's statistical yearbook. 2012. Available: http://faostat.fao.org/site/.
18. Chapman HD. The mineral nutrition of citrus. In: The citrus industry. Reuther W, Barchelor LD and Webber MJ (eds.), Univ. Calif. Div. Agr. Sci., Berkeley, California. Rev. ed. II: 1968. 127–289.
19. Wild A. Rusell´s soil conditions and plant growth. Mundi Prensa Ed. (Madrid). 1992. 1025 pp.
20. Bingham FT, Davis S and Shade E. Water relations, salt balance and nitrate leaching losses of a 960 acre citrus watershed. Soil Sci. 1971. 112(6), 410–418
21. Quiñones A, Bañuls J, Primo-Millo E, Legaz F. Recovery of the 15N-labelled fertiliser in citrus trees in relation with timing of application and irrigation system. Plant Soil. 2005;268: 367–376
22. Ramos C, Agut A, Lidón AL. Nitrate leaching in important crops of the Valencian Community region (Spain). Environ Pollut 2002;118: 215–223. pmid:11939284
23. Martínez-Alcántara B, Quiñones A, Primo-Millo E, Legaz F. Nitrogen remobilization response to current supply in young Citrus trees. Plant Soil 2011;342(1–2): 433–443.
24. Martínez-Alcántara B, Quiñones A, Forner-Giner MA, Domingo JI, Primo-Millo E, Legaz F. Impact of fertilizer-water management on nitrogen use efficiency and potential nitrate leaching in citrus trees. Soil Sci Plant Nutr. 2012;58(5): 659–669. http://dx.doi.org/10.1080/00380768.2012.733678
25. World Health Organization. Nitrate and nitrite in drinking-water Background document for development of WHO Guidelines for Drinking-water Quality. 2004. Available: http://www.who.int/water_sanitation_health/dwq/chemicals /nitraten itrite2ndadd.pdf,
26. Muñoz G, Powell JM, Kelling KA. Nitrogen budget and soil N dynamics after multiple applications of unlabelled or 15Nitrogen-enriched dairy manure. Soil Sci Soc Am J. 2003;67: 817–825.

27. Sørensen P, Jensen ES, Nielsen NE. The fate of 15N-labelled organic nitrogen in sheep manure applied to soils of different texture under field conditions. Plant Soil 1994;162(1): 39–47.
28. Thomsen IK, Kjellerup V, Jensen B. Crop uptake and leaching of ^{15}N applied in ruminant slurry with selectively labelled faeces and urine fractions. Plant Soil 1997;197(2): 233–239.
29. Cusik PR, Kelling KA, Powell JM. Estimates of residual dairy manure nitrogen availability using various techniques. Journal of Environmental Quality 2006;35(6): 2170–2177. pmid:17071886
30. Quiñones A, Martínez-Alcántara B, Primo-Millo E, Legaz F. Fertigation: Concept and application in citrus. In: Srivastava AK, editors. Advances in citrus nutrition. 2012. pp 353–368.
31. Martínez-Alcántara B, Martínez-Cuenca MR, Fernández C, Legaz F, Quiñones A, Production of 15N-Labelled liquid organic fertilisers based on manure and crop residue for use in fertigation studies. Plos One, 2016. http:// dx.doi. org/ 10. 1371/journal.pone.0150851.
32. Parsons LR. Management of micro-irrigation systems for Florida citrus. In Fruit Crops Fact Sheet. Univ. of Florida Coop. Ext. Serv. FC-81. Univ. of Florida, Gainesville. 1989.
33. Smajstrala AG, Harrison DS, Zazutta FS, Parsons LR, Stone KC. Trickle irrigation scheduling for Florida citrus. In: Bull. 208. Florida Coop. Ext. Serv., Institute of Food and Agriculture Sciences. University of Florida, Gainesville. 1987.
34. Isaac RA, Johnson WC Jr.. Elemental determination by inductively coupled plasma atomic emission spectrometry. In: Kalra YP, editors. Handbook of Reference Methods for Plant Analysis. CRC Press, Boca Ratón, FL, USA. Chp. 1998. pp. 165–170.
35. Campbell CR, Plank CO. Preparation of plant tissue for laboratory analysis. In: Kalra YP, editors. Handbook of Reference Method for Plant Analysis. CRC Press, Boca Raton, FL. 1998. pp. 37–49.
36. Rivas F, Erner Y, Alós E, Juan M, Almela V, Agustí M. Girdling increases carbohydrate availability and fruti-set in citrus cultivars irrespective of parthenocarpic ability. Journal of Horticultural Science & Biotechnology 2006. 81, 289–295.
37. Raigón MD, Pérez-García M, Maquieira A, Puchades R. Determination of available nitrogen (nitric and ammoniacal) in soils by flow injection analysis. Analysis, 1992;20: 483–487.
38. Bremner JM. Nitrogen total. In Sparks DL et al. Eds Methods of Soil analysis, Part 3-Chemical Methods. Soil Science Society of America, American Society of Agronomy, Madison, WI. 1996. pp. 1085–1121.
39. Lea-Cox JD, Syvertsen JP, Graetz DA. Springtime 15N uptake, partitioning, and leaching losses from young bearing citrus trees of differing nitrogen status. J Am Soc Hortic Sci. 2001;126: 242–251.
40. IAEA, International Atomic Energy Agency. A guide to the use of Nitrogen-15 and radioisotopes in studies of plant nutrition: calculations and interpretation of data. IAEA, Vienna. 1983.
41. Hardarson G. Use of Nuclear Techniques in Studies of Soil-Plant Relationships. Training Course Series 2, IAEA, Vienna. 1990.
42. Gamal AM, Ragab MA. Effect of organic manure source and its rate on growth, nutritional status of the trees and productivity of Balady mandarin trees. Assiut J Agr Sci. 2003;54: 253–264.
43. Cacco G, Dell'Agnola G. Plant growth regulator activity of soluble humic complex. Can J Soil Sci. 1984; 64: 225–228.
44. Hassan HSA, Hagag LF, Rawash MA, El-Wakeel H, Abdel-Galel A. Response of Klamata olive young trees to mineral, organic nitrogen fertilization and some other treatments. Nature Sci. 2010;8(11): 59–65.
45. Baldi E, Toselli M, Marangoni B. Nutrient partitioning in potted peach (prunus persica L.) trees supplied with mineral and organic fertilizers. J Plant Nutr. 2010;33: 2050–2061.

46. Kato T. Nitrogen metabolism and utilization in citrus. In: Janick J (ed.). AVI Publishing Co. Horticultural Reviews. 1986;8: 81–216.
47. Serna MD, Borrás R, Legaz F, Primo-Millo E. The influence of nitrogen concentration and ammonium/nitrate ratio on N-uptake, mineral composition and yield citrus. Plant Soil. 1992;147: 13–23.
48. Rapisarda P, Calabretta ML, Romano G, Intrigliolo F. Nitrogen metabolism components as a tool to discriminate between organic and conventional citrus fruits. J Agr Food Chem. 2005;53: 2664–2669.
49. Bosshard C, Sørensen P, Frossard E, Dubois D, Ma¨Der P, Nanzer S, et al. Nitrogen use efficiency of 15N-labelled sheep manure and mineral fertiliser applied to microplots in long-term organic and conventional cropping systems. Nutr Cycl Agroecosys. 2009;83: 271–287,
50. Helail BM, Gobran YN, Mostafa MH. Study on the effect of organic manure application and bio-fertilizers on: a-Tree growth and leaf mineral content of Washington. J Appl Sci. 2003;18(44): 270–296.
51. Baldi E, Marcolini G, Quartieri Q, Sorrenti G, Toselli M. Effect of organic fertilization on nutrient concentration and accumulation in nectarine (Prunus persica var. nucipersica) trees: The effect of rate of application. Sci Hort. 2014;179: 174–179.
52. Canali S, Trinchera A, Intrigliolo F, Pompili L, Nisini L, Mocali S, et al. Effect of long term addition of composts and poultry manure on soil quality of citrus orchards in southern Italy. Biol Fert Soils. 2004;40: 206–210
53. Gasparatos D, Roussos PA, Christofilopoulou E, Haidouti C. Comparative effects of organic and conventional apple orchard management on soil chemical properties and plant mineral content under Mediterranean climate conditions. J Soil Sci Plant Nutr, 2011;11(4): 105–117 http://dx.doi.org/10.4067/S0718-951620110 00400008.
54. Rosen CJ, Bierman PM. Using manure and compost as nutrient sources for fruit and vegetable crops. Department of Soil, Water, and Climate. University of Minnesota. 2005.
55. Hakkinen SH, Torronen AR. Content of flavonols and selected phenolic acids in strawberries and Vaccinium species: Influence of cultivar, cultivation site and technique. Food Res Int. 2000;33: 517–524.
56. Asami DK, Hong YJ, Barrett DM, Mitchell AE. Comparison of the total phenolic and ascorbic acid content of freeze-dried and air-dried marionberry, strawberry, and corn using conventional, organic, and sustainable agricultural practices. J Agri Food Chem. 2003;51: 1237–1241.
57. Weibel FP, Bickel R, Leuthold S, Alfoldi T. Are organically grown apples tastier and healthier? A comparative field study using conventional and alternative methods to measure fruit quality. Acta Hort. 2000;517: 417–426.
58. Aminifard MH, Aroiee H, Azizi M, Nemati N, Jaafar HZE. Effect of compost on antioxidant components and fruit quality of sweet pepper (capsicum annuum L.). J Central European Agric 2013;14(2): 47–56.
59. Zekri M, Obreza T. Magnesium (Mg) for Citrus Trees. University of Florida. IFAS Extension. 2013. SL380.
60. Berry PM, Stockdale EA, Sylvesterbradley R, Phillips L, Smith KA, Lord EI, et al. N, P and K budgets for crop rotations on nine organic farms in the UK. Soil Use Manage. 2003;19: 112–118.
61. Gosling P, Shepherd M. Long—term changes in soil fertility in organic arable farming systems in England, with particular reference to phosphorus and potassium. Agriculture, Ecosystems and Environment 2005;105: 425–432.
62. Shiralipour A, Mcconnell D, Smith W. Physical and chemical properties of soils as affected by municipal solid waste compost application. Biomass Bioenergy. 1992;3: 261–266.
63. Canali S, Di Bartolomeo E, Trinchera A, Nisini L, Tittarelli F, Intrigliolo F, et al. Effect of different management strategies on soil quality of citrus orchards in southern Italy. Soil Use Manage. 2009;25: 34–42.

64. Diacono M, Montemurro F. Long-term effects of organic amendments on soil fertility. A review. Agron Sustain Dev. 2010;30(2): 401–422.
65. Intrigliolo F, Montemurro N, Roccuzzo G, Giuffrida A, Canali S, Calabretta ML, et al. Field survey on soil fertility and plant nutritional status in organic and conventional citrus orchards. In: El-Fouly MM, editors. Book of Abstracts. X International Colloquium for the Optimization of Plant Nutrition, Cairo, Egypt.2000. pp. 106–107.
66. Burkart MR, Stoner JD. Nitrate in Aquifers beneath agricultural systems. Water Sci Technol. 2002;45(9): 19–29. pmid:12079102
67. de Paz JM, Ramos C. Simulation of nitrate leaching for different nitrogen fertilization rates in a region of Valencia (Spain) using a GIS-GLEAMS system. Agr Ecosyst Environ. 2004;103(1): 59–73.
68. Davidson EA, Hart SC, Shanks CA, Firestone MK. Measuring gross nitrogen mineralization, immobilization, and nitrification by 15N isotopic pool dilution in intact soil cores. J Soil Sci. 1991;42. 335–349.

CHAPTER 8

Analysis of The Genetic Diversity and Structure Across a Wide Range of Germplasm Reveals Prominent Gene Flow in Apple at The European Level

Jorge Urrestarazu[1,2,]

[1]IRHS, INRA, AGROCAMPUS-Ouest, Université d'Angers, SFR 4207 QUASAV, 42 rue Georges Morel, 49071 Beaucouzé cedex, France

ABSTRACT

Background

The amount and structure of genetic diversity in dessert apple germplasm conserved at a European level is mostly unknown, since all diversity studies conducted in Europe until now have been performed on regional or national collections. Here, we applied a common set of 16 SSR markers to genotype more than 2,400 accessions across 14 collections representing three broad European geographic regions (North + East, West and South) with the aim to analyze the extent, distribution and structure of variation in the apple genetic resources in Europe.

Results

A Bayesian model-based clustering approach showed that diversity was organized in three groups, although these were only moderately differentiated ($F_{ST} = 0.031$). A nested Bayesian clustering approach allowed identification of subgroups which revealed internal patterns of substructure within the groups, allowing a finer delineation of the variation into eight subgroups ($F_{ST} = 0.044$). The first level of stratification revealed an asymmetric division of the germplasm among the three groups, and a clear association was found with the geographical regions of origin of the cultivars. The substructure revealed clear partitioning of genetic groups among countries, but also interesting associations between subgroups and breeding purposes of recent cultivars or particular usage such as cider production. Additional parentage analyses allowed us to identify both putative parents of more than 40 old and/or local cultivars giving interesting insights in the pedigree of some emblematic cultivars.

Conclusions

The variation found at group and subgroup levels may reflect a combination of historical processes of migration/selection and adaptive factors to diverse agricultural environments that, together with genetic drift, have resulted in extensive genetic variation but limited population structure. The European dessert apple germplasm represents an important source of genetic diversity with a strong historical and patrimonial value. The present work thus constitutes a decisive step in the field of conservation genetics. Moreover, the obtained data can be used for defining a European apple core collection useful for further identification of genomic regions associated with commercially important horticultural traits in apple through genome-wide association studies.

Keywords: *Malus* x *domestica* Borkh. Genetic resources Population structure Variability SSR markers Differentiation Parentage analysis

BACKGROUND

Cultivated apple (*Malus* x *domestica* Borkh.) is one of the most important fruit crops grown in temperate zones and the most important in the *Rosaceae* family [1]. Although there are more than 10,000 documented apple cultivars worldwide and the apple production area is widespread geographically, the global production is dominated by relatively few cultivars, many of which are closely related [2, 3]. Moreover, in the last century, despite the existence of a large number of apple breeding programs worldwide, only a few well-adapted genotypes (e.g., 'Red Delicious', 'Golden Delicious', 'Jonathan', 'McIntosh' or 'Cox´s Orange Pippin') were extensively used in apple breeding to release new varieties with desirable traits [2, 4, 5]. The additional release of clonal selections of the most popular and widely grown varieties has further contributed towards the uniformity of commercial apple orchards [6, 7, 8]. The gradual replacement of the traditional and locally well-adapted cultivars by a few wide-spread modern varieties has led to a dramatic loss of genetic diversity in the orchards and may also hamper future plant breeding.

The recognition of this situation has encouraged the establishment of action towards the preservation of apple genetic resources worldwide. Multiple apple collections are presently maintained in Europe, preserving mainly old cultivars which have been grown traditionally in their respective regions, but also other cultivars with diverse geographic origins introduced a long time ago, that represent elite selections from before the time of formal breeding. Most of these existing collections were established before molecular identification became available, and in the absence of marker data, the criteria used in the past for selecting the germplasm to be preserved in collections focused mainly on morphology (pomology), eco-geography and/or passport information [9]. The effectiveness of these conservation approaches depends upon the criteria used for selecting germplasm and it has been suggested that genetic diversity may not always be optimal in these, or equivalent collections in other crops [10, 11], and therefore, unintended internal redundancies are expected. Assessment of the genetic diversity in fruit tree species is nowadays mainly performed by marker genotyping techniques [12]. Molecular markers have therefore become an indispensable tool in the management of germplasm collections, and their use is widely applied in characterization to assist and complement phenotypic assessments and to re-examine the composition of the collections [11, 13, 14, 15, 16]. The use of molecular markers has not only important implications with regard to the efficiency of the management of the genetic resources, but constitutes a key instrument to evaluate diversity, to elucidate the underlying genetic structure of the germplasm and to quantify relatedness and differentiation between populations among other multiple applications [17, 18, 19, 20]. Such knowledge is of high relevance since the conservation of plant genetic resources only fulfills its full potential when they are used effectively, which requires knowledge of the extent and structure of the variation occurring within the material preserved [21].

Until now, the studies of diversity and genetic structure conducted in European apple have been based on the analyses of material from limited geographic areas (mostly nation-scale) [11, 14, 22, 23, 24, 25, 26]. By contrast, the extent and structure of the apple genetic diversity conserved at a European level have remained largely unknown. The main obstacle is the different sets of SSR markers used in the different European collections preventing an overall comparison [27]. Thus, in the frame of the EU-FruitBreedomics project [28] a single set of 16 SSR markers was used in a very broad set of apple germplasm (~2440 accessions, mostly of dessert use) preserved in collections located in eleven countries and representing three broad European geographical regions (North + East, West and South) in order to determine the diversity in apple collections at a European scale, to evaluate gene flow in cultivated apple across Europe, as well as to elucidate the stratification of germplasm into population subdivisions and finally, to perform parentage analysis. This is the largest study of apple genetic resources at the pan-European level.

RESULTS

SSR polymorphism – identification and redundancy

Among the 2,446 accessions, ten accessions did not show clear PCR amplifications and were discarded from the analysis. Pairwise comparison of multilocus profiles revealed 219 groups of redundancies (Additional file 1), leading to the removal of 405 redundant accessions before further analyses (16 % of redundancy). The number of accessions in each of these identical SSR profile groups varied from two to nine. The cumulative probability of identity (P_{ID}) was extremely low: $P_{ID} = 1.3 \times 10^{-22}$, thus highlighting the low risk of erroneous attribution of accessions to duplicate groups. Redundancies were found both within and between collections, leading to the confirmation of numerous previously documented synonyms (e.g., 'Papirovka' and 'White Transparent', 'London Pippin' and 'Calville du Roi', or 'Président van Dievoet' and 'Cabarette') and allowing the putative identification of numerous unknown synonyms or mutant groups (e.g., 'Gloria Mundi' = 'Mela Zamboni' = 'Audiena de Oroz' = 'Belle Louronnaise', 'Court-Pendu Plat/Doux/Gris' = 'Krátkostopka královská', 'Reinette de Champagne' = 'Maestro Sagarra' or 'Reinette Simirenko' = 'Renetta Walder' = 'Burdinche'). Redundancy groups also supported the notion of several national/local name translations such as the English cultivar 'Cornish Gilliflower' translated into 'Cornwallské hřebíčkové' (i.e., 'Cornish clove'), or 'White Transparent' and 'Skleněné žluté' (i.e., 'yellow glass') in Czech and 'Transparente Blanca' in Spanish, the Russian cultivar 'Korichnoe polosatoe' translated into 'Kaneläpple' in Swedish (i.e., 'cinnamon apple'), or the cultivar 'La Paix' translated into 'Matčino' (i.e., 'Mother', a synonym of 'La Paix') in Czech. Several cases of homonymy (i.e., accessions with the same name but different SSR profiles) were also found, e.g., three different SSR profiles for the same accession names 'Pomme Citron' or 'Charles Ross'. Data allowed identifying some obvious labeling errors, e.g., X2698 'Court Pendu Plat' which was shown to be the rootstock 'MM106', or CRAW-0362 'Transparente de Croncels' which was found likely to actually be 'Filippa' (Additional file 1). Following these observations, the apple germplasm dataset was reduced to 2,031 unique genotypes (i.e., exhibiting distinct SSR profiles). Among these individuals, 162 (8 % of the different genotypes) were removed since they had a putative triploid profile, while another ten were discarded because of too much missing SSR data, or because further identified as rootstock or outliers in a preliminary Principal Coordinate Analysis. The final number of unique diploid genotypes further analyzed was therefore 1,859. Using passport data and other accessible information, it was possible to attribute geographical regions of origin (either for three broad designated European regions or, when possible, specific countries) for a large part of the unique genotypes. Roughly 89 % (1,653) of these genotypes could be geographically assigned, with 261, 1,074 and 318 genotypes assigned to Northern + Eastern, Western and Southern historical regions of origin, respectively (Additional file 1). In brief, the Northern + Eastern region was composed of germplasm originating in Nordic European countries plus Russia, the Western region was composed of germplasm originating in Western and Central European countries and the Southern region was composed of germplasm from Spain and Italy (see Methods for more details). The remaining 11 % consisted of either genotypes lacking passport information or genotypes with contradictory information in passport data from different origins. Similarly, the specific country of origin could be attributed to 1,550 genotypes out of the 1,653 geographically assigned (Additional file 1). It is important to note that the European region or country of origin assigned to a genotype was independent from the location of the collection where the sampled accession was maintained, since many collections contained accessions from various origins.

Genetic diversity across and within European regional groups

The 16 SSR markers amplified a total of 369 alleles across the 1,859 apple accessions used for diversity analysis, ranging from 17 (CH02c09 and CH05f06) to 35 (CH02c06) alleles per locus. The average number of alleles per locus was 23.06, whereas the mean effective number of alleles per locus was 6.59 (Table 1). High average number of alleles per locus and almost identical mean effective number of alleles per locus were noted for the three geographical regions of origin of the germplasm. Allelic richness was normalized to the smallest group (i.e., North + East) to avoid a group size-dependent bias of results. Overall, the results obtained for the material of the three designated regions of origin suggested the existence of a high and relatively homogeneous allelic diversity across Europe (Table 1). Within the 369 alleles identified in the overall set (i.e., across Europe), 73.4 % and 52.0 % were found at frequencies below 5 % and 1 %, respectively (Table 1; data not shown for 1 %). A similar proportion of rare alleles was obtained for the material from the three designated geographical regions of origin, with the exception of alleles detected at a frequency < 1 % with Northern + Eastern and Southern European origins, for which slightly lower percentages were identified (≈38 %). Almost identical mean *He* values were obtained for the overall dataset (0.83) and for the germplasm from each of the three geographical groups (Table 1). Cross-comparison of the allelic composition for the accessions classified into geographic categories showed that 221 out of the 362 alleles (seven alleles appeared only in accessions that could not be classified into geographic groups) were detected in all three geographical groups, 59 alleles (16.3 %) were identified in two geographic groups only, whereas 82 alleles (22.6 %) were specifically found only in one geographic group (i.e., private alleles). At the national level (i.e., countries of origin of the unique genotypes), some countries exhibited a higher rate of private alleles than others: especially, genotypes assigned to Switzerland, Italy and Russia harboured 15, 14 and 14 private alleles (respectively), genotypes from Spain and France harboured 7 private alleles each, whereas genotypes from the Netherlands, Belgium, Great Britain or Sweden had a maximum of one private allele. The pattern of distribution of the frequent alleles (frequency > 0.05) between Southern, Northern + Eastern and Western germplasm was analyzed for each locus separately using Chi^2 tests. Highly significant differences in the allelic distributions ($P < 0.001$) were found between all the geographic groups for all markers except for the CH-Vf1 locus when comparing Southern and Western germplasm (data not shown).

Genetic structure and differentiation

A Bayesian model-based clustering method was applied to the 1,859 unique diploid genotypes in order to elucidate the underlying genetic structure at a European scale. The analysis of Evanno's ΔK statistic indicated unambiguously $K = 3$ as the most likely level of population stratification (Fig. 1 a1). The mean proportion of ancestry of the genotypes to the inferred groups was 0.81. Using the threshold of $qI \geq 0.80$ to define strong assignments to groups, 1,175 genotypes (63 %) were identified as strongly associated to a group. This partitioning level corresponded to an asymmetric division of the material into three groups: K1 composed of 506 genotypes, K2 containing 401 genotypes, and K3, the largest group, comprising 952 genotypes. Diversity estimates revealed high levels of

Table 1 Average measures of genetic diversity at two different levels: overall set of accessions and according to the three geographical regions of origin (Northern + Eastern, Southern and Western)

Material	N_A	N_B [a]	N_E	A_R [b]	Ho	He
Overall set (1859 genotypes)	23.06	16.94	6.59	-	0.81	0.83
European regions of origin						
Northern + Eastern Europe	16.75	10.87	6.24	16.57	0.83	0,82
Southern Europe	17.50	11.87	6.29	16.95	0.81	0.82
Western Europe	20.31	13.94	6.18	16.36	0.81	0.82

[a]Rare alleles were considered if they appeared in a frequency below 5 %

[b]For the geographical European regions of origin, allelic richness was computed after normalization according to the smallest population size (i.e., Northern + Eastern Europe). Number of alleles per locus (N_A), number of rare alleles (N_B), effective number of alleles (N_E), allelic richness (A_R), and observed (Ho) and expected (He) heterozygosity are included

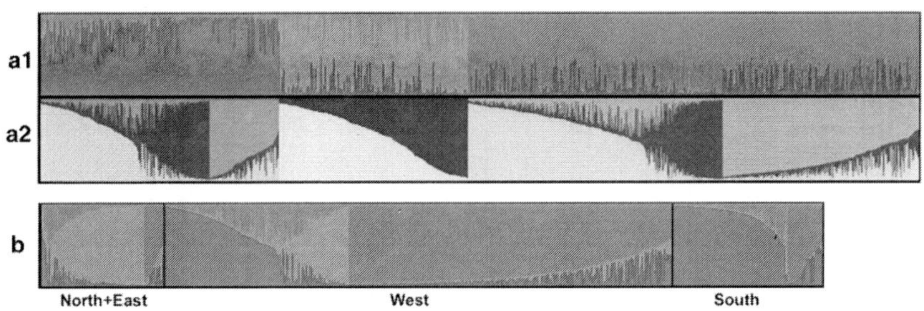

Figure. 1 Graphical display of the results of the Structure analyses. **a1**) Proportions of ancestry of 1859 unique diploid apple genotypes for $K = 3$ groups inferred with Structure v.2.3.4 software [67]. Each genotype is represented by a vertical bar partitioned into $K = 3$ segments representing the estimated membership fraction in three groups. The three groups are depicted using the following color codes: Red = group K1; Blue = group K2; Green = group K3. **a2**) Proportions of ancestry of the same 1859 genotypes following a nested Structure analysis within each previously defined group. For K1 and K3 three subgroups are shown and for K2 two subgroups are

shown. Each genotype is represented by a vertical bar partitioned into $K = 2$ or 3 subgroups representing the estimated membership fraction in each subgroup. Genotypes are presented in the same order than in a1. The subgroups are depicted using the following color codes: light Pink = K1.1; Purple = K1.2; dark Pink = K1.3; light Blue = K2.1; dark Blue = K2.2; fluorescent Green = K3.1; dark Green = K3.2; light Green = K3.3. **b**) Proportions of ancestry of 1653 unique diploid apple genotypes with known European region of origin for $K = 3$ groups inferred with the same Structure analysis as in a. The genotypes are sorted according to their European region of origin (North + East, West, and South)

allelic variation within each group, with allelic richness ranging between 16.0 (K3) and 18.6 (K1) (Table 2). Genetic discrimination between the three groups was confirmed through a multivariate Principal Coordinate Analysis (PCoA) (Fig. 2). In the bi-dimensional plot, K1 was located mostly to the left of the Y axis, and K2 mostly below the X axis, while K3 occurred to the right of the Y axis and mostly above the X axis. A Neighbor-joining tree also showed three different main clusters (Fig. 3), supporting the identification of the three groups by the Bayesian method.

Table 2 Descriptive information for each of the three major groups and eight subgroups of genotypes identified by the Bayesian model-based clustering method

	Number of genotypes in the group/subgroup			**Number of alleles**				
Group/Subgroup	**Number Genotypes**	**Frequency of genotypes with $qI \geq 0.8$**	**He**	**Total**	**Private**	**Unique**	**A**	**Allelic richness**
K1	506	60 %	0.823	307	34	16	19.19	18.63
K2	401	57 %	0.816	287	23	15	17.94	17.76
K3	952	67 %	0.801	294	22	14	18.36	15.99
K1.1	209	42 %	0.842	282	17	12	17.63	16.38
K1.2	149	54 %	0.789	215	3	1	13.44	13.2
K1.3	148	36 %	0.761	228	6	3	14.25	13.86
K2.1	244	48 %	0.818	268	14	11	16.75	14.73
K2.2	157	53 %	0.778	211	5	4	13.19	12.67
K3.1	375	41 %	0.775	242	7	6	15.13	12.32
K3.2	162	57 %	0.76	171	0	0	10.69	10.31
K3.3	415	51 %	0.809	255	14	8	15.94	13.43

Summary statistics include the partitioning of number of individuals in each group, expected heterozygosity (He), total, private, unique, and average number of alleles (A). Allelic richness is scaled to the smallest group (K2; $N = 401$) or subgroup (K1.3; $N = 148$)

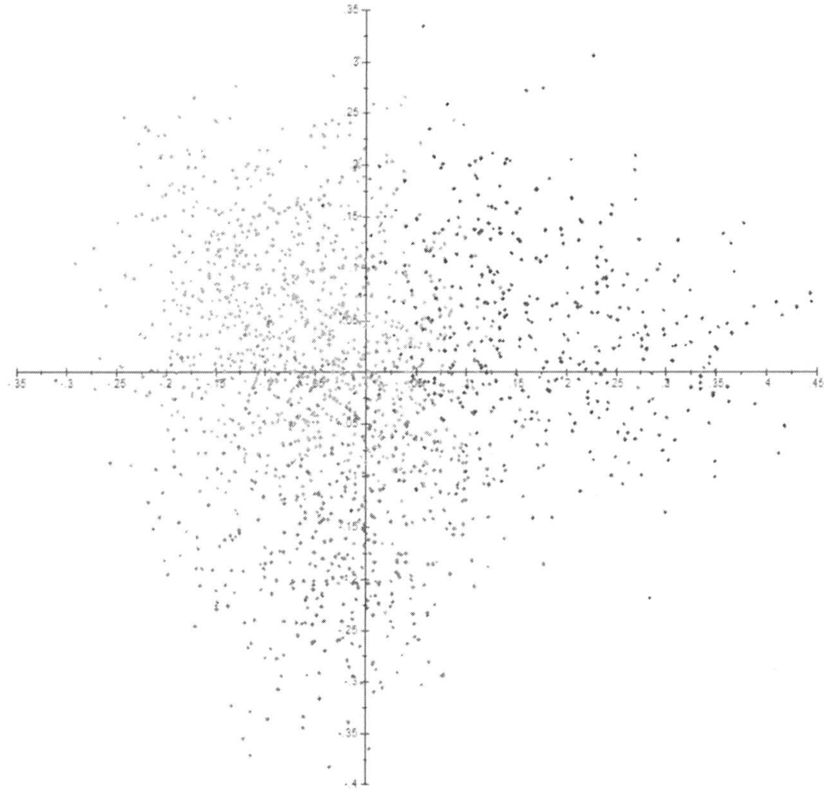

Figure. 2 Scatter plot of the Principal Coordinate Analysis (PCoA) of the 1859 apple accessions based on the 16 SSR data. The three groups are depicted using the following color codes: Red = group K1; Blue = group K2; Green = group K3

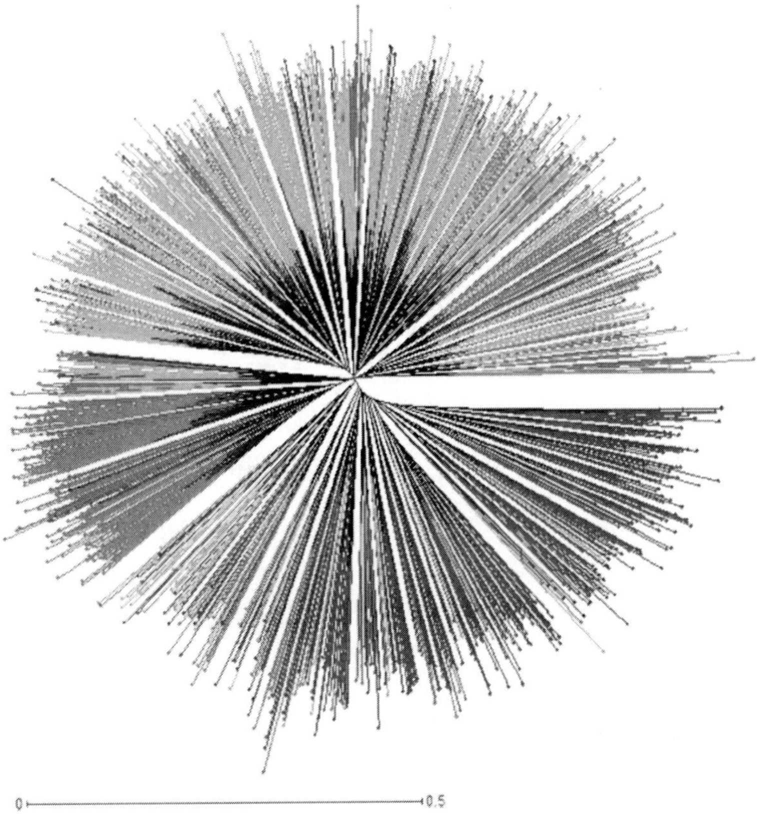

Figure. 3 Neighbor-joining dendrogram based on simple matching dissimilarity matrix calculated from the dataset of 16 SSR markers for the 1859 genotypes clustered in the three groups revealed by the Bayesian model-based clustering method. The three groups are depicted using the following color codes: Red = group K1; Blue = group K2; Green = group K3

The genetic differentiation between the three designated geographic regions of origin was low ($F_{ST} = 0.021$, $P < 0.001$, Table 3), suggesting a weak genetic structure for this crop at a European scale in terms of geographical origin. The level of genetic differentiation between the three groups inferred by Structure was only slightly higher ($F_{ST} = 0.031$, $P < 0.001$). The largest differentiation between pairs of groups was found between Northern + Eastern and Southern germplasm ($F_{ST} = 0.042$, $P < 0.001$), whereas much lower F_{ST} values were found between the Western and each of the Northern + Eastern ($F_{ST} = 0.023$, $P < 0.001$) and Southern ($F_{ST} = 0.015$, $P < 0.001$) materials.

Table 3 Analysis of molecular variance (AMOVA) based on the 16 SSR loci of the apple germplasm evaluated in this study corresponding to three regions of origin (Northern + Eastern, Southern and Western Europe) and groups and subgroups defined by Structure analysis

Populations	***df*** [a]		**Variance components (%)**		
	W [b]	***A*** [c]	***W***	***A***	***p* value**
3, geographic origins	1653	2	97.9	2.1	0.001
3, groups defined by Structure	1859	2	96.9	3.1	0.001
3, subgroups of K1	506	2	96.3	3.7	0.001
2, subgroups of K2	401	1	96.6	3.4	0.001
3, subgroups of K3	952	2	97.3	2.7	0.001
8, subgroups (K1+ K2+ K3)	1859	7	95.6	4.4	0.001

[a] *df* degrees of freedom

[b] *W* within populations

[c] *A* among populations

The relationship between membership of accessions within the three groups defined by Structure and their geographical regions of origin was also analyzed. 80 % and 75 % of the accessions from Northern + Eastern and Southern Europe clustered in K2 and K1 respectively. The relationship between the material with Western European origin and the third group (K3) was less evident (63 %), but still visible by comparison (Fig. 1b). Although the genetic differentiation revealed between the three groups defined by Structure was not very high, the existence of a relationship between the grouping by geographical regions of origin of the accessions and the three inferred groups is noteworthy. Furthermore, when considering the specific country of origin attributed to the cultivars, the distribution within the three Structure-defined groups appears to follow a clear gradient from North(East) to South of Europe (Fig. 4); the cultivars from Northern Europe and Russia were mainly assigned to the K2 group and the Spanish and Italian cultivars were mainly assigned to the K1 group, with intermediate patterns found for those countries located at the interfaces of the broad regions.

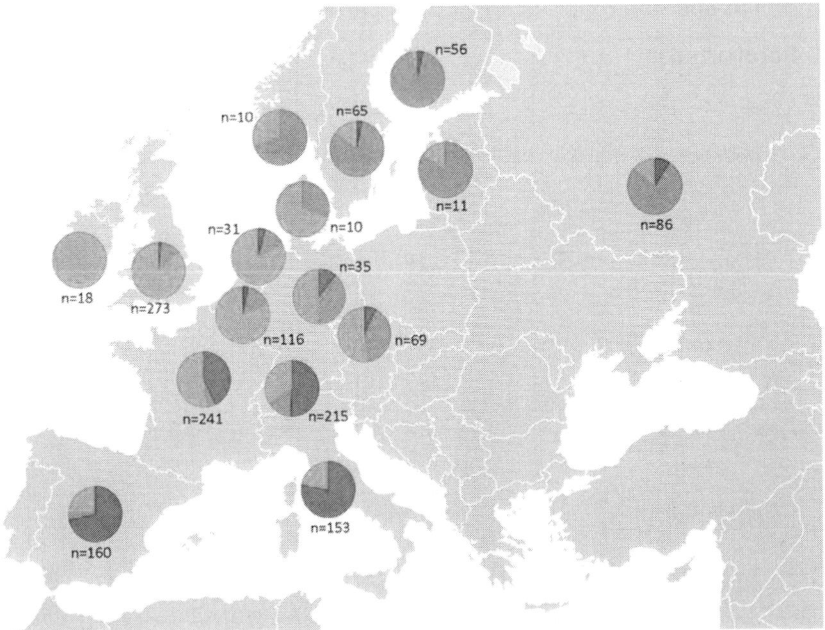

Figure. 4 Genetic composition of the groups of cultivars clustered by country of origin for $K = 3$ groups inferred with Structure. For the detailed country list, see Additional file 1. The pies represent the proportion of each group in each country; color codes are as per Fig. 1 a1

Within the admixed accessions (i.e., $qI < 0.8$) for which the geographical regions of origin (Northern + Eastern, Southern and Western) was known, we defined a membership coefficient threshold ($qI < 0.55$) with the aim of identifying genotypes unambiguously in admixis, in order to examine whether a supplemental relationship could be found between geographical region and grouping by Structure for the admixed material. For the unambiguously admixed material (i.e., $qI < 0.55$) of Southern European origin, the average proportion of ancestry (qI) was 0.45 to K1 (the group mostly associated with material from Southern Europe), followed by 0.42 to K3 and 0.13 to K2, the groups mostly composed by material from Western and Northern + Eastern European origins, respectively (data not shown); a slightly less pronounced, but complementary, pattern was observed for the unambiguously admixed germplasm (i.e., $qI < 0.55$) of Northern + Eastern Europe with average proportions of ancestry of 0.43, 0.35 and 0.22 to K2,

K3 and K1, respectively. For the unambiguously admixed material (i.e., $qI < 0.55$) of Western origin the average proportion of ancestry to each of these three groups was almost identical (approximately 1/3). This result was in line with the lower F_{ST} values found between the groups K1/K3 ($F_{ST} = 0.024$, $P < 0.001$) in comparison with the slightly higher differentiation between the groups K1/K2 ($F_{ST} = 0.039$, $P < 0.001$) and K2/K3 ($F_{ST} = 0.036$, $P < 0.001$). The dispersion of the three groups in the PCoA plot was also in agreement with these results, showing the highest overlap between K1 and K3 followed by K2 and K3.

Nested-Bayesian clustering approach: substructuring of the diversity

In order to investigate the substructuring of the diversity within each of the three groups identified in the initial analysis we used a nested application of the Structure software. To do this, the three groups were analyzed independently. To evaluate the strength of the hypothetical subdivisions (i.e., subgroups) within each group, simulations for each K value were examined, paying attention to the internal consistency between the runs, the mean proportion of ancestry of accessions within each subgroup, and the proportion of accessions unequivocally assigned ($qI \geq 0.80$).

The analysis of the relationships between K and ΔK for K1 suggested a probable subdivision of this material into three subgroups and the assignment of genotypes was well correlated between runs. The average proportion of ancestry for the accessions clustered in the three subgroups of K1 was 0.75, with 44 % of the accessions showing strong assignments. Two subgroups for K2 and three for K3 were similarly established. In both cases, the assignment of genotypes was well correlated between runs, and almost identical average proportions of ancestry to those for the subgroups of K1 were obtained with slightly higher proportions of strongly assigned accessions (47 % and 50 % respectively). Secondary peaks at other K values were also explored but these subdivisions had less statistical support (data not shown). Therefore, we adopted eight subgroups as the most suitable partitioning degree of substructuring (Fig. 1 a2). For these eight subgroups the affinity of almost half of the individuals (47 %) to their respective subgroups was strong and the assignment of admixed accessions was consistent between runs. The examination of the eight subgroups showed considerable differences in size, ranging from 148 (K1.3) to 415 (K3.3) genotypes, and variable proportion of accessions strongly assigned to the inferred subgroups (Table 2). K3.2 was the subgroup with the highest proportion of strongly assigned genotypes (57 %), whereas K1.3 had the highest proportion of admixed accessions. The proportion of accessions unambiguously assigned for the remaining six subgroups ranged from 41 %

to 54 %, whereas the mean proportion of ancestry for the accessions clustered in each one of the eight subgroups was very stable (≈0.75).

The analysis of the relationship between the different subgroups and the putative countries of origin of the germplasm indicated potentially interesting correlations, especially for groups K1 and K3. About 70 % of the subgroup K1.2 consisted of germplasm originating from Spain. Similarly, 46 % of the subgroup K1.1 and 50 % of the subgroup K1.3 consisted of germplasm originating from Switzerland and Italy, respectively (Additional file 2); the latter subgroup was also composed of a further 39 % of the cultivars with a French origin and interestingly, a significant proportion of these were attributed to Southeastern France (data not shown). The disentangling of the substructuring pattern therefore allowed not only the dissection of the internal distribution of the diversity within group K1, but also the detection of three subgroups strongly associated with some particular countries of origin. With respect to the collections from the Northern + Eastern part of Europe (Sweden, Finland and Russia), no clear differentiation of the germplasm in the two subgroups of K2 was observed (Additional file 2). For the subgroup K3.1, about half of the germplasm consisted of cultivars from either the United Kingdom or France. All of the 40 cultivars selected in the French collection as being recently bred, clustered in a single small subgroup (K3.2) which was mostly composed of English, US and, perhaps more surprisingly, Spanish cultivars. Major standard cultivars such as 'Golden Delicious', 'Red Delicious', 'Jonathan' and 'Ingrid Marie' were also assigned to this subgroup, as well as 'Cox's Orange Pippin' and 'James Grieve'. Interestingly, most of the 40 cider apple cultivars (87 %) were assigned to one subgroup (K3.3) which was mostly composed of French, English, and Swiss cultivars. The other standard cultivars were assigned to the latter subgroup and to subgroup K3.1.

Genetic diversity estimates were calculated for all the subgroups obtained by the nested Bayesian model-based clustering (Table 2). While *He* ranged from 0.76 (K1.3 and K3.2) to 0.84 (K1.1), indicating a high level of heterozygosity contained in all the subgroups, the percentage of alleles represented in each one of the eight subgroups was very variable, ranging from 46 % (K3.2) to 76 % (K3.1). Some private alleles were identified in all subgroups except for K3.2. They were most abundant in K1.1, but a considerable number of them were found also in K3.3 and K2.1. Most of the private alleles (approx. 72 %) were also unique as they were identified in only one accession. To properly evaluate the allelic diversity between the eight subgroups, we applied a rarefaction approach to compensate for the differences in subgroup size. The allelic richness obtained for the eight subgroups supported the previous results, confirming the highest diversity in K1.1 and the lowest diversity in K3.2.

Estimates of genetic differentiation showed that only 3.7 % (K1) and 3.4 % (K2) accounted for variation among subgroups within groups (Table 3). The genetic differentiation between the subgroups into which K3 was subdivided was considerably lower (Table 3). Considering the eight subgroups obtained by the overall Nested Bayesian approach, the results showed that variation among subgroups accounted for 4.4 % of the total variation. Regarding the F_{ST} pairwise tests between subgroups (Table 4), irrespective of whether they belonged to the same group or not, the highest F_{ST} corresponded to the pair K1.3/K2.2 ($F_{ST} = 0.087$, $P < 0.001$), followed by K1.3/K2.1 ($F_{ST} = 0.077$, $P < 0.001$), and the lowest to the pairs K1.1/K3.3 ($F_{ST} = 0.016$, $P < 0.001$) and K3.1/K3.3 ($F_{ST} = 0.023$, $P < 0.001$).

Table 4 Pairwise estimates of F_{ST} among the eight subgroups obtained by the nested Bayesian clustering approach

Subgroup	**K1.1**	**K1.2**	**K1.3**	**K2.1**	**K2.2**	**K3.1**	**K3.2**	**K3.3**
K1.1	—							
K1.2	0.030	—						
K1.3	0.035	0.051	—					
K2.1	0.028	0.067	0.077	—				
K2.2	0.049	0.076	0.087	0.035	—			
K3.1	0.034	0.051	0.061	0.061	0.055	—		
K3.2	0.051	0.065	0.070	0.070	0.058	0.029	—	
K3.3	0.016	0.042	0.060	0.038	0.051	0.023	0.038	—

All the estimates were highly significant ($P < 0.001$)

Parentage reconstruction

Two-parents-offspring relationships within the 1,859 diploid genotypes were explored using CERVUS software. A total of 46 putative trios (offspring and two inferred parents) were identified with high (95 %) confidence level. These consisted of two already documented trios, ('Calville Rouge du Mont Dore' and 'Belle de Mleiev' and their parents; [23]), as well as another 10 recent and 34 old cultivars (Table 5). The two parents of the 10 modern cultivars, for which full parentage was already documented were correctly inferred (e.g., 'Heta', 'Jaspi' and 'Pirkko' = 'Lobo' x 'Huvitus', 'Pirja' = 'Huvitus' x 'Melba', or 'Mio' = 'Worcester Pearmain' x 'Oranie'). In most cases, the two parents of the older cultivars were not known and thus newly inferred (Table 5). Inferred parentage was found for old cultivars from various European countries (6x for Italy; 4x for Great Britain, Switzerland, Czech Republic, and Sweden; 3x for Germany and Spain; 2x for Belgium). Perhaps unsurprisingly, some accessions were more frequently inferred as parents, such as the two French cultivars 'Reine des Reinettes' (= 'King of the Pippins') or 'Transparente de Croncels' which were each identified three times. Geographic convergence of parentage was frequently observed (e.g., 'Kramforsäpple' = 'Sävstaholm' x 'Åkerö', all three from Sweden; 'Beauty of Moray' = 'Keswick Codlin' x 'Stirling Castle', all three from Great Britain; 'Roja de Guipuzcoa' = 'Urte Sagarra' x 'Maxel Gorri', all three from Spain; or 'Scodellino' = 'Abbondanza' x 'Decio', all three from Italy). But hybridizations between cultivars from distant countries were also observed (e.g., 'Rotwiler' presumably from Switzerland = 'King of the Pippins' x 'Alexander' from France and Ukraine, respectively; or 'Godelieve Hegmans' from Belgium = 'Red Astrakan' x 'Transparente de Croncels' from Russia and France, respectively). It should be noted that the female and male status of the inferred parents could not be specified from the available SSR markers.

DISCUSSION

Identification and redundancy

The exchange of genotyping data between research units has increased considerably in recent years, with the aim to investigate the extent and distribution of diversity for specific crops at a wide geographic scale. In this study, the application of a common set of 16 SSR markers on a wide set of

Table 5 Full parentages of 46 apple cultivars inferred within the set of the 1859 apple unique accessions using 16 SSR markers with their accession codes, accession names (AcceNumber), their duplicate codes according to the SSR profile (FBUNQ) and their putative country of origin (OriginHist)

Offspring ID	Accename	FBUNQ	OriginHist	First candidate ID	Accename	FBUNQ	OriginHist	Second candidate ID	Accename	FBUNQ	OriginHist	Status[j]
X1618	Calville Rouge du Mont Dore	963	FRA	BAL086	Alexander	30	UKR	DCA_I05	Mele Ubriache[a]	361	FRA	doc.
X1846	Belle de Mleiev	1563	-	X0557	Mc Intosh	508	CAN	1957218	King of the Pippins	37	FRA	doc.
BAL035	Hета	1774	FIN	CRAW-0433	Lobo	788	CAN	FIN09	Huvitus	4922	FIN	recent
BAL039	Jaspi	1776	FIN	CRAW-0433	Lobo	788	CAN	FIN09	Huvitus	4922	FIN	recent
FIN18	Pirkko	4930	FIN	CRAW-0433	Lobo	788	CAN	FIN09	Huvitus	4922	FIN	recent
BAL010	Rödluvan	107	SWE	CRAW-0433	Lobo	788	CAN	BAL023	Barchatnoje	1768	RUS	recent
BAL109	Arona	1819	LVA	CRAW-0433	Lobo	788	CAN	BAL112	Iedzenu	1822	LVA	recent
BAL176	Nyckelby	1861	SWE?	CRAW-0433	Lobo	788	CAN	1957168	Cox's Pomona	2033	GBR	recent?
BAL059	Pirja	444	FIN	FIN09	Huvitus	4922	FIN	CRAW-0836	Melba	167	CAN	recent
FIN43	Pirkkala	4949	FIN	BAL042	Kaneläpple	512	RUS	FIN14	Lavia	4926	FIN	recent
BAL154	Mio	543	SWE	CZ_G2D_0045	Worcester parména	550	GBR	BAL056	Oranie	48	SWE	recent
BAL052	Oberle	1784	CAN	BAL027	Early Red Bird	236	CAN	CRAW-0266	Stark Earliest	468	USA	old
BAL091	Forlovningsäpple	1804	SWE	CHE0893	Heuapfel	1248	CHE	X1646	Saint Germain	31		old
BAL167	Vallidaäpple	1853	SWE	CHE0893	Heuapfel	1248	CHE	BAL179	Göteborgs Flickäpple	1863	SWE	old
BAL099	Kramforsäpple	1811	SWE	BAL161	Säystaholm	573	SWE	BAL195	Åkerö	308	SWE	old
BAL158	Stenkyrke	463	SWE	BAL171	Fulleró	1857	SWE	CZ_LJ_0045	Malinové podzimní[b]	722	POL	old
FIN07	Finne	4920	FIN	BAL161	Säystaholm	573	SWE	FIN08	Grenman	4921	FIN	old
1942035	Beauty of Moray	1925	GBR	2000053	Keswick Codlin	1438	GBR	2000090	Stirling Castle	2103	GBR	old
1951242	Brighton	2011	NZL?	X4915	Red Dougherty	939	NZL	CZ_LC_0411	Hlaváčkovo[c]	23	USA	old
1957208	Ben's Red	2035	GBR	CRAW-0020	Devonshire Quarrenden	622	GBR	1965077	Box Apple	2025	GBR	old
1965004	Fred Webb	2054	GBR	1946088	Winter Marigold	324	GBR	1957181	Gascoyne's Scarlet	45	GBR	old
2000083	Rivers' Early Peach	2099	GBR	2000051	Irish Peach	2093	IRL	BAL169	Aspa	1855	SWE	old
BMN0011	Roja de Guipuzcoa	3854	ESP	BMN0017	Urte Sagarra	956	ESP	BMN0171	Maxel Gorri	3896	ESP	old
BMZ016	Cella	3935	ESP	BMN0022	Erreka Sagarra	957	ESP	X5102	Bisquet	535	FRA	old
BMN0070	Madotz 01	3869	ESP	1957218	King of the Pippins	37	FRA	X7201	Transparente de Croncels	62	FRA	old
CHE1322	Rotwiler	1271	CHE?	1957218	King of the Pippins	37	FRA	BAL086	Alexander	30	UKR	old
CHE1288	Roseneggler	3718	CHE	1957218	King of the Pippins	37	FRA	CZ_BoN_0429	Trat. Laze	2281	CZE	old
CHE0032	Ernst Bosch	1003	DEU	1947074	Ananas Reinette	69	NLD	CZ_GF_0415	Evino[d]	7	GBR	old
CHE0168	Elbner	3258	CHE	CRAW-0836	Melba	167	CAN	CZ_BoN_0424	Trevinské červené[e]	71	USA	old
CHE1390	Nefeler	3589	CHE	KRAS123	Papirovka	25	RUS	X7199	Rose de Berne	63	CHE	old
CRAW-0226	Laubain n°1	2126	BEL	CRAW-0086	Bismarck	3	AUS	CZ_GS_0478	Ušlechtilé žluté[f]	90	GBR	old
CRAW-0105	Godelieve Hegmans	2116	BEL	BAL175	Röd Astrakan	82	RUS	X7201	Transparente de Croncels	62	FRA	old
CZ_BB_0442	Nathusiovo	2268	DEU	CZ_GL_0461	Bláhovo Libovické	2311	CZE	X7201	Transparente de Croncels	62	FRA	old
CZ_BB_0434	Panenske veliké	2265	CZE	CZ_GP_0469	Panenské české	1529	CZE	X1344	Reinette de Landsberg	61	DEU	old
CZ_GL_0412	Proche	2308	CZE	CRAW-0425	Calville Rouge d'Automne	13	FRA	X1344	Reinette de Landsberg	61	DEU	old
CZ_BoN_0421	Moravcovo	2283	CZE	CZ_GP_0469	Panenské české	1529	CZE	CHE0269	Pomme Bölleöpfel	1377	-	old
CZ_GL_0456	Bláhův poklad	694	-	CZ_GL_0461	Bláhovo Libovické	2311	CZE	CZ_GG_0438	Malinové hornokrajské[g]	47	NLD	old
CZ_GP_0473	Petr Broich	2321	DEU	1957175	Annie Elizabeth	15	GBR	2000075	Peasgood's Nonsuch	51	GBR	old
CZ_BB_0458	Šarlatová parmena	2269	CZE	CZ_GG_0442	Malinové holovouské	452	CZE	X8233	Petite Madeleine	24	-	old
CZ_BB_0466	Podzvičinské[h]	231	-	X0691	Boiken	106	DEU	X1071	Reinette de Caux	629	NLD	old
DCA_017	S.Giuseppe	1646	ITA	DCA_D30	Abbondanza	327	ITA	DCA_C44	Rambour Frank (MI)	493	FRA	old
DCA_H03	Scodellino	1642	ITA	DCA_D30	Abbondanza	327	ITA	DCA_E52	Decio	397	ITA	old
DCA_E72	Gelato Cola	330	ITA	DCA_E69	Gelato (CT)	780	-	DCA_F74	Limoncella (TN)[i]	708	ITA	old
DCA_H62	Liscio di Cumiana	1713	ITA	DCA_H29	Carla	114	-	DCA_C21	Renetta di Grenoble	263	ITA	old
DCA_I96	Res Magior	1658	ITA	DCA_I60	Rus d' Musiot	321	-	X1115	Rome Beauty	334	USA	old
DCA_F47	Mela Golden Simile di Villa Collemandina	1692	ITA	DCA_A20	Rosa Mantovana (TN)	101	ITA	CRAW-0025	Yellow Bellflower	77	USA	old

[a]DCA_I05 'Mele Ubriache' duplicate with 'Calville Rouge d'Hiver' [23]
[b]based on 11 SSR [64] the accession CZ_LJ_0045 'Malinové podzimní' was shown to be duplicated with 'Danziger Kantapfel'
[c]based on 11 SSR [64] the accession CZ_LC_0411 'Hlaváčkovo' duplicate with 'Northen Spy'
[d]based on 11 SSR [64] the accession CZ_GF_0415 'Evino' duplicate with 'Mank's Codlin'
[e]based on 11 SSR [64] the accession CZ_BoN_0424 'Trevinské červené' duplicate with 'King David'
[f]based on 11 SSR [64] the accession CZ_GS_0478 'Ušlechtilé žluté' duplicate with 'Golden Noble'
[g]based on 11 SSR [64] the accession CZ_GG_0438 'Malinové hornokrajské' duplicate with 'Framboise'
[h]based on 11 SSR [64] and on 13 SSR [14] the accession CZ_BB_0466 'Podzvičinské' duplicate with 'Altlander Pfannkuchenapfel' and 'Thurgauer Kent'
[i]based on 11 SSR [64] the accession DCA_F74 'Limoncella' (TN) duplicate with 'Cola'
[j]recent or old cultivars ; doc. = inferred parentage already documented in [23]

dessert apple cultivars distributed across three broad European regions allowed the detection of redundant accessions and duplicated genotypes between and within collections, and the description of the structuration of a significant part of the European apple diversity. Cross-comparison of SSR data in attempts to combine datasets from multiple sources has often been problematic due to challenges in harmonizing the allelic sizes between different laboratories [18, 29, 30]. By combining existing data over numerous shared reference accessions in our collections with the re-genotyping of a subset of the accessions, we were able to strongly secure the SSR allele adjustment over sites. This dataset represents a highly valuable resource for the comparison of apple germplasm collections throughout Europe and the rest of the world. Taking into consideration the rich allelic diversity present in the European apple germplasm, it would be useful to identify a relatively small set of varieties that offer a good representation of the allelic variability identified in this germplasm to act as an internal control (i.e., a reference set) between laboratories for future use.

Interestingly, duplicate groups involving accessions from different collections underlined some putative drift in the cultivar denomination. Some good examples were 'Pott's seedling' and 'Pottovo' (FBUNQ14), or 'Signe Tillish' and 'Signatillis' (FBUNQ34). In addition, 'sports' are often given derivative names (e.g., 'Crimson Peasgood' as a sport of 'Peasgood's Nonsuch') but the current analysis was not set up to distinguish between clones and 'sports' of cultivars with potential morphological differences. Many likely errors in denomination of genotypes were also detected when multiple representatives of a given cultivar were detected within a group, but a single supposed representative was obviously outside of the group and was often associated with representatives of a different cultivar. For example, 'Drap d'Or' and 'Chailleux' (FBUNQ92) are known to be synonyms used in France for the same cultivar, and accession DCA_D35 'Drap Dore', which was found to belong to the group FBUNQ50, was most likely a denomination error since almost all other members of this group were 'Winter Banana'. In other cases, accessions with uncertain denomination could be resolved, such as CRAW-1858 'Reinette Baumann?' (FBUNQ21) and accession CRAW-1108 'Peasgood Nonsuch?' (FBUNQ51) for which the molecular analyses confirmed that they were most likely 'true-to-type' cultivars. The question of 'trueness-to-type' is a major issue in apple germplasm management where extensive budwood exchange between regions and countries has occurred for centuries. Indeed, an erroneously denominated accession can be transmitted from collection to collection for years, such that a large number of representatives within a duplicate group (as per the present study) should not always be considered definitive proof of the trueness-to-type of accessions but this objective evidence is extremely valuable in highlighting issues to resolve. Since genebank curators have often collected material of old cultivars from private gardens or from tree pasture orchards, unidentified or misidentified material can later be detected either by classical phenotypic characters and/or by using genetic markers. As an example, this study showed that an old so called local cultivar 'Madame Colard' (CRAW-0365 – FBUNQ72), described to have been raised in 1910 by the nurseryman Joseph Colart at Bastogne (Belgium), exhibited the same SSR profile as the old English apple cultivar called 'Royal Jubilee' (UK-NFC 2000085) raised already in 1888. Further comparison with historical descriptions could conclude that they are the same cultivar. Additional insights from the passport data of accessions would be needed to help in tracing the transmission of the material from collection to collection and pomological characterization will be required to compare accessions to published descriptions of the variety. This will remain a task for the curators of collections, in order to improve curation of germplasm in a coordinated way.

It is important to note that the criteria used to select the accessions at the country-level were not always the same. For instance, the INRA and UNIBO material corresponded to former "core collections" built to encompass a large variability not restricted to the national/local accessions [23, 24]. Conversely, the UK-NFC and FRUCTUS material was restricted to older diploid accessions considered to derive from UK and Switzerland, respectively. A similar, despite less stringent situation was applied also for CRA-W, RBIPH, SLU, and the Spanish accessions (UPNA, UDL and EEAD). For MTT, NCRRI, VNIISPK, and KNAU, the national representativeness was more limited and strictly restricted to accessions considered to be emblematic landrace cultivars. The germplasm was thus somewhat heterogeneous in nature, but still allowed a broad examination of the European dessert apple diversity. In the future, it will be useful to enlarge the dataset to include additional accessions from the collections considered here as well as other European collections [11, 31] or collections from other regions worldwide [32, 33, 34] to provide a wider perspective on genetic resource conservation of apple worldwide.

Genetic diversity

The high level of diversity and heterozygosity in apple germplasm at a European level agreed with previous results obtained at collection-scale in several European countries, e.g., Italy [24], Spain [26], France [23], Sweden [22], Czech Republic [25] or Switzerland [14]. The large diversity found is consistent with the weak bottleneck effect reported in connection with the domestication of this species [35, 36, 37]. Probably a combination of factors are involved: i) vegetative propagation methods that have been adopted since ancient times favoring the dispersal of cultivars across geographic regions [38, 39], ii) forced allogamy due to the self-incompatibility system of *Malus* × *domestica* [40], iii) multiple hybridization events at each geographical region combined with human activities, e.g., selection and breeding [36, 37] and, iv) diversifying selection associated with adaptive criteria for the subsistence in diverse agricultural environments [41, 42]. Interestingly, the distribution of private SSR alleles over the countries of origin of the unique genotypes was somewhat unbalanced at the European level with much higher occurrences in genotypes assigned to Switzerland, Italy or Russia than in genotypes originating from Northern-Western Europe. Whilst these findings should be considered with caution because of possible biases linked to the initial sampling or to the size differences of the genotype sets, this study underlines that accessions originating from Southern Europe and Russia could be expected to bring original genetic diversity into modern breeding programs especially for traits related to more extreme climate adaptation. Overall, the highly diverse germplasm studied here contains much more genetic variation than do modern apple cultivars, many of which having been selected for optimal performance within a narrow range of environmental conditions [5, 37, 42].

Coordinated actions: a key point for better knowledge of the resources conserved

This large-scale analysis in apple germplasm constitutes a good example of the efficiency and value of coordinated international actions to enhance the knowledge of diversity conserved at a European level. The results obtained offer a valuable step to undertake actions to coordinate European resources towards optimizing the management of apple germplasm across Europe in line with the aspirations of the European Cooperative program on Plant Genetic Resources (ECPGR). The results also offer a potential starting point that may open new opportunities for apple breeding in the near future. All breeding advances are built upon the diversity available, and a key role of the germplasm collections is to help safeguard natural forms of genetic variation and to make them accessible to plant biologists, breeders, and other key users [15]. The extensive germplasm evaluated in this study consisted mainly of old and/or locally grown accessions across Europe, many of which remain underutilized in cultivation or breeding programs. The preservation of traditional cultivars in living germplasm collections must be regarded as an invaluable reservoir of insufficiently explored genetic diversity that may become useful for apple breeding in a near future, and the establishment of coordinated genetic data is hoped to increase the accessibility of this material to breeding programs. From the perspective of modern-day fruit production, most of these old varieties would now be considered as obsolete since they are not particularly well-adapted to current agricultural practices and marketing. Nevertheless, this material should be considered as a reservoir of potentially interesting genes to be used for further improvement. This is particularly relevant in a crop like apple, for which the current production is highly dependent on a very limited number of cultivars with a narrow genetic basis for the bulk of current production [5]. As an example, it can be mentioned that 50 % of the commercially marketed apple production in the European Union consists of only four cultivars, 'Golden Delicious', 'Gala', 'Idared' and 'Red Delicious' [43]. The low diversity of the subset of elite cultivars used for commercial production during recent decades is likely to result in a bottleneck hampering future genetic improvement [37]. The recognition of this situation should encourage the establishment of coordinated actions across different levels (regional, national and international scales) to define strategies for the efficient conservation of the genetic resources of this species.

Genetic structure: major divisions and substructuring of the diversity

The attribution of country of origin to traditional cultivars can be a matter of endless debate, especially for those dating back two-three centuries or more. Initial descriptions in pomologies and booklets can be subject to errors in denomination confused by historical distribution and renaming, resulting in synonymy, as well as the re-use of old names for more recent findings or misidentifications. This is less problematic for the better known old cultivars as many of them have been widely documented and monitored over years in several countries. However, for local cultivars and/or landraces where less information is available, the correct attribution can be complicated, especially between neighboring countries. It is also worthy of note that the 'country of origin' relies on a political construct, which can be prone to significant change within the potential lifetime of many varieties of apple (and other long lived perennial crops). Therefore, we first used a conservative approach and discussed our findings in terms of three broad European regions of origin. Then, we analyzed the structuration at a country-scale, but noting that the exact attribution of a given country to a genotype was not always unanimously agreed so that this finer level of analysis should be considered with an element of caution.

Using a Bayesian model-based clustering method we were able to initially discern the existence of three robust groups reflecting major divisions of the germplasm. These groups were linked with the three geographical regions of origin, although differentiated only to a low degree. This would reflect a situation whereby the cultivars from a given region were more frequently derived from crosses between parental cultivars from the same region than from cultivars from elsewhere. Nevertheless, the migration of the plant material associated to human movement together with hundreds of years of empirical selection may have caused a significant gene flow across Europe. This is clearly indicated by the low genetic differentiation between groups and has shaped the overall pattern of genetic diversity. A spatially and temporally dynamic process where seeds and mainly graftwood were exchanged between geographically distinct populations has contributed to the increase of the genetic diversity in each area through unintentional gene flow or human-mediated intentional crosses [35, 36, 44]. The background common to other long lived tree fruits, including factors such as multiple origins of cultivated populations, ongoing crop-wild species gene flow and clonal distribution of genotypes together with the features associated with fruit tree species (lengthy juvenile phase, extensive outcrossing, widespread hybridization or mechanisms to avoid selfing) has defined the way they evolve in nature and resulted in extensive population genetic variation, but limited population structure [44]. A possible cause of divergence between the three identified groups could be the differential adaptation to distinct environmental conditions as are the case between Southern, Western and Northern + Eastern Europe. A similar situation was postulated for grapevine cultivars where the genetic structure appeared to be strongly shaped by geographic origin and intentional selection [13]. But since selection causes differentiation in particular regions of the genome on which selection pressure is acting [45], another likely cause of the population structure is genetic drift (i.e., changes in allelic frequencies caused by chance events) as also shown in e.g., apricot [46]. Together with selection, migration and drift can shape the local adaptation of species [47].

Although there may have been some mistakes in attributing cultivars to country of origin, the genetic makeup of the cultivars at the European level clearly appeared to show a North-East to South gradient. Interestingly, some countries exhibit intermediate marker data patterns in consistency with their intermediate geographic positions. This was clearly manifested at the national scale for the German and Czech cultivars which were shared between K2 and K3 groups. Similarly, the French and Swiss cultivars were shared between K1 and K3 groups. By contrast, cultivars from Southern Europe (Spain and Italy), from Northwestern Europe (United Kingdom and Ireland, Belgium, the Netherlands), and from North + Eastern Europe (Sweden and Finland) and Russia were mostly assigned to a single group (K1, K3, and K2, respectively). For the admixed germplasm from Southern and Northern + Eastern European geographical regions of origin, a certain degree of introgression with the Western germplasm was also indicated in contrast to the low contribution of the Northern + Eastern germplasm into the Southern germplasm and viceversa. Thus, in agreement with the correspondence between clustering and regions or countries of origin of the germplasm, the geographical proximity appears to align with the patterning observed in the admixed accessions.

In cases demonstrating the presence of a significant hierarchical population structure as this study suggests, this method preferentially detects the uppermost level of structure [26, 48, 49, 50]. As a consequence, when large datasets in species with a complex background are analyzed, it is possible for an underlying substructure to remain undetected within the major divisions of the germplasm. In this context, the "*nested (or two-steps) Structure*" clustering method has been shown to be an efficient tool to delineate further levels of substructure in both apple and other plant species [10, 24, 26, 49, 50, 51, 52]. In this study, the three groups inferred from the first round of Structure analysis were used as the starting point for revealing internal substructuring. Eight subgroups were identified with remarkable differences in both allelic composition and richness, as well as a considerable number of private alleles associated to particular subgroups. Nevertheless, the relationship between the placement of the genotypes in the subgroups and their country of origin varied considerably between subgroups in contraposition to the clearer and more consistent clustering trend within the three groups. As discussed earlier, this stratification may reflect historical processes of selection and adaptation to local conditions that might suggest a "*fine-delineation*" of the intra-variation within each main geographical region of origin. This is most probably the case for the K1.1 subgroup which mainly consists of Spanish cultivars and could reflect a process of both local adaption and isolation by distance related to the Pyrenean barrier. For the K1.3 subgroup, local adaptation to the Southern region could be inferred together with a potential for more intense commercial exchange between Italy and Southern France. For other subgroups, the relationship with particular countries or small regions was not obvious, but some interesting associations between subgroups of group K3 and recent cultivars and some of their founders or particular usage (cider apple cultivars) could be noticed.

Relatedness and family relationships

The previously reported parentage of 10 recent cultivars was correctly inferred in all cases. These results served as a control and validated the parentage assignment obtained with the CERVUS software [53] indicating that the number and informativeness of SSR markers were sufficient at least for these cultivars. The 16 SSR markers were nevertheless limited in their ability to infer parentages, and additional cases might have been detected with a larger number of SSR markers. In a recent paper [54], it was suggested that the number of 27 SSR loci used in that study was a minimum to be utilized for full parentage reconstruction. Basically, the LOD score tests used in the CERVUS software are computed according to the SSR allelic frequencies, and thus, parentages involving common alleles are more difficult to detect. By contrast, parentages involving low frequency and rare alleles are more easily detected. On that basis, it is worthy to note that the more frequently detected parents (i.e., 'Reine des Reinettes' = 'King of the Pippins', and 'Transparente de Croncels') are possibly representing a biased view of the frequently involved parents, as they most probably carry rare or low frequency alleles in at least some SSR loci. Putative parents present in the dataset but carrying more common alleles may have been hidden because of the statistical limits of their detection with 16 SSR markers. A similar situation was observed by [23] with the frequent appearance of 'Reine des Reinettes' as a parent of four old cultivars out of 28, using 21 SSR markers. In the near future, medium and high density SNP arrays [55, 56, 57] will provide much more power to infer parentages.

The parentage of some old cultivars was either confirmed, in the case of 'Ernst Bosch' = 'Ananas Reinette' x 'Mank's Codlin' (synonym: 'Evino') or augmented, in the case of 'Ben's Red' = 'Devonshire Quarrenden' x 'Box Apple' (Table 5) where the second parent was initially hypothesized to be 'Farleigh Pippin' [58]. Distances between the geographic origins of the inferred parents (when known), ranged from crosses between geographically close cultivars to crosses between very distant cultivars, reflecting the large gene flow across Europe caused by, e.g., extensive exchange of budwood over centuries.

Some traditional folklore about the origination of old apple cultivars could be either substantiated or refuted by the SSR-based parentage information. As one example, the old Swedish cultivar 'Förlovningsäpple' is said to derive from a locally acquired seed in Northern Sweden where only a few cold-hardy apples can be grown. The two unknown parents were here inferred to be the Swiss cultivar 'Heuapfel' and the wide-spread cultivar 'Saint Germain' (X1646) also known as 'Vitgylling' in Sweden, a name used for a group of more or less similar, white-fruited, early-ripening and winter-hardy cultivars. Interestingly, the 'Vitgylling' accession included in the present study (BAL072) did not have the same SSR profile as 'Saint Germain', but they share one allele for all 16 SSR loci and may therefore be related. In two other cases, traditional Swedish folklore indicated that a sailor brought an exotic seed to the island of Gotland and to Kramfors in Northern Sweden, respectively, resulting in 'Stenkyrke' and 'Kramforsäpple'. For 'Stenkyrke', one parent is the Swedish 'Fullerö' and the second is the German cultivar 'Danziger Kantapfel' which has been much grown in Sweden. The origin of 'Stenkyrke' is thus probably much more local than anticipated. Similarly, the surmised American sailor origin of the seed giving rise to 'Kramforsäpple' is refuted by the fact that the parents of this cultivar are the Swedish 'Sävstaholm' and 'Åkerö'.

It is important to keep in mind that trueness-to-type of the accessions is not guaranteed, thus the labeling of the offspring or the parents can be erroneous in some cases. Conversely, the inferred parentages are robustly established so that the genetic relationships between the accessions are valid independently of their names. Crosses between the two inferred parents could be performed to reproduce the cross which gave birth to the offspring cultivar, especially if genetic analysis of some particular traits of the latter genotype indicates an interesting application in plant breeding.

CONCLUSIONS

The analysis of a large and representative set of *Malus* x *domestica* genotypes indicated that apple germplasm diversity reflects its origination within three main geographic regions of Europe, and that a weak genetic structure exists at the European level. This structuring of genetic variation in European dessert apple is caused by evolutionary processes relevant to the domestication of perennial fruit species with factors such as gene flow created by, e.g., ancient roads of commerce across the continent, other human activities like intentional selection and later breeding, and genetic drift. The remarkable differences in the allelic variation found at group and subgroup levels of germplasm stratification constitute a strong indication of that the diversity is hierarchically organized into three *genepools*, with consistent evidence of a pattern of internal substructure. The potential value for modern fruit production is mostly unknown since a majority of the accessions are poorly evaluated from an agronomic point of view. Thus, phenotypic data obtained with standardized methods is required to determine the commercial potential of the preserved material and to enable its use in new crosses to increase the genetic basis of the cultivated apple.

The integration of data for collections from different European geographic regions using standardized methods will undoubtedly form an important step in developing the European strategy for conservation of apple germplasm and constitute the starting point to define a European "apple core collection". This will constitute a decisive step in the field of conservation genetics, and may also have direct implications on the improvement of our understanding of the species, including i) the identification of genomic regions associated with commercially important horticultural traits, ii) the discovery of new germplasm features that may be taken advantage of for efficient breeding and iii) the analysis of genotype x environmental interactions for studying the stability of the most economically important traits for this species.

METHODS

Plant material

Apple germplasm collections from nine European countries, plus Western part of Russia and Kyrgyzstan, were available for this study (Additional file 1): France (INRA, Institut National de la Recherche Agronomique, 399 accessions), Italy (UNIBO, University of Bologna, 216 acc.), Belgium (CRA-W, Centre Wallon de Recherche Agronomique, 408 acc.), Czech Republic (RBIPH, Research and Breeding Institute of Pomology Holovousy, 263 acc.), United Kingdom (UK-NFC, University of Reading, 310 acc.), Sweden (SLU, Swedish University of Agricultural Sciences, 199 acc.), Finland (MTT Agrifood Research, 50 acc.), Spain (UPNA, Public University of Navarre, UDL, University of Lleida, and EEAD, Aula Dei Experimental Station, 269 acc.), Switzerland (FRUCTUS, Agroscope, 237 acc.), Russia (NCRRIHV, North Caucasian Regional Research Institute of Horticulture and Viticulture, and VNIISPK, The All Russian Research Institute of Horticultural Breeding, 83 acc.), and Kyrgyzstan (KNAU, Kyrgyz National Agrarian University, 12 acc.). In all countries, the accessions were mostly chosen as old local/national dessert cultivars (registered or at least known before 1950), but 12 standard dessert cultivars were also included to strengthen comparisons between collections, namely 'Golden Delicious', 'Red Delicious', 'McIntosh', 'Rome Beauty', 'Granny Smith', 'Jonathan', 'Winter Banana', 'Ingrid Marie', 'Ananas Reinette', 'Reinette de Champagne', 'Discovery' and 'Alkmene'. Moreover, 40 old cider apple cultivars and 40 recently-bred dessert cultivars were sampled in the INRA collection in order to investigate particular patterns. Altogether, 2,446 accessions were thus considered (Additional file 1). Available collections were somewhat heterogeneous in nature as some of them corresponded to already established core collections (INRA and UNIBO) whereas others were selected for the present study thanks to available SSR marker data (UK-NFC and FRUCTUS, see below), or were chosen as a subset of mainly local cultivars (CRA-W, RBIPH, SLU, MTT, UPNA, UDL, EEAD, NCRRIHV, VNIISPK and KNAU). Cultivars that were known to be triploid or duplicated were avoided since this analysis was performed with an aim to subsequently use a major part of the material in a Genome Wide Association Study to be carried out within the EU FruitBreedomics project [28].

SSR genotyping

A set of 16 SSR markers developed by different groups [59, 60, 61, 62] was used to genotype the 2,446 accessions (Additional file 3). These SSR markers are distributed over 15 out of the 17 apple linkage groups, and 15 of them are included in a former list recommended by the ECPGR *Malus/Pyrus* working group [63]. The 16th marker of this list, NZ05g08, was replaced by the marker CH-Vf1 because the former showed either complex scoring pattern or low level of polymorphism in previous studies [23, 26]. SSR marker data were fully available for the collection from INRA [23]. SSR data were available (i.e., for some, but not all of the 16 SSR markers) for collections from UK-NFC [64], FRUCTUS [14], UPNA, UDL and EEAD [26], and UNIBO [24], so that only the missing SSR marker data were generated in the present study. Fully new SSR datasets were generated for collections from CRA-W, RBIPH, SLU, MTT, NCRRIHV, VNIISPK, and KNAU.

Forward primers were labeled with four different fluorescent dyes (6-FAM, VIC, NED, or PET) in order to be combined into four different multiplexed reactions (Additional file 3). Polymerase chain reactions (PCR) for the four multiplex PCRs were performed in a final volume of 11 μL using 10 ng of DNA template, 0.18 μM of each primer (with the exception of some markers as described in Additional file 3), and 1× PCR Master mix of QIAGEN kit multiplex PCR (Qiagen, Hilden, Germany). PCR cycling conditions were as follows: pre-incubation for 15 min at 94°C, followed by 4 cycles using a touchdown amplification program with an annealing temperature reduced by 1°C per cycle from 60°C to 55°C, followed by 34 cycles, each consisting of 30 s denaturing at 94°C, 90 s annealing at 55°C, and 60 s elongation at 72°C, the last cycle ending with a final 15-min extension at 72°C. SSR amplification products were analyzed with an ABI3730 XL sequencing system (Applied Biosystems, Foster City, CA, USA). Fragment analysis and sizing were carried out using GeneMapper v.4.0 software (Applied Biosystems, Foster City, CA, USA); chromatograms were independently read by two operators. When SSR marker data were already available and obtained at different sites, SSR allele sizes were carefully adjusted between collections, both by use of reference accessions known to be in common between collections and by re-genotyping a subset of each collection with the full set of 16 SSR markers to confirm the allele adjustment.

Diversity assessments

The multilocus SSR profiles were compared pairwise in order to establish the genetic uniqueness of each accession. Accessions were considered as duplicates if they had identical SSR fingerprints, or if they had one allelic difference for a maximum of two SSR loci thus making room for some genotyping errors and/or spontaneous SSR mutations. On this basis, redundant profiles were removed from the dataset to avoid bias in genetic analyses and duplicate groups were labeled with unique group ID codes (FBUNQ codes). An accession was declared as a putative triploid when at least three of the 16 SSR loci exhibited three distinct alleles. Analyses of descriptive diversity statistics were conducted at locus level. For each SSR marker, SPAGeDi v.1.3 software [65] was used to estimate the number of alleles (N_A), the number of alleles with a frequency below 5 % (N_B), the number of effective alleles (N_E), and the observed (*Ho*) and expected (*He*) heterozygosity. The probability of identity (P_{ID}) was calculated as follows [66]:

$$P_{ID} = \sum p_i^4 + \sum\sum \left(2p_ip_j\right)^2$$

where *pi* and *pj* are the frequencies of the ith and jth alleles and $i \neq j$. The cumulative P_{ID} over the 16 SSR was computed as the product of the P_{ID} of each individual marker.

Determination of the geographical regions of origin of the unique genotypes

Using passport data along with reviewing published records with a focus on old literature (national compilations/varietal catalogues/reports) and specialized websites we were able to discern the geographical regions of origin for a large part of the unique genotypes analyzed. This was further helped by the resolution of identified duplicates and comparison of accessions against additional SSR data of the whole UK-NFC apple collection kindly made available from the UK-NFC database [64] and of the whole FRUCTUS collection kindly made available by Agroscope [14]. We first decided to define three broad historical European regions of origin of the germplasm according to geographical proximity and traditional agricultural relations between them: North + East (Sweden, Norway, Finland, Denmark, Baltic countries, plus Russia, Ukraine and Kyrgyzstan), West (Ireland, United Kingdom, France, Belgium, the Netherlands, Switzerland, Germany, Czech Republic) and South (Spain and Italy). When available, countries of origin of the cultivars were also documented although, this information should be considered with caution since the information on the countries of origin was not always fully consistent within duplicates groups.

Analysis of the genetic structure

The software Structure v.2.3.4 [67] was used to estimate the number of hypothetical subpopulations (*K*) and to quantify the proportion of ancestry of each genotype to the inferred subpopulations. No prior information about the geographical origin of the accessions was considered in the analysis. Ten independent runs were carried out for *K* values ranging from two to 10 using 500,000 Markov Chain Monte Carlo (MCMC) iterations after a burn-in of 200,000 steps assuming an admixture model and allelic frequencies correlated. In order to assess the best *K* value supported for our dataset, the Δ*K* method [68] was used through the Structure harvester v.0.6.93 website [69] to examine the rate of change in successive posterior probabilities over the range of *K* values. When the results described above suggested additional substructuring of the diversity in subgroups, a second-level (nested) application of the Structure clustering method was carried out analyzing separately each of the *K* major groups previously obtained [10, 24, 26, 50, 51]. Genotypes were assigned to the group (or subgroup) for which they showed the highest membership coefficient, considering an accession strongly assigned to each partitioning level if its proportion of ancestry (*qI*) was ≥0.80 [70, 71, 72]; otherwise they were considered as "admixed". The placement of genotypes on groups (or subgroups) was determined using CLUMPP v.1.1 [73], which evaluates the similarity of outcomes between population structure runs. CLUMPP output was used directly as input for Distruct v1.1 [74] in order to graphically display the results.

To validate the genetic structure revealed by the Bayesian model-based clustering two complementary approaches using the Darwin software package v6.0.10 [75] were considered: i) an unweighted neighbor-joining tree constructed based on dissimilarities between the unique genotypes (using a Simple Matching coefficient), and ii) a multivariate Principal Coordinate Analysis (PCoA).

Genetic differentiation

Population differentiation was estimated by analyses of molecular variance (AMOVA) through Genodive [76] under two scenarios: i) three broad European geographic regions of origin of the material (North + East, West and South); and ii) the major groups (and subgroups) defined by Structure. Pairwise F_{ST} estimates for the different partitioning levels considered in each case were also obtained using Genodive [76]. Descriptive statistics were calculated for the material clustered according to geographical regions of origin as well as for each group (or subgroup) identified by the Bayesian model-based clustering method, including *Ho* and *He*, number of total alleles, number of private alleles, i.e., those only found in one (sub)division level, and number of unique alleles, i.e., those only detected in one unique accession. The software FSTAT v.2.9.3.2 [77] was applied to compute the allelic richness after scaling down to the smallest partitioning level in the different scenarios considered.

Parentage reconstruction and relatedness between the accessions

On the basis of the SSR profiles of the unique genotypes, accessions were analyzed to infer possible parent-offspring relationships using Cervus v.3.0 software [53]. In order to reveal only robust parentages, we limited the study to the inferences of 'two-parents offspring' relationships and did not consider inferences of 'one-parent offspring' relationships where the lacking parent offers more flexibility but more speculative assignments as well, especially with only 16 SSR markers. Two criteria were considered to establish strict parentage relationships: i) a confidence level of the LOD score and ii) the Delta LOD value (defined as the difference in LOD scores between the first and second most likely two-candidate parents inferred) both higher than 95 %. Finally, an additional constraint was added to strengthen the results by limiting the maximum number of tolerated locus mismatches to only one in any inferred two-parents offspring trio, assuming that such a slight difference may be attributable to possible scoring errors, occurrence of null alleles or occasional mutational events [54, 78].

ABBREVIATIONS

ECPGR, European cooperative programme for plant genetic resources; F_{ST}, fixation index 'F-statistics'; LOD, logarithm of odds ratio; PCR, polymerase Chain reaction; SSR, simple sequence repeat.

ACKNOWLEDGMENTS

The staff at the ANAN genotyping platform of the SFR 149 QUASAV (Angers, France), at the GENTYANE genotyping platform (INRA, Clermont-Ferrand, France) and at the Unité Expérimentale Horticole (INRA, Angers, France) are warmly acknowledged for their help in producing the genotyping data and for maintaining the apple germplasm collection, respectively. The INRA MIGALE bioinformatic platform is also greatly acknowledged for having given support to intense data analyses. Natural Resources Institute, Finland, is gratefully acknowledged for providing access to 50 Finnish apple cultivars in their germplasm collections. UDL, University of Lleida, and EEAD, Aula Dei Experimental Station, are gratefully acknowledged for providing access to apple cultivars in their germplasm collections. Acknowledgements to Slepkov S. from Maykop experimental station of

Vavilov's N.I. Vavilov Institute of Plant Industry (MOSVIR) for making available plant material of Russian apple cultivars included in the present analysis.

REFERENCES

1. Food and Agriculture Organization of the United Nations. FAO statistics database on the World Wide Web http://faostat.fao.org/site/567/ Desktop Default.aspx ?PageID0567#ancor. Accessed 27 July 2015.
2. Hokanson SC, Lamboy WF, Szewc-McFadden AK, McFerson JR. Microsatellite (SSR) variation in a collection of *Malus*(apple) species and hybrids. Euphytica. 2001;118:281–94.
3. Janick J, Moore JN. Fruit breeding. Volume I: tree and tropical fruits. New York: Wiley; 1996.
4. Laurens F, Durel CE, Patocchi A, Peil A, Salvi S, Tartarini S, Velasco R, van de Weg WE. Review on apple genetics and breeding programs and presentation of a new initiative of a news European initiative to increase fruit breeding efficiency. J Fruit Sci. 2010;27:102–7.
5. Noiton DAM, Alspach PA. Founding clones, inbreeding, coancestry, and status number of modern apple cultivars. J Am Soc Hortic Sci. 1996;121:773–82.
6. Brooks RM, Olmo HP. Register of new fruit and nut varieties list 35. HortScience. 1991;26:951–78.
7. Brooks RM, Olmo HP. Register of new fruit and nut varieties list 36. HortScience. 1994;29:942–69.
8. Brooks RM, Olmo HP. Register of fruit and nut varieties. Alexandria: ASHS; 1997.
9. de Vicente. The evolving role of genebanks in the fast-developing field of molecular genetics. Ed. Issues in genetic resources No XI, August 2004. Rome: International Plant Genetic Resources Institute-IPGRI; 2004.
10. Jing RC, Vershinin A, Grzebyta J, Shaw P, Smykal P, Marshall D, Ambrose MJ, Ellis THN, Flavell AJ. The genetic diversity and evolution of field pea (*Pisum*) studied by high throughput retrotransposon based insertion polymorphism (RBIP) marker analysis. BMC Evol Biol. 2010;10:44.
11. van Treuren R, Kemp H, Ernsting G, Jongejans B, Houtman H, Visser L. Microsatellite genotyping of apple (*Malus* x *domestica* Borkh.) genetic resources in the Netherlands: application in collection management and variety identification. Genet Resour Crop Evol. 2010;57:853–65.
12. Nybom H, Weising K, Rotter B. DNA fingerprinting in botany: past, present, future. Investig Genet. 2014;5:1.
13. Bacilieri R, Lacombe T, Le Cunff L, Di Vecchi-Staraz M, Laucou V, Genna B, Péros JP, This P, Boursiquot JM. Genetic structure in cultivated grapevines is linked to geography and human selection. BMC Plant Biol. 2013;13:25.
14. Bühlmann A, Gassmann J, Ingenfeld A, Hunziker K, Kellerhals M, Frey JE. Molecular characterization of the Swiss fruit genetic resources. Erwerbs-Obstbau. 2015;57:29–34.
15. McCouch SR, McNally KL, Wang W, Sackville HR. Genomics of gene banks: a case study in rice. Am J Bot. 2012;99:407–23.
16. Wünsch A, Hormaza JI. Cultivar identification and genetic fingerprinting of temperate fruit tree species using DNA markers. Euphytica. 2002;125:59–67.
17. Ellegren H. Microsatellites: simple sequences with complex evolution. Nat Rev Genet. 2004;5:435–45.

18. Guichoux E, Lagache L, Wagner S, Chaumeil P, Léger P, Lepais O, Lepoittevin C, Malausa T, Revardel E, Salin F, Petit RJ. Current trends in microsatellite genotyping. Mol Ecol Resour. 2011;11:591–611.
19. Kalia RK, Rai MK, Kalia S, Singh R, Dhawan AK. Microsatellite markers: an overview of the recent progress in plants. Euphytica. 2011;177:309–34.
20. Sunnucks P. Efficient genetic markers for population biology. Trends Ecol Evol. 2000;15:199–03.
21. Urrestarazu J, Royo JB, Santesteban LG, Miranda CM. Evaluating the influence of the microsatellite marker set on the genetic structure inferred in *Pyrus communis* L. PLoS One. 2015;10:e0138417.
22. Garkava-Gustavsson L, Kolodinska Brantestam A, Sehic J, Nybom H. Molecular characterisation of indigenous Swedish apple cultivars based on SSR and S-allele analysis. Hereditas. 2008;145:99–112.
23. Lassois L, Denancé C, Ravon E, Guyader A, Guisnel R, Hibrand-Saint-Oyant L, Poncet C, Lasserre-Zuber P, Feugey L, Durel CE. Genetic diversity, population structure, parentage analysis and construction of core collections in the French apple germplasm based on SSR markers. Plant Mol Biol Rep. 2015. doi:10.1007/s11105-015-0966-7.
24. Liang W, Dondini L, De Franceschi P, Paris R, Sansavini S, Tartarini S. Genetic diversity, population structure and construction of a core collection of apple cultivars from Italian germplasm. Plant Mol Biol Rep. 2015;33:458–73.
25. Patzak J, Paprštein F, Henychová A, Sedlák J. Comparison of genetic diversity structure analyses of SSR molecular marker data within apple (*Malus* × *domestica*) genetic resources. Genome. 2012;55:647–65.
26. Urrestarazu J, Miranda C, Santesteban LG, Royo JB. Genetic diversity and structure of local apple cultivars from Northeastern Spain assessed by microsatellite markers. Tree Genet Genomes. 2012;8:1163–80.
27. Sehic J, Garkava-Gustavsson L, Nybom H. More harmonization needed for DNA-based identification of apple germplsm. Acta Hortic. 2013;976:277–83.
28. Laurens F, Aranzana MJ, Arús P, Bassi D, Bonany J, Corelli L, Durel CE, Mes J, Pascal T, Patocchi A, Peil A, Quilot B, Salvi S, Tartarini S, Troggio M, Vecchietti A, Velasco R, van de Weg WE. Review of fruit genetics and breeding programmes and a new European initiative to increase fruit breeding efficiency. Acta Hortic. 2012;929:95–102.
29. Morin PA, Manaster C, Mesnick SL, Holland R. Normalization and binning of historical and multi-source microsatellite data: overcoming the problems of allele size shift with ALLELOGRAM. Mol Ecol Resour. 2009;9:1451–5.
30. Putman AI, Carbone I. Challenges in analysis and interpretation of microsatellite data for population genetic studies. Ecol Evol. 2014;4:4399–28.
31. Gasi F, Simon S, Pojskic N, Kurtovic M, Pejic I. Genetic assessment of apple germplasm in Bosnia and Herzegovina using microsatellite and morphologic markers. Sci Hortic. 2010;126:164–71.
32. Gharghani A, Zamani Z, Talaie A, Oraguzie NC, Fatahi R, Hajnajari H, Wiedow C, Gardiner SE. Genetic identity and relationships of Iranian apple (*Malus* x *domestica* Borkh.) cultivars and landraces, wild *Malus* species and representative old apple cultivars based on simple sequence repeat (SSR) marker analysis. Genet Resour Crop Evol. 2009;56:829–42.
33. Gao Y, Liu F, Wang K, Wang D, Gong X, Liu L, Richards CM, Henk AD, Volk GM. Genetic diversity of *Malus* cultivars and wild relatives in the Chinese National Repository of Apple Germplasm Resources. Tree Genet Genomes. 2015. doi:10.1007/s11295-015-0913-7.
34. Gross BL, Volk GM, Richards CM, Forsline CL, Fazio G, Chao CT. Identification of "duplicate" accessions within the USDA-ARS National Plant Germplasm System *Malus* Collection. J Am Soc Hortic Sci. 2012;137:333–42.
35. Cornille A, Gladieux P, Smulders MJM, Roldán-Ruiz I, Laurens F, Le Cam B, Nerseyan A, Clavel J, Olonova M, Feugey L, Gabrielyan I, Zhang XG, Tenaillon MI, Giraud. New

insight into the history of domesticated apple: secondary contribution of the European wild apple to the genome of cultivated varieties. PLoS Genet. 2012;8:e1002703.**View ArticlePubMedPubMed CentralGoogle Scholar**

36. Cornille A, Giraud T, Smulders MJM, Roldán-Ruiz I, Gladieux P. The domestication and evolutionary ecology of apples. Trends Genet. 2014;30:57–65.
37. Gross BL, Henk AD, Richards CM, Fazio G, Volk GM. Genetic diversity in *Malus* x *domestica* (*Rosaceae*) through time in response to domestication. Am J Bot. 2014;101:1770–9.
38. Hartmann HT, Kester DE, Davies FT, Geneve RL. Plant propagation: principles and practices. 7th ed. Upper Saddle River: Prentice Hall; 2002.
39. Zohary D, Hopf D. Domestication of plants in the Old World: the origin and spread of cultivated plants in West Asia, Europe and the Nile Valley. Oxford: Oxford University Press; 2000.
40. De Franceschi P, Dondini L, Sanzol J. Molecular bases and evolutionary dynamics of self-incompatibility in the *Pyrinae*(*Rosaceae*). J Exp Bot. 2012;63:4015–32.
41. Knutson L, Stoner AK. Biotic diversity and germplasm preservation: global imperative. Boston: Kluwer Academic Publishers; 1989.
42. McCouch S. Diversifying selection in plant breeding. PLoS Biol. 2004;2:e347.
43. World Apple and Pear Association. European apple and pear crop forecast. Brussels: World Apple and Pear Association; 2013.
44. Miller AJ, Gross BL. Forest to field: perennial fruit crop domestication. Am J Bot. 2011;98:1389–414.
45. Biswas S, Akey JM. Genomic insights into positive selection. Trends Genet. 2006;22:437–46.
46. Bourguiba H, Audergon JM, Krichen L, Trifi-Farah N, Mamouni A, Trabelsi S, D'Onofrio C, Asma BM, Santoni S, Khadari B. Loss of genetic diversity as a signature of apricot domestication and diffusion into the Mediterranean Basin. BMC Plant Biol. 2012;12:49.
47. Blanquart F, Gandon S, Nuismer SL. The effects of migration and drift on local adaptation to a heterogeneous environment. J Evol Biol. 2012;25:1351–63.
48. Emanuelli F, Lorenzi S, Grzeskowiak L, Catalano V, Stefanini M, Troggio M, Myles S, Martinez-Zapater JM, Zyprian E, Moreira FM, Grando MS. Genetic diversity and population structure assessed by SSR and SNP markers in a large germplasm collection of grape. BMC Plant Biol. 2013;13:39.
49. Lia VV, Poggio L, Confalonieri VA. Microsatellite variation in maize landraces from Northwestern Argentina: genetic diversity, population structure and racial affiliations. Theor Appl Genet. 2009;119:1053–67.
50. Li XW, Meng XQ, Jia HJ, Yu ML, Ma RJ, Wang LR, Cao K, Shen ZJ, Niu L, Tian JB, Chen MJ, Xie M, Arus P, Gao ZS, Aranzana MJ. Peach genetic resources: diversity, population structure and linkage disequilibrium. BMC Genet. 2013;14:84.
51. Jacobs MMJ, Smulders MJM, van den Berg RG, Vosman B. What's in a name; genetic structure in *Solanum* section *Petota*studied using population-genetic tools. BMC Evol Biol. 2011;11:42.
52. Jing R, Ambrose MA, Knox MR, Smykal P, Hybl M, Ramos A, Caminero C, Burstin J, Duc G, van Soest LJM, Swiecicki WK, Pereira MG, Vishnyakova M, Davenport GF, Flavell AJ, Ellis THN. Genetic diversity in European *Pisum* germplasm collections. Theor Appl Genet. 2012;125:365–80.
53. Kalinowski ST, Taper ML, Marshall TC. Revising how the computer program cervus accommodates genotyping error increases success in paternity assignment. Mol Ecol. 2007;16:1099–2006. Salvi S, Micheletti D, Magnago P, Fontanari M, Viola R, Pindo M, Velasco R. One-step reconstruction of multi-generation pedigree networks in apple (*Malus* × *domestica* Borkh.) and the parentage of Golden Delicious. Mol Breed. 2014;34:511–24.
54. Bianco L, Cestaro A, Sargent DJ, Banchi E, Derdak S, Di Guardo M, Salvi S, Jansen J, Viola R, Gut I, et al. Development and validation of a 20K single nucleotide

polymorphism (SNP) whole genome genotyping array for apple (*Malus* x *domestica*Borkh). PloS ONE. 2014;9:e110377.

55. Bianco L, Cestaro A, Linsmith G, Muranty H, Denancé C, Théron A, Poncet C, Micheletti D, Kersshbamer E, Di Pierro EA, Larger S, Pindo M, van de Weg WE, Davassi A, Laurens F, Velasco R, Durel CE, Troggio M. Development and validation of the Axiom®Apple480K SNP genotyping array. Plant J. 2016;86:62–74.
56. Chagné D, Crowhurst RN, Troggio M, Davey MW, Gilmore B, Lawley C, Vanderzande S, Hellens RP, Kumar S, Cestaro A et al. Genome-wide SNP detection, validation, and development of an 8K SNP array for apple. PloS ONE. 2012;7:e31745.
57. Morgan J, Richards A, Dowle E. The new book of Apples: the definitive guide to apples, including over 2000 varieties. London: Ebury Press; 2002.
58. Hokanson SC, Szewc-McFadden AK, Lamboy WF, McFerson JR. Microsatellite (SSR) markers reveal genetic identities, genetic diversity and relationships in a *Malus* x *domestica* Borkh. core subset collection. Theor Appl Genet. 1998;97:671–83.
59. Liebhard R, Gianfranceschi L, Koller B, Ryder CD, Tarchini R, van de Weg E, Gessler C. Development and characterisation of 140 new microsatellites in apple (*Malus* x *domestica* Borkh.). Mol Breed. 2002;10:217–41.
60. Silfverberg-Dilworth E, Matasci CL, van de Weg WE, van Kaauwen MPW, Walser M, Kodde LP, Soglio V, Gianfranceschi L, Durel CE, Costa F, Yamamoto T, Koller B, Gessler C, Patocchi A. Microsatellite markers spanning the apple (*Malus* x *domestica* Borkh.) genome. Tree Genet Genomes. 2006;2:202–24.
61. Vinatzer BA, Patocchi A, Tartarini S, Gianfranceschi L, Sansavini S, Gessler C. Isolation of two microsatellite markers from BAC clones of the Vf scab resistance region and molecular characterization of scab-resistant accessions in *Malus*germplasm. Plant Breed. 2004;123:321–6.
62. Evans KM, Fernández F, Govan C. Harmonising fingerprinting protocols to allow comparisons between germplasm collections - *Pyrus*. Acta Hortic. 2009;814:103–6.
63. Fernández-Fernández F. Fingerprinting the National apple and pear collections. Final report of DEFRA research project GC0140. 2010. p. 1–18. http://randd.defra.gov.uk/Document.aspx?Document=GC0140SID5FingerprintingFernandez.pdf.
64. Hardy OJ, Vekemans X. SPAGEDi: a versatile computer program to analyse spatial genetic structure at the individual or population levels. Mol Ecol Notes. 2002;2:618–20.
65. Waits LP, Luikart G, Taberlet P. Estimating the probability of identity among genotypes in natural populations: cautions and guidelines. Mol Ecol. 2001;10:249–56.
66. Pritchard JK, Stephens M, Donnelly P. Inference of population structure using multilocus genotype data. Genetics. 2000;155:945–59.
67. Evanno G, Regnaut S, Goudet J. Detecting the number of clusters of individuals using the software STRUCTURE: a simulation study. Mol Ecol. 2005;14:2611–20.
68. Earl DA, von Holdt BM. Structure harvester: a website and program for visualizing STRUCTURE output and implementing the Evanno method. Cons Genet Resour. 2012;4:359–61.
69. Breton C, Pinatel C, Médail F, Bonhomme F, Bervillé A. Comparison between classical and Bayesian methods to investigate the history of olive cultivars using SSR-polymorphisms. Plant Sci. 2008;175:524–32.
70. Miranda C, Urrestarazu J, Santesteban LG, Royo JB, Urbina V. Genetic diversity and structure in a collection of ancient Spanish pear cultivars assessed by microsatellite markers. J Am Soc Hortic Sci. 2010;135:428–37.
71. Marra FP, Caruso T, Costa F, Di Vaio C, Mafrica R, Marchese A. Genetic relationships, structure and parentage simulation among the olive tree (*Olea europaea* L. subsp. europaea) cultivated in Southern Italy revealed by SSR markers. Tree Genet Genomes. 2013;9:961–73.

72. Jakobsson M, Rosenberg NA. CLUMPP: a cluster matching and permutation program for dealing with label switching and multimodality in analysis of population structure. Bioinformatics. 2007;23:801–6.
73. Rosenberg NA. Distruct: a program for the graphical display of population structure. Mol Ecol Notes. 2004;4:137–8.
74. Perrier X, Jacquemoud-Collet JP. DARwin Software. 2006. http:// darwin. cirad.fr/ darwin.
75. Meirmans PG, van Tienderen PH. Genotype and Genodive: two programs for the analysis of genetic diversity of asexual organisms. Mol Ecol Notes. 2004;4:792–4.
76. Goudet J. FSTAT 2.9.3.2. 2002. http://www2.unil.ch/popgen/softwares/fstat.htm.
77. Lacombe T, Boursiquot JM, Laucou V, Di Vecchi-Staraz M, Péros JP. This P. Large-scale parentage analysis in an extended set of grapevine cultivars (*Vitis vinifera* L.). Theor Appl Genet. 2013;126:401–14.

CHAPTER 9

Indigenous Fruit Trees of Tropical Africa: Status, Opportunity for Development and Biodiversity Management

Rasheed Olufemi Awodoyin[1], Oluseun Sunday Olubode[1], Justin Ugochukwu Ogbu[2*], Rasheed Babatunde Balogun[2], Jonah U. Nwawuisi[3], Kalu O. Orji[4]

[1]*Department of Crop Protection and Environmental Biology, University of Ibadan, Ibadan, Nigeria*

[2]*Department of Horticulture and Landscape Technology, Federal College of Agriculture (FCA), Ishiagu, Nigeria*

[3]*Department of Pest Management Technology, Federal College of Agriculture (FCA), Ishiagu, Nigeria*

[4]*Department of Agronomy, Michael Okpara University of Agriculture (MOUA), Umuahia, Nigeria*

ABSTRACT

Tropical fruit trees constitute important biological resources in the global agrobiodiversity context. Unlike the tropical fruit trees of American and Asian origin, indigenous fruit trees (IFT) of tropical Africa have scarcely achieved the status of international recognition in commodity markets and research arena outside Africa. This paper presented a critical review of the status of IFT in the Tropical African sub-regions (of West Africa, Central Africa, East Africa, Southern Africa and the Indian Ocean Islands) in relation to the introduced naturalised fruit trees from tropical America and Asia, threats to the diversity and sustainable use of IFT, analysis of the opportunities and challenges of developing IFT, as well as targets for crop improvement of the rich IFT of Tropical Africa. Domestication programme via relevant vegetative propagation techniques for priority IFT of the sub-regions was examined and advocated, in addition to the adoption of complementary conservation strategies, including Field GeneBanks in the management of the continent's IFT diversity.

Keywords: Biodiversity, Conservation, Crop Improvement, Domestication, Indigenous Fruit Trees, Introduced Species, Tropical Africa

1. INTRODUCTION

Fruit trees constitute important biological resources in many agroecological systems and forest ecosystems all over the world. The assertion is evident by the fact of these tree species long time economic and ecological impacts in nature. Fruits are full of nature's rich essential nutrients, antioxidants and health benefits for ready use by humans (and other animals) without alternation in most cases, unlike vegetables and other edible agricultural/horticultural produce that may require necessary pre-treatments, like heating, sometimes before consumption [1] -[3] . The tropics, more than any other region of the world, is endowed with great diversity of fruit tree species that have provided humans with basic food and nourishment for ages since the domestication of beneficial wild plants (crops species). Tropical continents of the world possess rich variety of fruit trees with about 1000 species identified in Americas, 1200 species in Africa and 500 species in Asia [4] [5] . Although only relatively few fraction of these diversities are marketed worldwide, the diversities are nature's inestimable assets for the livelihoods of local people throughout the tropical regions.

In contrast to the noted diversity of fruit trees, only 10 annual cereal grains, pulses and oil seeds dominate 80% of the world agronomic fields. At present, wheat, rice and maize cover half of the world's croplands, while adding other annual gains and pulses accounts for up to two-third of all arable land in the world [5] [6] . Also despite the fact that more than 90% of fresh fruit produce are consumed locally, the importation demand for tropical fruits has been on steady increase globally in the past few decades. According to FAO [7] , there is a great potential for growth of the fruit produce in the international markets due to the increasingly health consciousness and much hyped nutritional awareness about the consumption of more fresh fruits as well as vegetables in people's diets.

1.1. Indigenous Fruit Trees (IFT) of Tropical Africa

According to the authors [6] in PROTA report, Tropical Africa comprised of 48 countries which are distributed across the various sub-regions of the continent, namely West Africa (16 states), East Africa (9 states), Central Africa (10), some part of Southern Africa (7) and the Indian Ocean islands (6) including Madagascar (Table 1). Others described Tropical Africa as Sub-Saharan Africa excluding South African republic and the few countries enmeshed in it [9] .

These sections of the continent are diverse in their climate, soil, topography and vegetation, which invariably influence the array and distribution of African floristic diversity. Tropical Africa sub-regions are home to many valuable fruit tree species whose potentials have not been fully realized. A good number of these fruit species have remained rather of local importance, and are yet to be domesticated. However, tangible economic produce are been harvested from their wild and or protected volunteer stands in home gardens, farmlands and forest reserves [10] [11] . Following PROTA reports [12] [13] , Tropical Africa has 477 edible fruit and nut species (including both indigenous and introduced naturalised ones) grown across its landscape. Attempts to identify and compile some of these local fruit commodity species that had their roots in Tropical Africa sub-regions, with regards to the species origin and natural distributions are shown in Table 2. The list in Table 2 should not be seen as exhaustive, since some of the IFT also yield other valuable products like timber and medicine for which they are sometimes treated as more of forest trees or medicinal plants than orchard species.

1.2. Status of IFT in Tropical Africa

According to the National Academy of Science [14] , fruit production in Africa at present has been dominated by such species introduced from tropical Americas and Asia. These introduced species, including Banana, Citrus spp, Mango, Papaya, and Pineapple among others (Table 3) have also constituted bulk of international trade on tropical fresh fruit produce and processed fruit products. These and other adapted exotic tropical fruit crops which were already improved through horticultural selection and breeding, arrived on the African continent cen- turies ago and increasingly displaced the indigenous fruit species that had fed African for millennia. Moreover these introduced species received the support of colonial powers who wanted familiar crops that were highly economical to cultivate. Consequently, most of these introduced fruit species rather than the indigenous species were grown and established in orchards and plantations for large scale production and distribution. Thus, the IFT known to the local

Table 1. Countries of Tropical Africa and their respective sub-regions.

Sub-regions	Member countries
West Africa	Benin, Burkina Faso, Cape Verde, Cote d'Ivoire, Gambia, Ghana, Guinea, Guinea Bissau, Liberia, Niger, Nigeria, Mali, Mauritania, Senegal, Sierra Leone, Togo
Central Africa	Burundi, Cameroon, Central Africa Republic, Chad, Congo, Sao Tome and Principe, Democratic Republic of Congo, Equatorial Guinea, Gabon, Rwanda
East Africa	Djibouti, Eritrea, Ethiopia, Kenya, Somalia, Sudan, South Sudan, Tanzania, Uganda
Southern Africa	Angola, Botswana, Malawi, Mozambique, Namibia, Zambia, Zimbabwe
Indian Ocean Islands	Comoros, Madagascar, Mauritius, Mayotte, Reunion, Seychelles

people continued their downward spiral of dwindling cultivation and knowledge without research investment and improvement. In fact, until recently, most of the researchers in horticultural crops production in Nigeria for example still inadvertently show preferences for detailed research work on introduced fruit species over the indigenous species, according to a recent survey [15] .

The fact about Tropical Africa indigenous fruit trees (IFT) diversity and economic potential is something one would never readily recognized by looking at fresh fruits local markets in most parts of the continent, or even consulting most college recommended text books on tropical agriculture and horticulture, for instance: [1] [5] [16] -[23] . The reasons for this insidious neglect of the continent's IFT economic resources may not be far-fetched. Most African native fruit species have not been brought up to their full potential in terms of quality breeding and selection, scale of production and distribution, value addition and availability. In terms of geographical spread and distribution, relatively few of them have moved beyond Africa's shores, unlike similar tropical fruit species from other continents.

2. EXPLORING AND EXPLOITING THE UNDER-TAPPED TREASURIES OF IFT

2.1. Threats to Diversity and Sustainable Use of IFT

Historically, the protection of IFT/wild fruit species in many countries of Africa has been carried out by local farmers in their community forest reserves, traditional home gardens, protected volunteer stands of such important plant species in farmlands, market squares and village squares. Until recently, organised orchards and plantations of IFT were rather rare and only few exceptional ways used in the perpetuation of IFT kinds in some African communities. Due to little or nil formal research development and investments for improvement, many of the IFT still retain their natural massive size that would make group planting and management in orchards/ plantations setting difficult [14] [15] [24] [25] .

Various other factors have contributed as threats to survival and sustainability of IFT in the Tropical African sub-regions. In the first place, local populations in the sub-regions are heavily dependent on forests, which provide them with firewood, timber and food (nuts, fruit, mushrooms and honey). They use wild fruit and nut species as food, as well as for their medicinal properties, in handicrafts and to dye home-made products. Indiscrimi- nate and illegal logging of IFT stands in the remaining forested areas and farmlands for varied reasons have in no small measure contributed to the depletion IFT local population and distribution in many Tropical African rural communities. Such uncontrolled tree felling may be due to domesticate needs as fuel woods, charcoal production, livestock feeds, land clearing of site for building construction and or intensive agricultural activities, especially crop production.

As earlier pointed out, the insidious domination and preference of other introduced tropical fruit tree species over the African IFT by most researchers and investors have impeded the expected developmental needs of the continent's many IFT. For most of the IFT, standardised agrotechniques for the propagation, protection, production, value added products and improved processing are nearly lacking or at best just at rudimentary stage when contrast to similar crop species from Tropical America and Asia. However, where such standardised agrotechniques might have been locally developed, they often remained as prototypes in relevant national research centres and government ministries yet to be commercialized. A SWOT (Strength, Weakness, Opportunity and Threat) analysis (Table 4) of Tropical African indigenous fruit trees' (IFT) opportunities, potentials, limitations and threats captured and summarized the prevailing situation.

2.2. Domestication of Indigenous Fruit Tree (IFT)

Tree domestication has been explained by various authorities in agroecology, agroforestry, agronomy and plant breeding [26] -[30] . However, common among the varied definitions is that domestication entails settling a species (whether plant or animal) as a member of a household; to cause to feel at home or naturalized. Thus, domestication in this context is an anthropogenic change in the genetics of a plant to conform to human needs and agro eco system [6] . By extension therefore, domesticating underutilized IFT/wild fruit trees involves accelerated and human-induced evolution to bring species into wider cultivation

Table 2. List of some scantly researched indigenous fruit trees/shrubs of Tropical African origin.

Common name	Botanical name[1]	Family name	Species origin
1. African breadfruit	*Treculia africana*	Moraceae	W, C, E
2. Ackee	*Blighia sapida*	Sapindaceae	W, C
3. Butter fruit	*Dacryodis edulis*	Burseraceae	W, C, E
4. Baobab	*Adansonia digitata*	Bombacaceae	W, C, E
5. Monkey kola (yellow)	*Cola lepidota*	Malvaceae	W, C
6. Monkey kola (white)	*C. pachycarpa*	,,	,,
7. Monkey kola (red)	*C. lateritia*	,,	,,
8. Magic plant	*Synsepalum dulcificum*	Sapotaceae	W, C
9. African walnut	*Plukenetia conophora*	Euphorbiaceae	W, C
10. Black plum/Mbembe	*Vitex doniana*	Labiatae	W, C, E, S
11. -	*V. payos*	,,	C, E, S
12. Tamarind	*Tamarindus indica*	Fabaceae	W, C, E, S
13.Velvet tamarind	*Dialiumgu ineense*	Fabaceae	W, C, E
14. African oil bean	*Pentaclethra macrophylla*	Fabaceae	W, C, E, S, I
15. Sheabutter	*Vitellaria paradoxa*	Sapotaceae	W, C, E
16. Irvingia nut	*Irvingia garbonensis*	Irvingiaceae	W, C, E
17. African star apple	*Chrysophyllum albidum*	Sapotaceae	WC, E, S
18. Incense tree	*Canarium schweifurthii*	Burseraceae	W, C, E
19. Gum vine	*Landolphia kirkii*	Apocynaceae	W, C
20. -	*Uvariodendron connivens*	Annonaceae	E, S
21. -	*Sacoglottis gabonesis*	Humiriaceae	E, S
22. -	*Maesobortyabarteri*	Euphorbiaceae	E, C, S
23. Ebony	*Diospyros mespiliformis*	Ebenaceae	WC, E, S
24. -	*Chytranthus talbotii*	Sapindaceae	W, C
25. -	*Salvadora persica*	Salvadoraceae	W, C
26. -	*S. australis*	,,	W, C
27. Marula	*Sclerocarya birrea*	Anacardiaceae	W, C, E, S
28. African walnut	*Coulae dulis*	Olacaceae	W, C, S
29. Imbe	*Garcinia livingstonii*	Gultiferae	W, C, E, S
30. Medlar	*Vangueria apiculata*	Rubiaceae	C, E, S
31. -	*V. rotundata*	,,	C, E, S
32. Monkey orange	*Strychnos cocculoides*	Strychnaceae	W, C, E, S
33. Wild loquat	*Uapaca kirkiana*	Euphorbiaceae	W, C, E, S
34. Sugar plum	*U. heudelotii*	,,	W, C, E, S
35. Sweet Detar	*Detarium senegalense*	Fabaceae	W
36. Detar	*D. microcarpum*	,,	W
37. Tree grape	*Lannea microcarpa*	Anacardiaceae	W, C, E, S
38. Aizen	*Boscia senegalensis*	Capparaceae	W, E
39. Balanite	*Balanites aegyptiaca*	Balanitaceae	W, E
40. Carissa	*Carissa macrocarpa*	Apocynaceae	W, C, E
41. Kei apple	*Dovyalis caffra*	Flacourtiaceae	S
42. Custard apple	*Annona senegalensis*	Annonaceae	W, C, E, S
43. Ginger bread plum	*Parina riexcelsa*	Chrysobalanaceae	W, C, E, S
44. Gum vine	*Saba lanceolata*	Apocynaceae	W, C
45. Icacina	*Icacina oliviformis*	Icacinaceae	W, C
46. -	*Trichoscypha ferruginea*	Anacardiaceae	W, C, S
47. -	*Azanza garckeana*	Malvaceae	E, S
48. Bird plum/wild date	*Berchemia discolor*	Rhamnaceae	E, S
49. -	*Ximenia caffra*	Olacaceae	W, C, E, S
50. -	*Grewia tembensis*	Tiliaceae	W, C, S
51. -	*G. caffra*	Tiliaceae	W, C, S

[1]Note: some genus have more than one species of primary use as fruit, although not all are listed; [2]W—West Africa; C—Central Africa; E—East Africa; S—Southern Africa; I—Indian Ocean Islands.

Table 3. Some adapted introduced tropical fruit species from outside Africa.

Common name	Botanical name	Family name	Species origin
Avocado pear	*Persea americana*	Lauraceae	Central America
Banana	*Musa* spp	Musaceae	South East (SE) Asia
Cashew	*Anacardium occidentalis*	Anacardiaceae	South America
Coconut	*Cocosnu cifera*	Arecaceae	Pacific South/Central America
Carambola	*Averrhoa carambola*	Oxalidaceae	Indonesia
Custard apple	*Annona squamosal*	Annonaceae	South America
Date palm	*Phoenix dactylifera*	Arecaceae	Persian Gulf
Fig	*Ficus caria*	Moraceae	SE Asia
Grape citrus	*Citrus paradise*	Rutaceae	West Indies
Lemon	*C. limon*	"	SE Asia
Lime	*C. aurantifolia*	"	"
Orange	*C. sinensis*	"	"
Shaddock	*C. grantis*	"	"
Tangerine	*C. reticulate*	"	"
Tangelo	*C. reticulate × C. paradisi*	"	"
Tangor	*C. reticulata × C. sinensis*	"	"
Guava	*Psidium guajava*	Myrtaceae	West Indies America
Grapevine	*Vitis vinifera*	Vitaceae	Russia (Asia)
Passion fruit	*Passiflora edulis*	Passifloraceae	South America
Mango	*Mangifera indica*	Anacardiaceae	South Asia
Pawpaw/Papaya	*Carica papaya*	Caricaceae	Central America
Pineapple	*Ananas comosus*	Bromeliaceae	South America
Sour sop	*Annona muricata*	Annonaceae	Central America
Sweet sop	*Annona squamosal*	"	South America
Star apple	*Chrysophyllum cainito*	Sapotaceae	Central America
Breadfruit	*Artocarpus atilis*	Moraceae	Polynesia (Asia)
Jackfruit	*A. heterophyllus*	"	S. Asia
Mulberry	*Morus alba*	"	China-Japan
Hog plum	*Spondia mombin*	Anacardiaceae	Tropical America

through a farmer-driven and or market-led process. This is an iterative procedure involving the identification, production, management and adoption of desirable germplasm. Strategies for individual species vary according to their functional use, biology and target environment. Remarkably, domestication of any plant species consists of selection and management by humans, and

Table 4. SWOT analysis of opportunities and challenges of Tropical African IFT.

Internal factors	Strength (+ve)	Weakness (−ve)
	- Area of origin/natural distribution - Local preference/consumption of produce - Form part of traditional home garden farming system - Hold high cultural value among the local people - Viable revenue generator for rural resource poor farmers - Available local/regional markets	- Grossly under-researched - Relatively poor and inefficient production technology - Near absence of improved propagation agrotechniques - Most harvest still derived from wild/volunteer stands - Slow rate of natural regeneration - IFT information are poorly documented or in grey literatures
External factors	**Opportunity (+ve)**	**Threat (−ve)**
	- International market/export prospects as novel crops - Potentials diversification of products and markets - Recognition as candidates for agroforestry multipurpose trees for the sub-regions - Recent recognition by Bioversity International—NUS, ICUC, ICRAF - Recent works of PROTA documentation/Publications - Satisfy needs for crop diversification	- Over exploitation of available *in situ* stands on farms and in forests - Depleting tree stands in the wild due to deforestation - Inadequate deliberate replanting scheme and organised orchards - Poor marketing outlets and low pricing of IFT produce - Preference of African researchers for introduced naturalized species over IFT

Note: NUS—Neglected and Underutilised Species/Bioversity International Rome; ICUC—International Centre for Underutilised Crops, Colombo Sri Lanka; ICRAF—International Centre for Research in Agroforestry (now known as World Agroforestry Centre), Nairobi Kenya; PROTA—Plant Resources of Tropical Africa, Wageningen, the Netherlands.

is not only about breeding *per se*, but definitely an aspect of whole species improvement. Domestication, in this perspective, aims at promoting the cultivation of IFT with economic potentials as new cash or novel crops, and provides incentive to subsistence farmers to grow such trees that contribute toward achieving poverty reduction, enhancement of food and nutritional security [25] [31] -[33] .

Deliberate tree improvement programme (involving specific domestication interventions) for useful characters has been advocated for some highly promising IFT of Sub-Saharan sub-regions. This is to encourage local farmers to continue their age long practice of on-farm conservation and sustainable exploitation of IFT alongside their field crops [8] [34] -[36] . The long period that it takes before enjoying tangible benefits from tree-based agriculture is one of the reasons why some farmers see it as deterrent to planned cultivation of such economic IFT species. However, domestication via standardised vegetative propagation (like budding, stem cuttings and layering) can play significant role in ensuring that trees start producing quality fruits in a shorter period [28] [37] -[39] . Farmers and nursery managers must be able to have access to superior stocks for multiplication and distribution, if they are to capitalise on the benefit of such IFT cultivation.

2.3. Vegetative Propagation for Domestication of Indigenous Fruit Trees (IFT)

Vegetative propagation refers to the regeneration of selected plants from vegetative organs such as roots, stems, leaves, buds and even single cells/tissues; and it offers a wide range of benefits for any domestication programme of IFT, as well as in general Plant Genetic Resources (PGR) conservation efforts. By capturing the genetic variation of trees in natural stands, researchers are able to select for desirable traits found in wild tree populations. Ultimately, the aim is to produce large number of improved propagules for resource poor farmers and reforestation programme [40] [41] . The concept of vegetative propagation is that an exact copy of the genome of a mother plant is made and continued in new individuals (clones). This is possible in plants unlike animals or humans, for several reasons. Firstly, every living cell of a plant is initially totipotent, meaning that it contains all the genetic information needed to regenerate the entire plant. Furthermore, cell division continues to occur during the normal growth and development of most plants. Thirdly, cells can reform a meristem, which divides and produces the missing part [42] . Thus clones (i.e. all individuals produced by vegetative propagation from a single or original stock plant), are genetically identical, unless rare somatic mutations (chimeras) occur and are perpetuated. Whereas propagation by seeds provides opportunity for variation and evolutionary advancement, vegetative propagation aims at the identical reproduction of plants with desirable features such as high productivity, superior quality or high tolerance to biotic and or abiotic stresses, and as such plays a very vital role in perpetuating a preferred trait [36] [43] .

According to author [37] , the most important vegetative propagation techniques for trees are the regeneration by stem or root cuttings, grafting and budding, as well as various methods of marcotting (layering) and micropropagation procedures. Rooting of stem cuttings is the most common vegetative propagation technique used by gardeners and nursery managers. For this purpose, a portion of a stem with a leaf and axillary bud is cut from the parent plant and then set in an environment where humidity is high. After few weeks in such humid environment, roots may form on the cutting, and a new independent plant can then develop from the stem section. Most high success record regeneration of tree plants by cutting is done with materials from seedlings or from coppice shoots, because these parts of the plants are young with less lignification, and tend to grow vigorously. On the other hand, grafting and budding are techniques that join parts of plants together in such a way that they unite and continue their growth as a single plant. These techniques are used to perpetuate clones that cannot be conveniently reproduced by other vegetative propagation methods. Grafting is often used to multiply mature material that is difficult to root as cuttings. It is mainly applied in multiplication of high value fruit and ornamental woody species [38] [42] .

Marcotting, a form of layering, is one of the oldest methods of vegetative propagation, although it still has relevance in present day plant multiplication, domestication and conservation efforts. It involves the development of roots on a stem while the stem is still attached to the parent plant. The rooted stem is then detached to be new plant growing from its own roots. Unlike in grafting, the need of compatible seedling rootstock is unnecessary in layering, which is an added advantage. As rooting from cuttings of mature tree plant materials is known to be difficult, layering and grafting are preferred for the vegetative propagation plan in IFT domestication programme [37] [41] .

Finally, micropropagation, also referred to as tissue or in vitro culture, is a relatively new vegetative propagation technique, which uses a plant's potential to regenerate a complete new plant from single cells or small amounts of living tissue through the cultivation of these in aseptic and controlled environments. Since this technique requires a substantial investment in infrastructure, equipment and materials, its application is mostly justified in the case of high value plants where traditional vegetative propagation methods are considered unsuccessful. The method also allows for the production of virus-free plant materials and large amounts of new plants issued from a limited amount of initial material [42] [44] .

Evidently, vegetative propagation leading to the domestication and conservation of IFT offers exciting opportunities for research and sustainable development of Plant Genetic Resources (PGR). Reasons for vegetative propagation options include: maintenance of superior genotype, problematic seed germination and storage behavior, shortening time to first flowering and fruiting, controlling phases of tree development, combining desirable characteristics of more than one genotype into a single plant stand, uniformity of plantations/orchards, among others [28] [41] . For instance, time to first fruiting have been reduced through vegetative propagation in Bush mango [Irvingia gabonensis and I. wombolu] from 7 years to 3 years; Bush butter tree [Dacryodis edulis] from 5 years to 2 years; Baobab [Adansonia digitata] from 10 years to 4 years; and inSheanut [Vitellaria paradoxa] from 20 years to less than 5 years [37] [45] .

2.4. Major Targets in the Improvement of Tropical African IFT

Most Tropical African IFT have so far been grown in traditional ways as homestead shade trees, nurse trees; in community protected sacred forests, wild and volunteer stands on farms, market squares, village squares and forest areas. A good number of these species are scarcely grown deliberately by the farmers, who most times depended on wild saplings they came across for transplanting and regeneration of their choice local fruit trees. Knowledge and technical know-how of propagation and nursery management of the IFT are more or less lacking. Scientifically planned selection and breeding programmes have not been largely undertaken to select promising pomological traits (such as fruit size, shape, flavor, taste, suitability for processing into juice, wine and other beverages, resistance to biotic and abiotic stresses) among the various landraces of species. High production technologies such as high density planting system, pruning, efficient rootstocks cultivars and plant protection schedules have not been standardized. Faced with these several challenges, there is need to initiate focused tree improvement programme that would address essential pomological needs of the continent's IFT to bring it them at par with similar commodity crops from tropical America and Asia. Hence, the following major general targets for improvement are enumerated, although there may be other species specific needs relevant to the consumers and or markets.

Screening and selection of priority species for domestication: Specific screening plots for priority species identified by local farmers in the sub-regions should be established as well as on-farm trials to determine farmers' preferred niches for IFT.

IFT phenology, physiology and propagation study: Initiation of detailed research works on the IFT phenology, physiology and profitable propagation options for priority species in the various localities.

Varietal selection for superior fruit quality among the landraces: Quality of fruit will always be a major objective of fruit tree improvement programme. Pulp taste, aroma and texture, fruit size and shape are main pomological attributes of fruit quality.

Development of robust rootstock that is resistant or tolerant to biotic stress: Efficient and resistant cultivars and rootstocks are primary requirements for modern fruit production which can be achieved by undertaking sound science based improvement programme. Common objectives of rootstock improvement for fruit trees include ease of propagation, graft compatibility, high yield, control of scion vigour, longevity, adaptation to wide environmental conditions, resistance to diseases and pests.

Enhanced productivity: This involves selection and breeding of promising cultivars with greater economic yield per unit area and cultivars with reduced tree size for ease of management.

Value addition of IFT produce: Involves development and commercialization of enhanced processing techniques and value added products from fresh produce for market diversification of IFT. This would boost the resource poor farmers' expected revenue and encourage them to embark on deliberate cultivation of IFT.

3. MANAGING TROPICAL AFRICAN IFT GENETIC RESOURCES

In Situ and ex Situ Conservation Approaches

There are two complementary approaches for conservation of IFT, namely in situ and ex situ. The in situ conservation involves maintaining genetic resources in the natural habitats where they occur, whether as wild and uncultivated plant communities or crop cultivars in farmers' fields as components of traditional agricultural systems. The aim of in situ conservation is to protect habitats of target species so that a population of that species can steadily persist [46] . Its target is for the protected areas to allow for multiple uses, and to allow the system to preserve rare, endangered and threatened species. In these systems, there is need to increase the geographic distribution of target species, improve population structure, and influence the dynamics and genetic variability within and between populations. Also, identification of threats to targets species in the wild and suitable mitigation actions as well as species recovery programmes need to be implemented. In situ conservation of cultivated and exploited wild species is mostly concerned with the on-farm maintenance of landraces, forage, fruit trees and agroforestry species [47] [48] .

The ex situ conservation on the other hand involves conservation outside the native habitat and is generally used to safeguard populations in danger of destruction, replacement or deterioration. Samples from such species are stored in centralized banks away from the origin. Methods employed in ex situ conservation include field gene banks, seed storage banks, botanical gardens, arboretum, research centres and laboratories, DNA and pollen storage banks, and world heritage sites [49] . Other advanced biotechnological approaches include in vitro conservation and cryopreservation. These biotechnological approaches have been succinctly discussed by the authors [50] .

According to author [46] , the current level of development in conservation options for tropical fruit tree genetic resources, including African IFT, had necessitated the need for Field GeneBanks (FGBs) as important tool in the conservation and use of fruit tree genetic resources. Many important varieties of agronomic crops, horticultural crops and forestry trees are either difficult or impossible to conserve as seeds, due to inability of some species to naturally set seeds and as such are propagated vegetatively, or that the seeds are recalcitrant. Most of the IFT species are not exempted from this tendency, and hence are preferably conserved in FGBs along with other complementary conservation strategies [51] . The FGBs offer easy and ready access to conserve bulky plant genetic materials for research and use.

Moreover a number of plant species, including many IFT of Tropical Africa, have not had alternative methods of conservation fully developed for them. However, FGBs have its own challenges too: requires more space due to natural large size and evergreen nature of IFT species, relatively expensive to maintain in the long run, risk of damage by natural disasters, pestilences, neglects among others.

4. CONCLUSION

Rasheed Olufemi Awodoyin, Oluseun Sunday Olubode, Justin Ugochukwu Ogbu, Rasheed Babatunde Balogun, Jonah U. Nwawuisi, Kalu O. Orji (NARCs) list of mandate crops in Tropical African sub-regions should be re-considered for rather increasing focus and investment on exploiting the untapped rich treasuries of African IFT vis-à-vis developing improved agrotechniques and varietal selection for the species. The Plant Resources of Tropical Africa (PROTA) programme, World Agroforestry Centre and International Conference on Neglected and Underutilised Species (NUS) under the auspices of Bioversity International are by their various activities championing international campaign on re-awakening interests of all relevant stakeholders, including the academia, researchers, policy makers and industrialists on NUS generally and IFT in particular.

REFERENCES

1. Rathore, D.S. (2003) Role of Genetic Resources in Improvement of Tropical Fruit Species. In: Chaudhury, R., Panday, R., Malik, S.K. and Mal, B., Eds., In Vitro Conservation and Croypreservation of Tropical Fruits Species, IPGRI Office for South, Asia New Delhi/NBPGR, New Delhi, 17-25
2. Kulnlein, H., Erasmus, B. and Spigelski, D., Eds. (2009) Indigenous People Food Systems. FAO and Centre for Indigenous People's Nutrition and Environment, Rome, 251-281
3. Lapeña, I., Turdieva, M., Noriega, I.L. and Ayad, W.G. (2014) Conservation of Fruit Tree Diversity in Central Asia: Policy Options and Challenges. Bioversity International, Rome, 251 p
4. Paull, R.E. and Duarte, O., Eds. (2011) Tropical Fruits. 2nd Edition, CAB International, London, 1-10
5. Sthapit, B., Rao, V.R. and Sthapit, S., Eds. (2012) Tropical Fruit Trees Species and Climate Change. Bioversity International, New Delhi, 15-26, 97-125
6. Jaramillo, E.H., Sensi, A., Brandenberg, O., Ghosh, K. and Sonnino, A. (2011) Biosafety Resourcebook; Module (b): Ecological Aspects. Food and Agricultural Organisation (FAO), Rome, 10-49
7. FAO (2010) Statistical Yearbook 2010. http://www.fao.org/fileadmin/templates/ess/ess_test_folder/publications/yearbook_2010/c11.xls

8. Grace, O.M., Borus, D.J. and Bosch, C.H., Eds. (2008) Vegetable Oils of Topical Africa. Conclusions and Recommendations Based on PROTA 14: Vegetable Oils. Plant Resources of Tropical Africa (PROTA) Foundation, Nairobi, 84 p
9. Newman, J.L., Mehretu, A., Shillington, K. and Stock, R. (2009) Africa. Microsoft® Encarta® 2009 [DVD]. Microsoft Corporation, Redmond, 2008
10. Okigbo B.N. (1997) Neglected Plants of Horticultural and Nutritional Importance in Traditional Farming Systems of Africa. Acta Horticulturae, 53, 131-150
11. Meregini, A.O.A. (2005) Some Endangered Plants Producing Edible Fruits and Seed in Southeastern Nigeria. Fruits, 60, 211-220
12. Bosch, C.H., Siemonsma, J.S., Lemmens, R.H.M.J. and Oyen, L.P.A. (2002) Plant Resources of Tropical Africa Basic List of Species and Commodity Grouping. PROTA Programme, Wageningen, 7-12
13. Siemonsma, J.S., Schmelzer, G.H. and Rodrigues, W. (2004) PROTA 2004 Annual Report. Plant Resources of Tropical Africa (PROTA), Wageningen University and Research, Wageningen, 5-7
14. The National Academies (2008) Lost Crops of Africa: Volume 3. Fruits. National Academies Press, Washington DC, 381 p. http://www.nap.edu/catalog/11879.html
15. Ogbu, J.U. and Ibekwe, H.N. (2013) Characterization and Distribution of Indigenous Plants Research in Horticultural Society of Nigeria (HORTSON) Publications. Nigerian Journal of Horticultural Science, 18, 35-41
16. Cobley, L.S. and Steele, W.M. (1976) Introduction to the Botany of Tropical Crops. 2nd Edition, Longman, London
17. Purseglove, J.W. (1978) Tropical Crops: Monocotyledons Vol. 1 & 2. Longman Group, London, 607 p
18. Purseglove, J.W. (1984) Tropical Crops: Dicotyledons Vol. 1 & 2. Longman Group, London, 719 p
19. Youdeowei, A., Ezedinma, F.O.C. and Onazi, O.C. (1986) Introduction to Tropical Agriculture. Longman, London, 344 p
20. Opeke, L.K. (1987) Tropical Tree Crops. Spectrum Books, Ibadan, 327 p
21. Yayock, J.Y., Lombin, G. and Owonubi, J.J. (1988) Crop Science and Production in Warm Climates. Macmillan Publishers, London, 307 p
22. MacDonald, I. and Low, J. (1990) Fruits and Vegetables. Evans Brothers Ltd., Nairobi, 137 p
23. Matthew, I.P. and Karikari, S.K. (1990) Horticulture: Principles and Practices. Macmillan Publishers, London, 202 p
24. Muok, B.O., Owuor, B., Dawson, I. and Were, J. (2000) The Potential of Indigenous Fruit Trees: Results of a Survey in Kitui District, Kenya. Agroforestry Today, 12, 13-6
25. Mithofer, D. (2005) Economics of Indigenous Fruit Tree Crops in Zimbabwe. Ph.D. Thesis, Department of Economics and Business Administration, University of Hannover, Hannover, 250 p
26. Simon, T. (1997) Tree Domestication—Better Trees for Rural Prosperity. Agroforestry Today, 9, 4-5
27. Altieri, M.A. (2000) Agroecology: Principles and Strategies for Designing Sustainable Farming Systems. Hayworth Press, New York.
28. Leakey, R.R.B. (2000) Tree Domestication. The Overstory #31. http://www.agroforestry.net/overstory/overstory31.html.
29. Pauku, R.L. (2005) Domestication of Indigenous Nuts for Agroforestry in the Solomon's Islands. Ph.D. Thesis, James Cook University, Cairns, 381 p
30. Akinnifesi, F.K., Leakey, R.R.B., Ajayi, O.C., Sileshi, G., Tchoundjeu, Z., Matakala, P. and Kwesiga, F. (2007) Indigenous Fruit Trees in the Tropics: Domestication, Utilization and Commercialization. CAB International, Wallingford, 28-49. http://dx.doi.org/10.1079/9781845931100.0000
31. Jaenicke, H. and Hoschle-Zeledon, I. (2006) Strategic Framework for Underutilized Plant Species Research and Development with Special Reference to Asia and the Pacific, and to Sub-Saharan Africa. International Centre for Underutilized Crops, Rome, 33 p

32. Dawson, I.K., Guarino, L. and Jaenicke, H. (2007) Underutilized Plant Species: Impact of Promotion on Biodiversity. Position Paper No. 2. International Centre for Underutilized Crops (ICUC), Colombo.
33. Leakey, R.R.B., Nevenimo, T., Moxon, J., Pauku, R., Tate, H., Page, T. and Cornelius, J. (2010) Domestication and Improvement of Tropical Crops for Multi-Functional Farming Systems. Contemporary Crop Improvement: A Tropical View. 14th Australasian Plant Breeding Conference (APBC) and the 11th Congress of the Society for the Advancement of Breeding Research in Asia and Oceania (SABRAO), 10-14 August 2009, Cairns, 57
34. O'Neil, G.A., Dawson, I., Sotelo-Montes, C., Guarino, L. and Weber, J.C. (2001) Genetic Conservation of Tropical Trees. Biodiversity and Conservation, 10, 837-850. http://dx.doi.org/10.1023/A:1016644706237
35. Dawson, I., Harwood, C., Jamnadass, R. and Beniest, J. (2012) Agroforestry Tree Domestication: A Primer. The World Agroforestry Centre, Nairobi, 148 p
36. Halewood, M., Baidu-Forson, J.J., Clancy, E. and Vodouhe, R.S. (2014) Cooperating to Make the Best Use of Plant Genetic Resources in West and Central Africa: A Regional Imperative. Bioversity International, Rome, and CORAF/ WECARD, Dakar
37. Jaenicke, H. and Beniest, J. (2002) Vegetative Tree Propagation in Agroforestry. ICRAF, Nairobi, 1-30, 75-82
38. Verheij, E. (2006) Agrodok 5 Fruit Growing in the Tropics. 3rd Edition, Agromisa Foundation & CTA, Wageningen, 35-40
39. Scrase, R. (2009) Focus on—Domesticating Wild Trees in Botswana. New Agriculturist Online Journal. file:///E:/newag/focus/focusitem795.html
40. Tchoundjeu, Z., De Wolf, J. and Jaenicke, H. (1997) Vegetative Propagation for Domestication of Agroforestry Trees. Agroforestry Today, 9, 10-12
41. Verheij, E. (2004) Agrodok-19 Propagating and Planting Trees. 2nd Edition, Agromisa Foundation, Wageningen, 15-53
42. Hartmann, H.T., Kester, D.E., Davies Jr., F.T. and Geneve, R.L. (2007) Plant Propagation: Principles and Practices. 7th Edition, Prentice-Hall Inc., New Delhi, p 293-63
43. Acquoah, G. (2004) Horticulture: Principles and Practices. 2nd Edition, Prentice-Hall Press, New Delhi, 316-356
44. Chadha, K.L. (2009) Handbook of Horticulture. Indian Council of Agricultural Research (ICAR), New Delhi, 76-82
45. Jamnadass, R.H., Dawson, I.K., Franzel, S., Leakey, R.R.B., Mithöfer, D., Akinnifesi, F.K. and Tchoundjeu, Z. (2011) Improving Livelihoods and Nutrition in Sub-Saharan Africa through the Promotion of Indigenous and Exotic Fruit Production in Smallholders' Agroforestry Systems: A Review. International Forestry Review, 13, 338-354. http://dx.doi.org/10.1505/146554811798293836
46. Rao, V.R. and Sthapit, B.R. (2012) Tropical Fruit Tree Genetic Resources: Status and Effect of Climate Change. In: Sthapit, B., Rao, V.R. and Sthapit, S., Eds., Tropical Fruit Trees Species and Climate Change, Bioversity International, New Delhi, 97-128
47. Rao, N.K. (2004) Plant Genetic Resources: Advancing Conservation and Use through Biotechnology. African Journal of Biotechnology, 3, 136-145
48. Rathore, D.S., Srivastava, U. and Dhillon, B.S. (2005) Management of Genetic Resources of Horticultural Crops: Issues and Strategies. In: Dhillon, B.S., Tyagi, S., Saxena, S. and Randhawa, G.T., Eds., Plant Genetic Resources: Horticultural Crops, Narosa Publishers, New Delhi, 1-18
49. Oldfield, S. and Newton, A.C. (2012) Integrated Conservation of Tree Species by Botanic Gardens: A Reference Manual. Botanic Gardens Conservation International, Richmond, 56 p.
50. Ahuja, M.R. and Ramawat, K.G. (2014) Biotechnology and Biodiversity. Sustainable Development and Biodiversity. Vol. 4, Springer International Publishing Switzerland. http://www.springer.com/series/11920

51. Engelmann, F. and Engels, J.M.M. (2002) Technologies and Strategies for ex Situ Conservation. In: Brown, A. and Jackson, M., Eds., Managing Plant Genetic Diversity, CAB International /IPGRI, Wallingford, 89-104

CHAPTER 10

Characterization and Expression Analysis of Genes Encoding Ubiquitin Conjugating Domain-Containing Enzymes in *Carica Papaya*

Dengwei Jue[1], Xuelian Sang[1], Bo Shu[1], Liqin Liu[1], Yicheng Wang[1], Zhiwei Jia[1], Yu Zou[2], Shengyou Shi1*

[1]*Key Laboratory of Tropical Fruit Biology (Ministry of Agriculture), South Subtropical Crops Research Institute, Chinese Academy of Tropical Agricultural Sciences, Zhanjiang, China,*

[2] *Rice Research Institute, Anhui Academy of Agricultural Sciences, Hefei, China*

ABSTRACT

Background

Ripening affects the quality and nutritional contents of fleshy fruits and is a crucial process of fruit development. Although several studies have suggested that ubiquitin-conjugating enzyme (E2s or UBC enzymes) are involved in the regulation of fruit ripening, little is known about the function of E2s in papaya (*Carica papaya*).

Methodology/Principal findings

In the present study, we searched the papaya genome and identified 34 putative UBC genes, which were clustered into 17 phylogenetic subgroups. We also analyzed the nucleotide sequences of the papaya UBC (*CpUBC*) genes and found that both exon-intron junctions and sequence motifs were highly conserved among the phylogenetic subgroups. Using real-time PCR analysis, we also found that all the CpUBC genes were expressed in roots, stems, leaves, male and female flowers, and mature fruit, although the expression of some of the genes was increased or decreased in one or several specific organs. We also found that the expression of 13 and two CpUBC genes were incresesd or decreased during one and two ripening stages, respectively. Expression analyses indicates possible E2s playing a more significant role in fruit ripening for further studies.

Conclusions

To the best of our knowledge, this is the first reported genome-wide analysis of the papaya UBC gene family, and the results will facilitate further investigation of the roles of UBC genes in fruit ripening and will aide in the functional validation of UBC genes in papaya.

Funding: This work was supported by the Natural Science Foundation of Hainan Province (20163111), the Natural Science Foundation of China (31572087), the Natural Science Foundation of Guangdong Province of China (2014A030307031), the Basic Scientific Research Project of Non-profit Central Research Institutions (No. 1630062015004), the Germplasm Resource Protection of Ministry of Agriculture (2016NWB053), the Germplasm Resource Protection of Ministry of Agriculture Wild Plant (1251416305010), the Science and Technology Project of Guangdong Province (2016A020208002) and the Nature Science Foundation of Anhui Province (grant no. 1408085MKL63).

INTRODUCTION

Papaya (*Carica papaya*) is an economically important fruit crop that is widely grown in tropical and sub-tropical regions, and its fleshy fruits play an important role in the human food supply, as well as in human nutrition and health [1, 2]. Gaining in popularity among tropical fruits worldwide, papaya is now ranked fourth in total tropical fruit production after bananas, oranges), and mango [3]. Fruit ripening is an important and complex process, and, therefore, the regulatory mechanisms of fruit ripening have been intensively studied [4]. Fruit with different ripening mechanisms can be divided into two groups: climacteric fruit (e.g., bananas, tomatoes, pears, mangos, and papayas), in which ripening is accompanied by a peak in respiration and a concomitant burst of ethylene, and non-climacteric fruit (e.g., pineapple, carambola, borojo, and watermelon), in which respiration exhibits no dramatic change and ethylene production remains at a very low level [5].

Ethylene, in particular, has been extensively studied, owing to its crucial role in the ripening of climacteric fruits [6–9], and the biosynthesis, perception, and signal transduction of ethylene, as well as downstream gene regulation, are well known [10]. Tomato (*Solanum lycopersicum*) has long served as an excellent model for studies of fleshy fruit development and ripening [11, 12], and to date, many of the underlying genes have been cloned, thus providing insight into the regulation of ripening. The ripening-associated genes that have been reported in tomato include: *rin*, which encodes a MADS-box transcription factor [13]; *Nr*, which encodes an ethylene receptor [14]; *Cnr* (*Colorless nonripening*), which encodes an SBP-box transcription factor [15]; *Gr* (*Green ripe*), which encodes a still poorly defined component of the ethylene transduction pathway [6]; *SlNAC4*, a new tomato NAC transcription factor that positively regulates fruit ripening and carotenoid accumulation [16]; *APETALA2a* (*AP2a*), a transcription factor that regulates fruit ripening *via* regulation of ethylene biosynthesis and signaling [17]; *LeNCED1*, which initiates abscisic acid (ABA) biosynthesis at the onset of fruit ripening and might act as an original inducer in tomato [18]; and the tomato genes *SHATTERPROOF*, *TAGL1* [19], *TDR4/FUL1*, and *MBP7/FUL2* [20], which are also involved in regulating fruit ripening. However, in contrast to tomato, there have been only a few reports regarding fruit ripening in *C. papaya*.

Ubiquitination is an essential cellular process in all eukaryotes, from unicellular yeast to humans, and the ubiquitination-proteasomal pathway has been implicated in diverse aspects of eukaryotic cellular regulation, owing to its ability to degrade intracellular proteins [21, 22]. The process of protein ubiquitination is mediated through the action of three enzymes, which are known as ubiquitin-activating enzymes (E1s), ubiquitin-conjugating enzymes (E2s), and ubiquitin ligases (E3s) [23]. Ubiquitin-conjugating enzymes (E2s) function at an intermediate step in the protein ubiquitination pathway. The E2 family has expanded during evolution, and more ancestral eukaryotes possess fewer E2 enzymes (e.g., ≤20 in algae) than more derived ones (e.g., >40 in certain plants and animals) [24]. For example, 13 UBC domain proteins have been identified in *Saccharomyces cerevisiae* [25], 12, 18, and 19 E2 enzymes are encoded in the algae *Ostreococcus tauri*, *Micromonas sp. RCC299*, and *Chlamydomonas reinhardtii*, respectively (data not shown), 20 in *Caenorhabditis elegans* [26], 37 in both humans [24] and *Arabidopsis thaliana* [27], 48, 52, 74 and 75 UBC domain proteins in rice, tomato, banana and maize [28–31].

Previous studies have suggested that ubiquitination plays an important role in plant growth, development, and tolerance against various abiotic stresses, and several of the UBC genes in *A. thaliana* have been implicated specifically [32, 33]. For example, the *A. thaliana* ubiquitin-conjugating gene *AtUBC13* has been implicated in epidermal cell differentiation and iron deficiency responses [34, 35], and AtUBC32 is an endoplasmic reticulum-associated degradation (ERAD) component that functions in brassinosteroid-mediated salt stress tolerance [36], whereas AtUBC21 (AtPEX4) is specialized for ubiquitination in peroxisome maintenance [37]. In rice, 14 *OsUBCs* were differentially expressed under drought and salt stress conditions [38], and in *Zea mays*, 48 and 16 *ZmUBC* genes were significantly upregulated under salt and drought stress conditions, respectively [30].

Several studies suggest that E2s also participate in the fruit ripening process. In tomato, for example, *SlUBC32*, which is upregulated during tomato fruit ripening and downregulated in the rin mutant, plays an important role in the regulation of fruit ripening [31]; and in banana, 32 *MaUBC* genes are reportedly up- and downregulated during different ripening stages [29]. However, functional studies of E2s involved in fruit ripening are limited and have not been conducted in papaya. Therefore, in the present study, we performed a genome-wide analysis of E2 domains, analyses of CpUBC gene structures and phylogenetic relationships, and expression profiling of CpUBC genes in various *C. papaya* plant tissues and at different stages of fruit ripening. To the best of our knowledge, this is the first reported genome-wide analysis of the papaya UBC gene family, and the results will facilitate further investigation of the roles of the UBC genes in fruit ripening and will aide in the functional validation of UBC genes in papaya.

MATERIALS AND METHODS

2.1. Plant material and treatments

Papaya (*C. papaya* cv. 'sun up') plants were obtained from the South Subtropical Crops Research Institute of the Chinese Academy of Tropical Agricultural Science (Zhan-jiang, Guangdong Province, China), and root, stem, leaf, male flower, female flower, and fruit (mature green) organs were collected separately. For the fruit ripening experiment, we also collected fruit pulp from four development stages: immature green (IG), mature green (MG), breaker (Br), and mature fruit (MF). The experiments were performed in triplicate, and all samples were immediately frozen in liquid nitrogen and stored at −80°C for expression analysis.

2.2. Sequence retrieval and gene family member identification

To identify potential members of the papaya UBC protein family, published *Arabidopsis* and yeast (*S. cerevisiae*) UBC protein sequences were retrieved and used as queries in BLASTP searches against the *C. papaya* Genomics Database (CpGDB; http://www.plantgdb.org/CpGDB/) and PLAZA 3.0 (http://bioinformatics.psb.ugent.be/plaza/versions/plaza/) [39]. Amino acid sequences of all the UBC-encoding genes of *Arabidopsis* , *S. cerevisiae*, and *S. lycopersicum* were downloaded from the *Arabidopsis* Information Resource (http://www.arabidopsis.org/), *Saccharomyces* Genome Database (http://www.yeastgenome.org/), and Sol Genomics Network Tomato Database (http://solgenomics.net) websites, respectively. All putative papaya UBC protein sequences were then subject to domain analysis using Simple Modular Architecture Research Tool (SMART; http://smart.emblheidelberg.de/) [40], and information regarding the CpUBC genes, including location, open reading frame (ORF), amino acid (AA) length and intron number, were obtained from PLAZA 3.0 [39]. The molecular weight (MW) and theoretical isoelectric point (PI) of the candidate CpUBC proteins were investigated using ExPASy online tools (http://expasy.org/tools/) [41]. Finally, the structure of each gene was visualized using the Gene Structure Display Server (GSDS) (http:// gsds.cbi.pku.edu.cn/) [42].

2.3. Sequence alignment and phylogenetic analysis

To evaluate the evolutionary relationship of the CpUBC proteins, full-length amino acid sequences of 34 CpUBC proteins, 15 UBC proteins from *S. cerevisiae*, 48 UBC proteins from *Arabidopsis thaliana*, and 52 UBC proteins from *S. lycopersicum* were subject to multiple sequence alignment, using MUSCLE [43], with default parameters, and manual optimization. We excluded ambiguously aligned sequences, in order to produce an alignment of 34 amino acid characters for subsequent phylogenetic analyses. An unrooted phylogenetic tree was constructed using the neighbor-joining (NJ) method in MEGA (Version 6.0; http://www.megasoftware.net/), and the significance of nodes was assessed using a bootstrap test with 1,000 iterations [44]. Representations of the calculated trees were constructed using TreeView (Version 1.6.6; http://taxonomy.zoology. gla.ac.uk /rod/treeview).

2.4. RNA isolation and expression analysis

Total RNA was extracted for cDNA synthesis using the Super fast new plants of RNA extraction kit (Huayueyang Bio Co., Ltd, Beijing, China), following the manufacturer's instructions. Reverse transcription reactions were performed using the PrimeScript RT reagent kit with gDNA Eraser (Takara Bio, Inc., Kusatsu, Japan), according to the supplier's manual. Real-time PCR was performed using a Bio-Rad real-time thermal cycling system (LightCycler 480; Bio-Rad Laboratories, Inc., Hercules, CA, USA) and SYBR-green to assess the expression levels of 34 candidate CpUBC genes. The gene-specific primers were designed according to the CpUBC gene sequences using Primer 5 software and checked using Blast (S1 Table) In addition, the papaya actin 1 gene was used as an internal control for normalization. Each reaction consisted of 10 μL 2× SYBR Premix Ex Taq II (Takara Bio), 40 ng cDNA, and 250 nM of each primer, in a final volume of 20 μL. The following PCR program was used: 94°C for 10 min, followed by 40 cycles of 94°C for 10 s, 58–63°C for 20 s, and 72°C for 30 s. The relative mRNA levels of the genes were measured using the cycle threshold (Ct) $2^{(-\Delta Ct)}$ method. The analysis included cDNA from the three biological samples for each tissue, and all the reactions were run in triplicate. In the comparative expression analysis of *CpUBCs*, genes that were up- or downregulated by at least two-fold were considered differentially expressed.

RESULTS

Identification of UBC genes in papaya

In the present study, we used the published *Arabidopsis* and yeast (*S. cerevisiae*) UBC protein sequences as queries in BLASTP searches against the CpGDB and PLAZA 3.0[39], a total of 39 putative CpUBC genes were identified. After scanning of the 39 sequences for the UBC domain by motif scan using SMART search, we found only 34 sequences contain the UBC domain. According to their chromosome locations (Table 1), the 34 CpUBC genes were designated *CpUBC1* to *CpUBC34*. In these 34 CpUBCs, two were identified as Related to Ubiqutin (RUB) conjugating enzymes (CpUBC18 and CpUBC29), and one was identified as a SUMO-conjugating enzyme (CpUBC11). Six other UBC proteins (CpUBC2/14/16/19/24/25) lacked the Cys active site, identifying them as ubiquitin-conjugating enzyme variants (UEVs), which are not active by themselves, thus leaving 25 potential ubiquitin E2s. To better understand the papaya E2s, we constructed a recombinant E2 protein library of all 34 CpUBCs. The predicted proteins ranged from 98 (CpUBC15) to 668 amino acids (CpUBC20) in length, with corresponding molecular masses of 10.74 kDa and 74.03 kDa, respectively, and predicted isoelectric points of between 4.29 (CpUBC33) and 9.64 (CpUBC27).

Table 1. The information of *CpUBC* gene family.

Gene name	Gene locus	Chromosome Location	ORF(bp)	Size(aa)	PI	MW(KDa)	Intron
CpUBC1	CP00001G03780	supercontig_1: 4880545~4883588	504	167	5.04	18.72	4
CpUBC2	CP00002G01990	supercontig_2: 2835597~2840351	474	158	5.11	17.96	4
CpUBC3	CP00002G02590	supercontig_2: 3448167~3450541	447	148	7.72	16.48	3
CpUBC4	CP00003G00420	supercontig_3: 241676~245597	447	148	7.72	16.59	3
CpUBC5	CP00004G01190	supercontig_4: 2377968~2380915	585	194	4.69	21.28	4
CpUBC6	CP00006G01800	supercontig_6: 1485889~1488676	492	163	8.35	18.6	5
CpUBC7	CP00006G02820	supercontig_6: 2177997~2178675	447	148	8.44	16.55	3
CpUBC8	CP00009G00470	supercontig_9: 221437~222339	903	300	9.22	33.92	0
CpUBC9	CP00012G00790	supercontig_12: 606495~607675	148	447	7.72	16.46	3
CpUBC10	CP00017G01980	supercontig_17: 2605422~2607916	480	159	5.81	18.15	4
CpUBC11	CP00021G01620	supercontig_21: 1651878~1654594	483	160	8.42	18.03	4
CpUBC12	CP00023G00940	supercontig_23: 1591010~1594089	1593	530	6.33	58.62	5
CpUBC13	CP00027G00240	supercontig_27: 207634~210077	1170	389	4.99	43.83	4
CpUBC14	CP00033G01420	supercontig_33:1629710~1630930	1221	406	6.14	42.23	0
CpUBC15	CP00037G01340	supercontig_37:1138171~1138798	297	98	5.18	10.74	2
CpUBC16	CP00052G00760	supercontig_52: 720742~721809	936	311	5.41	35.21	1
CpUBC17	CP00053G00580	supercontig_53: 377001~380351	576	191	5.55	21.32	5
CpUBC18	CP00062G01080	supercontig_62: 722413~724323	612	203	8.53	23.25	5
CpUBC19	CP00073G00060	supercontig_73: 180777~182000	1224	407	6.68	45.56	0
CpUBC20	CP00080G00590	supercontig_80: 464531~468738	2007	668	4.66	74.03	7
CpUBC21	CP00084G00040	supercontig_84: 64510~66540	546	181	5.21	20.02	5
CpUBC22	CP00092G00550	supercontig_92: 491412~493626	780	259	8.69	28.16	4
CpUBC23	CP00097G00940	supercontig_97: 653771~662278	462	153	6.74	17.22	7
CpUBC24	CP00104G00590	supercontig_104: 426794~432223	441	146	6.2	16.62	3
CpUBC25	CP00119G00620	supercontig_119: 447043~449670	387	128	8.62	13.73	2
CpUBC26	CP00122G00250	supercontig_122: 616176~622862	447	148	7.72	16.55	3
CpUBC27	CP00124G00190	supercontig_124: 451236~454332	360	119	9.64	14.04	3
CpUBC28	CP00129G00500	supercontig_129: 420388~423075	465	154	4.91	17.22	5
CpUBC29	CP00158G00160	supercontig_158: 234845~236579	552	183	7.69	20.99	4
CpUBC30	CP00161G00220	supercontig_161: 289660~296412	459	152	5.37	17.32	4
CpUBC31	CP00183G00210	supercontig_183: 206758~210983	459	152	5.31	17.30	4
CpUBC32	CP00190G00050	supercontig_190: 25441~28352	426	141	5.58	16.14	4
CpUBC33	CP00214G00120	supercontig_214: 202216~221651	552	183	4.29	20.78	5
CpUBC34	CP01258G00030	supercontig_1258: 3731~7220	555	184	4.41	21.09	5

Structure and phylogenetic analysis of CpUBC genes

Using the GSDS website, we found that the number of introns in the 34 CpUBC genes ranged from zero (*CpUBC8*, *CpUBC14* and *CpUBC19*) to seven (*CpUBC20* and *CpUBC23*), with most of CpUBC genes containing three to five introns (n = 76.5%) (Table 1 and Fig 1). In addition, most CpUBC genes within the same subfamilies shared the same exon/intron structure. For example, in the UBC4/5 subfamily, *CpUBC3*, *CpUBC4*, *CpUBC7*, *CpUBC9*, and *CpUBC26*contained three introns, whereas in the UBC11 subfamily, *CpUBC17* and *CpUBC21* harbored five introns.

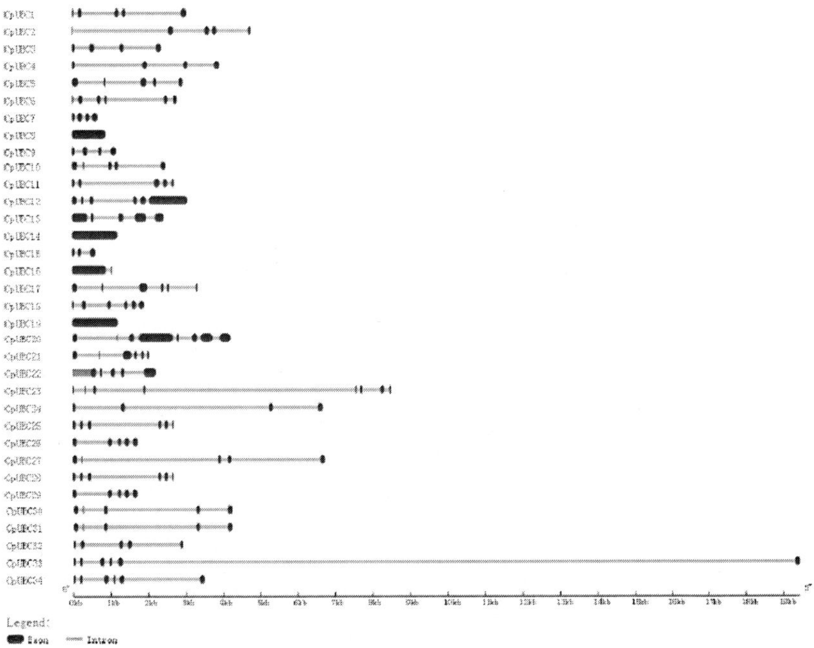

Figure 1. Structure of CpUBC genes.

Untranslated 5′ and 3′ regions, exons, and introns are indicated by red, blue, and gray, respectively. The scale bar represents 1000 bp.

Multiple sequence alignment of the predicted amino acid sequences revealed that the tryptophan (W) located at the C-terminal side of the active cysteine was conserved in most of the CpUBC genes (Fig 2A). The consensus active site motif "HPN" was found at six amino acids from the N-terminal site of the active cysteine, and the strongly conserved PxxPP motif was found at seven amino acids from the N-terminal side of the HPN motif (Fig 2A). In addition, the predicted amino acid sequences of the UBC domains of 34 CpUBC genes were analyzed using the MEME Suite website (http://meme-suite.org/index.html). The results indicated that the highly conserved sequence of the UBC domain is HPNINSNGSICLDILKEQWSP (Fig 2C).

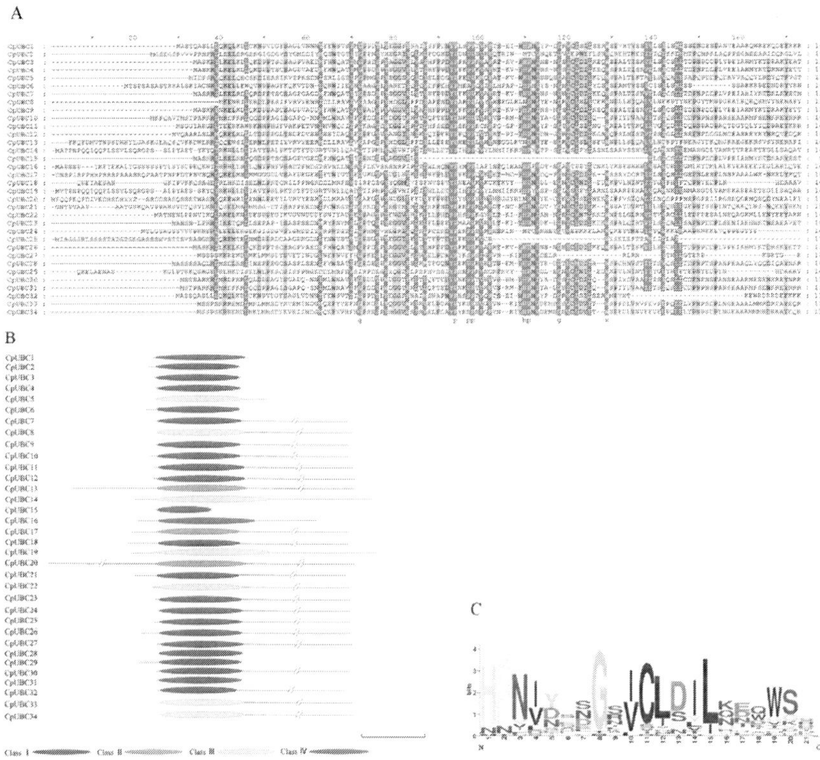

Figure 2. The multiple sequence alignment, structures and the architecture of conserved protein motifs of CpUBC proteins. (A) Multiple sequence alignment of CpUBC domains. Identical amino acids, conserved amino acids, and blocks of similar amino acid residues are shaded in black, charcoal gray, and gray, respectively. (B) Structures of the CpUBC proteins. The name of each corresponding protein is shown on the left. The position of the UBC domain is indicated in the figure. The length and order of the domains represent their actual location within each protein. The scale bar represents 100 amino acids. The different colors indicate the four E2 subtypes of UBC domains. (C) Conserved CpUBC domains. Logos of the protein alignment of the UBC domain is shown. The x-axis represents the conserved sequences, and the conservation of each residue is indicated by the height of each letter. The y-axis is a scale of the relative entropy, which reflects the conservation rate of each amino acid.

According to the UBC domain and the N- or C-terminal structure, E2 proteins are divided into four types. Class I E2s contain only the catalytic domain; Class II E2s contain N-terminal extensions; Class III E2s contain C-terminal extensions; and Class IV E2s have both N- and C-extensions [45, 46]. In the present study, we found that 23 of the CpUBCs belonged to Class I, and three, seven, and one belonged to Class II, Class III, and Class IV, respectively (Fig 2B).

Furthermore, phylogenetic analysis revealed that the UBC proteins could be divided into 14 E2 groups and three independent UEV groups (UBC17, UBC18, and UBC19, with UBC9 and UBC12 functioning in the SUMO and RUB1 conjugation pathways, based on >50% bootstrap support (Fig 3). Most of the CpUBC groups were named according to the identities of the *S. cerevisiae* proteins or proteins contained within the group (UBC1, UBC2, UBC3, UBC4/5, UBC6, UBC3/7, UBC8, UBC9, UBC10, UBC11, UBC12, UBC13, UBC17, and UBC18); however, there were no yeast proteins in groups UBC14, UBC15, UBC16, and UBC19, which indicated that the groups may be plant-specific or were lost in yeasts. Interestingly, the subfamilies UBC4/5 or UBC3/7 shared two highly identical paralogous yeast genes, respectively. For example, during the phylogenetic analysis of CpUBC genes using MEGA 5.0, UBC4 and UBC5 always clustered into the same subgroup, as did UBC3 and UBC7. According to previous reports [26, 27, 47], we designated the two clades as subfamilies UBC4/5 and UBC3/7.

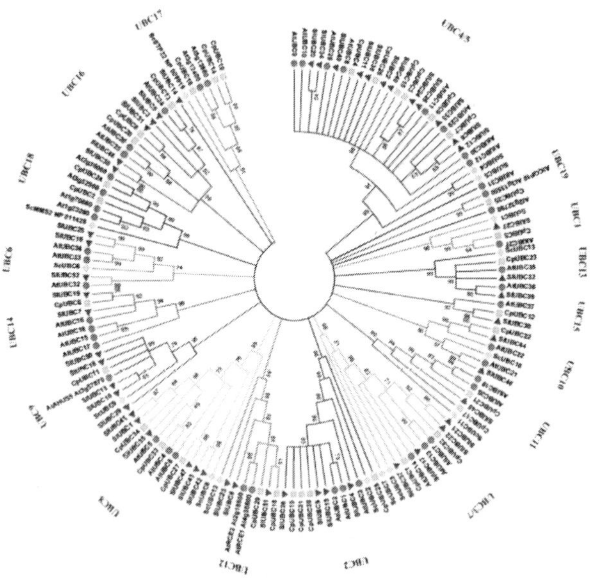

Figure 3. Phylogenetic tree of ubiquitin-conjugating (UBC) domains from *Carica papaya*, *Arabidopsis*, *Solanum lycopersicum*, and *Saccharomyces cerevisiae.*

The neighbor-joining tree includes 34 UBC proteins from papaya (squares), 15 from *S. cerevisiae* (rhombi), 48 from *A. thaliana* (circles), and 52 from *S. lycopersicum*(triangles). The different colors indicate different CpUBC subgroups.

Organ- and ripening stage-specific expression of CpUBC genes

All 34 of the candidate CpUBC genes were differentially expressed among the six papaya organs (Fig 4). In addition, we found that the expression of 13 (*CpUBC1/4/5/6/9/10/13/17/18/24/25/26/34*), 11 (*CpUBC2/3/15/21/23/27/28/29/30/31/33*), and seven (*CpUBC7/8/11/12/14/16/20*) genes were lowest in leaves, roots, and stems, respectively, and that the expression of only one gene was lowest in the male flowers (*CpUBC32*) and female flowers (*CpUBC19*), respectively. Meanwhile, 19 (*CpUBC1/2/3/5/ 6/9/10/11/12/15/17/20/23/24/26/30/31/33/34*) and two genes (*CpUBC21* and *CpUBC22*) were highly expressed in male flowers and female flowers, respectively, which implicated their involvement in the development of floral sex organs. Four (*CpUBC25/27/28/29*), five (*CpUBC7/8/14/16/19*), and two (*CpUBC4* and *CpUBC32*) genes were highly expressed in fruits, leaves, and stems, respectively, and a single gene (*CpUBC13*) was highly expressed in roots. Therefore, our results suggested that CpUBC genes play multiple roles in papaya development.

In the present study, we also analyzed the expression of the 34 CpUBC genes at different fruit ripening stages. We found that 13 (*CpUBC4/6/7/8/9/11/12/14/16/19/20/28/34*) and two (*CpUBC2* and *CpUBC10*) of the genes showed higher or lower expression during the progress of papaya fruit ripening, respectively, whereas the expression of the remaining genes was not significantly different (Fig 5). Of the 13 genes that were higher expressed in the fruit, the expression levels of eight genes (*CpUBC7/8/9/11/12/14/19/20*) were slightly increased at the IG, MG, and Br stages and most highly expressed at the MF stage. Meanwhile, the expression of four genes (*CpUBC6/16/28/34*) continuously increased during the first three stages of fruit ripening, before being decreased at the MF stage, and the expression of *CpUBC4* was increased at MG stage and then reduced at the two late stages. These results suggest that the papaya E2s maybe playing a significant role in fruit development and ripening.

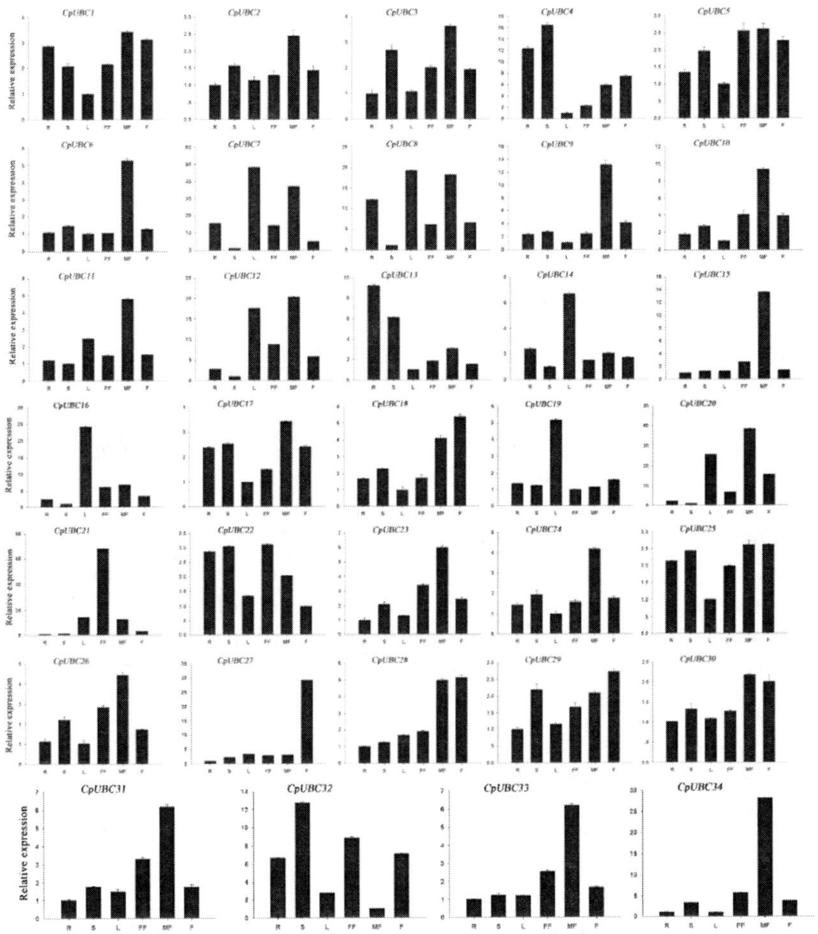

Figure 4. Tissue-specific expression of selected CpUBC genes. The x-axis indicates time course/treatment, and the y-axis indicates the relative expression level. Error bars indicate standard deviations of independent biological replicates (n ≥ 3). R, root; St, stem; L, leaf; T, tassel, Ys, young seed; and Si, silk.

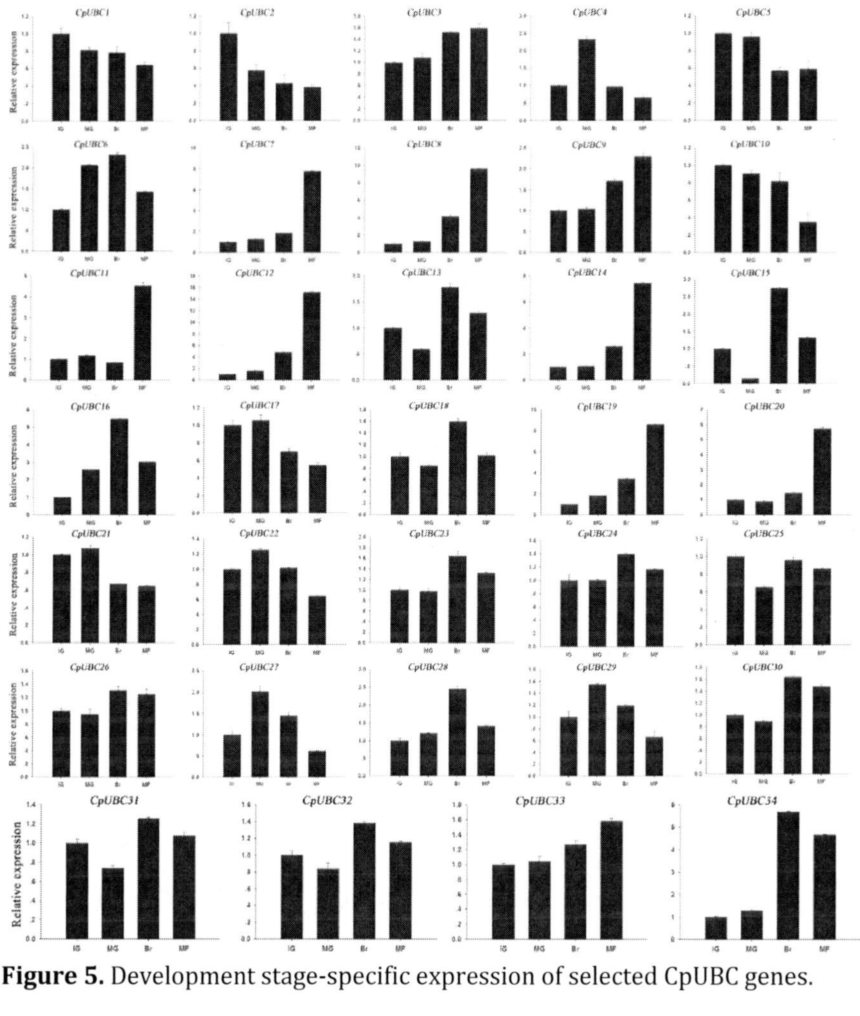

Figure 5. Development stage-specific expression of selected CpUBC genes.

The x-axis indicates time course/treatment, and the y-axis indicates the relative expression level. Error bars indicate standard deviations of independent biological replicates (n ≥ 3). IG, immature green; MG, mature green; Br, breaker; and MF, mature fruit.

DISCUSSION

Ripening affects the quality and nutritional contents of fleshy fruits and is a crucial process in the development of climacteric fruits [48]. In recent years, the study of genes involved in fruit ripening has rapidly progressed; however, most research has been conducted in tomato. In addition, E2 proteins are critically important in many aspects of plant growth and development; however, considering the potential functional significance of E2 proteins, only a few E2 family members have been described in higher plants[27, 31, 38], and only a few studies have investigated the role of the UBC gene family in fruit ripening [29, 31].

In the present study, we identified genes encoding 34 *C. papaya* UBC (CpUBC) proteins (Table 1), which includes two RUB-conjugating enzymes (CpUBC18 and CpUBC29), one SUMO-conjugating enzyme (CpUBC11), and six UEV proteins (CpUBC2/14/16/19/24/25). Similar results have also been reported by other studies. For example, 16 UBC proteins were identified in *S. cerevisiae*, including one RUB-conjugating enzyme (ScUBC9), one SUMO-conjugating enzyme (ScUBC12), and two UEV protein (ScMMS2, NP_011428 and ScSTP22, NP_009919) [49]. In addition, a total of 48 UBC domain-containing proteins have been identified in *Arabidopsis*, of which two are RUB-conjugating enzymes (RCE1, At4g36800 and RCE2, At2g18600), one is a SUMO-conjugating enzyme (AtSCE1, At3g57870), and eight others are UEVs, thus leaving 37 potential E2s [27]. Among 48 rice E2s, three proteins (OsUBC1/2/3) were identified as RUB E2s, three SUMO E2s (OsUBC4/5/6), and four UEVs (OsUBC28/29/30/31) [28]. Meanwhile, of the 52 SlUBC proteins in tomato, there are three SUMO E2s (SlUBC8/23/51) and five RUB E2s (SlUBC10/13/18/29/50) [31]; and of the ZmUBC proteins in maize, there are three RUB E2s (ZmUBC-37/40/43), eight SUMO E2s (ZmUBC-01/04/11/12/51/69/73/75), and six UEVs (ZmUBC-08/04/31/34/53/74) [30]. The number of CpUBC genes is far less than the 48, 48, 52, 75, and 74 UBC genes reported from the *Arabidopsis* [27], rice [28], tomato [31], maize [30], and banana genomes [29], respectively. The genome sizes of *Arabidopsis*, rice, tomato, maize, and banana are ~125, ~466, ~900, ~2300, and 523 Mb, respectively. However, although the genome sizes of rice and tomato are 3.7 and 7.2 times larger than that of *Arabidopsis*, respectively, the genomes of the three species contain a similar number of UBC genes (45, 48, and 52, respectively); and although the genome size of maize is 4.3 times larger than that of banana, the genomes of these two species also contain a similar number of UBC genes (75 and 74, respectively). Therefore, the difference in number of UBC genes among these species cannot be explained by genome size.

Alternatively, gene duplication events have been demonstrated to play a significant role in the expansion of gene family members in genomes [38], and an increasing number of studies has shown that segmental duplications are largely responsible for the expansion of gene families in maize, such as the CCCH, HD-Zip HSF, bZIP, and PRX gene families[50–53]. Research has also shown that segmental duplications are the main contributor to the expansion of UBC genes in maize and banana [29, 30]. However, owing to a lack of information regarding the chromosomes of papaya, we were unable to perform segmental and tandem duplication analysis in papaya.

Through systematic phylogenetic analyses, the present study provides a detailed classification and nomenclature of papaya E2s. The phylogenetic analysis categorized the 34 CpUBCs into 17 discrete groups, which was similar to the number of groups described in other species, such as tomato, maize, and banana [29–31]. However, there were still some minute differences between the topologies of the UBC genes in papaya and the other species. For example, there are 15 and 13 UBC groups in tomato and banana, respectively, not including the UEV subgroups [29, 31]. However, in *Arabidopsis*, rice, and maize, the UEV subgroups were taken into account [27]. In our study, to better understand the papaya E2s, the CpE2s were divided into 17 groups, which included three independent UEV groups. The corresponding homologs of AtUBC21 and AtUBC22 were grouped together in maize [30], banana [29], and papaya, whereas the corresponding homologs (and orthologs) were separated into two groups in tomato [31]. Additionally, homologs of AtUBC31, grouped together in the UBC4/5 group in maize and banana [29, 30], were not grouped in any groups in our study. These differences could have resulted from different parameter settings during the phylogenetic analyses.

In addition, the number of UBC genes was obviously different among the groups. The largest groups (UBC4/5) included eight members, whereas the UBC1, UBC2, UBC6, UBC9, UBC10, UBC13, UBC14, and UBC15 groups were only represented by one member each. The UBC4/5 groups was also the largest in other species, which indicates that, in plants, the group may possess more diverse functions than other groups. Phylogenetic data also suggested that the UBC9 and UBC12 groups were expanded in monocots but not in *Arabidopsis*, and previous research has shown that the genomes of maize and rice each contain eight and three members of the UBC9 and UBC12 groups, respectively [28, 30]. However, in papaya, we only identified one and two members in these groups. Furthermore, we also identified three UBC groups (UBC14, UBC15, and UBC16) in *Arabidopsis* and papaya for which no probable homologs exist in budding yeast; however, all three groups have potential homologs in animals [54], which indicates a possible gene loss during yeast evolution.

Although the genome sequence of papaya has been reported, the identification and functional studies of papaya genes have proceeded at a slow pace. Analysis of the temporal and spatial expression patterns of genes may provide useful information for characterizing their functions. Previous studies have shown that UBC genes have different expression patterns in different organs. In *Arabidopsis*, for example, *AtUBC1* and *AtUBC2* are ubiquitously expressed in roots, leaves, flowers, and seedlings, and the double mutant *atubc1-1atubc2-1* exhibits a dramatically reduced number of rosette leaves and an early-flowering phenotype [55]. In banana, *MaUBC10/11/33/34/61* are highly expressed in most organs, especially in roots, stems, leaves, and *MaUBC6/11/34/35/45/61* were highly expressed in stems. In addition, 12 *MaUBC* genes (*MaUBC13/ 18/29 /33/34 /36/43/46/48/53/67/70*) were predominantly expressed in roots [29]. Meanwhile, in the present study, we found that 19 and two genes were highly expressed in male and female flowers, respectively, which suggests that the genes may be involved in the development of floral sex organs, and various other CpUBC genes were specifically higher expressed in other tissues, which that CpUBC genes play multiple roles in the development of papaya.

Previous studies have demonstrated that ubiquitin-conjugating enzymes (E2s) are critically important in many aspects of plant growth and development, as well as in physiological processes, such as stress responses [56]. For example, *OgUBC1* from wild rice is involved in cellular responses to biotic and abiotic stresses [56], and three (*OsUBC2/5/18*) and five UBC genes (*AtUBC13/17/20/26/31*) from rice and *Arabidopsis*, respectively, are significantly reduced in response to salt and drought stress, whereas only three rice genes (*OsUBC13/15/45*) are significantly upregulated [38]. Although the function of E2s in plant development and stress responses have been well clarified, reports about the function of protein ubiquitination in fruit ripening remains extremely rare, except in tomato and banana [29, 31]. In tomato, a total of six UBC genes (*SlUBC6/8/24/32/41/42*) were directly regulated by the fruit-ripening regulator RIN, which suggests that specific UBC genes might be involved in fruit ripening, and in banana, five UBC genes (*MaUBC1/9/70/68/71*) exhibited approximately 10- to 40-fold higher levels of expression at the fifth stage than at other stages of fruit ripening, whereas seven other UBC genes (*MaUBC8/16/17/33/34/56/61*) exhibited continuously increasing expression throughout the fruit development. In the present study, 13 (*CpUBC4/6/7/8/9/11/12/14/16/19/20/28/34*) and two (*CpUBC2* and *CpUBC10*) of the 34 CpUBC genes were up- or downregulated during the progression of papaya fruit ripening, respectively (Fig 5), and our results indicated that the expression patterns of some of the CpUBC genes, such as *CpUBC17* and *CpUBC56*, was related to their phylogenetic relationships. By contrast, the expression patterns of some paralogs, such as *CpUBC8* and *CpUBC23*, were quite different. Meanwhile, we also found that the genes with close phylogenetic relationships that exhibited similar expression patterns in banana, tomato, and papaya. For example, our data indicated that the *CpUBC23* orthologs *MaUBC3/8* and *SlUBC32* were similarly expressed during fruit ripening. However, *CpUBC23*was not considered differentially expressed genes in the present study since its expression was upregulated by less than two-fold. *MaUBC1*, *MaUBC9*, *SlUBC41*, *SlUBC42*, and *CpUBC34*, which belong to the *UBC8* subgroup, also exhibited similar expression patterns during fruit ripening. Therefore, taken together, our results suggest that the papaya E2 family genes might be participated in the regulation of fruit development and ripening processes in papaya.

CONCLUSIONS

In the present study, we describe the genome-wide identification and analysis of UBC genes in papaya. A total of 34 putative CpUBC genes were identified, and phylogenetic analysis indicated that genes could be divided into 17 subfamilies. Analysis of exon-intron junctions and sequence motifs revealed high levels of conservation within and between phylogenetic groups. In addition, all the CpUBC genes were detected in roots, stems, leaves, male and female flowers, and mature fruit, although some genes showed higher or lower expression in one or several specific organs. Similar to the results in tomato and banana, we also found that the expression of 13 and two CpUBC genes were incresesd or decreased during one and two ripening stages, respectively. Therefore, the results of the present study suggest that CpUBC genes are involved in the regulation of fruit development and ripening processes. The results also provide novel insights into the function of plant UBC genes and will facilitate further investigation of the roles of the UBC genes in fruit ripening.

ACKNOWLEDGMENTS

This work was supported by the Natural Science Foundation of Hainan Province (20163111), the Natural Science Foundation of China (31572087), the Natural Science Foundation of Guangdong Province of China (2014A030307031), the Basic Scientific Research Project of Non-profit Central Research Institutions (No. 1630062015004), the Germplasm Resource Protection of Ministry of Agriculture (2016NWB053), the Germplasm Resource Protection of Ministry of Agriculture Wild Plant (1251416305010), the Science and Technology Project of Guangdong Province (2016A020208002) and the Nature Science Foundation of Anhui Province (grant no. 1408085MKL63).

REFERENCES

1. Devitt LC, Fanning K, Dietzgen RG, Holton TA. Isolation and functional characterization of a lycopene β-cyclase gene that controls fruit colour of papaya (Carica papaya L.). Journal of Experimental Botany. 2010;61(1):33–9. pmid:19887502
2. Ming R, Hou S, Feng Y, Yu Q, Dionnelaporte A, Saw JH, et al. The draft genome of the transgenic tropical fruit tree papaya (Carica papaya Linnaeus). Nature. 2008;452(7190):991–6. pmid:18432245
3. Evans EA, Ballen FH. An Overview of Global Papaya Production, Trade, and Consumption. Food & Resource Economics. 2015.
4. Giovannoni JJ. Genetic regulation of fruit development and ripening. The plant cell. 2004;16(suppl 1):S170–S80.
5. White PJ. Recent advances in fruit development and ripening: an overview. Journal of Experimental Botany. 2002;53(377):1995–2000.
6. Barry CS, Giovannoni JJ. Ripening in the tomato Green-ripe mutant is inhibited by ectopic expression of a protein that disrupts ethylene signaling. Proceedings of the National Academy of Sciences. 2006;103(20):7923–8.
7. Hackett RM, Ho C-W, Lin Z, Foote HC, Fray RG, Grierson D. Antisense inhibition of the Nr gene restores normal ripening to the tomato Never-ripe mutant, consistent with the ethylene receptor-inhibition model. Plant Physiology. 2000;124(3):1079–86. pmid:11080285
8. Kevany BM, Tieman DM, Taylor MG, Cin VD, Klee HJ. Ethylene receptor degradation controls the timing of ripening in tomato fruit. The Plant Journal. 2007;51(3):458–67. pmid:17655616
9. Oeller PW, Min-Wong L. Reversible inhibition of tomato fruit senescence by antisense RNA. Science. 1991;254(5030):437. pmid:1925603
10. Lin Z, Zhong S, Grierson D. Recent advances in ethylene research. Journal of experimental botany. 2009;60(12):3311–36. pmid:19567479
11. Consortium TG. The tomato genome sequence provides insights into fleshy fruit evolution. Nature. 2012;485(7400):635–41. pmid:22660326
12. Moore S, Vrebalov J, Payton P, Giovannoni J. Use of genomics tools to isolate key ripening genes and analyse fruit maturation in tomato. Journal of Experimental Botany. 2002;53(377):2023–30. pmid:12324526
13. Vrebalov J, Pan IL, Arroyo AJM, McQuinn R, Chung M, Poole M, et al. Fleshy fruit expansion and ripening are regulated by the tomato SHATTERPROOF gene TAGL1. The Plant Cell. 2009;21(10):3041–62. pmid:19880793
14. Wilkinson JQ, Lanahan MB, Yen H-C, Giovannoni JJ, Klee HJ. An ethylene-inducible component of signal transduction encoded by Never-ripe. Science. 1995;270(5243):1807. pmid:8525371
15. Manning K, Tör M, Poole M, Hong Y, Thompson AJ, King GJ, et al. A naturally occurring epigenetic mutation in a gene encoding an SBP-box transcription factor inhibits tomato fruit ripening. Nature genetics. 2006;38(8):948–52. pmid:16832354
16. Zhu M, Chen G, Zhou S, Tu Y, Wang Y, Dong T, et al. A new tomato NAC (NAM/ATAF1/2/CUC2) transcription factor, SlNAC4, functions as a positive regulator of fruit ripening and carotenoid accumulation. Plant and Cell Physiology. 2014;55(1):119–35. pmid:24265273
17. Karlova R, Rosin FM, Busscher-Lange J, Parapunova V, Do PT, Fernie AR, et al. Transcriptome and metabolite profiling show that APETALA2a is a major regulator of tomato fruit ripening. The Plant Cell. 2011;23(3):923–41. pmid:21398570
18. Zhang M, Yuan B, Leng P. The role of ABA in triggering ethylene biosynthesis and ripening of tomato fruit. Journal of Experimental Botany. 2009;60(6):1579–88.

19. Itkin M, Seybold H, Breitel D, Rogachev I, Meir S, Aharoni A. TOMATO AGAMOUS-LIKE 1 is a component of the fruit ripening regulatory network. The Plant Journal. 2009;60(6):1081–95. pmid:19891701
20. Bemer M, Karlova R, Ballester AR, Tikunov YM, Bovy AG, Wolters-Arts M, et al. The tomato FRUITFULL homologs TDR4/FUL1 and MBP7/FUL2 regulate ethylene-independent aspects of fruit ripening. The Plant Cell. 2012;24(11):4437–51. pmid:23136376
21. Callis J, Vierstra RD. Protein degradation in signaling. Current opinion in plant biology. 2000;3(5):381–6. pmid:11019805
22. Hershko A. The ubiquitin system. Ubiquitin and the Biology of the Cell: Springer; 1998. p. 1–17.
23. Mukhopadhyay D, Riezman H. Proteasome-independent functions of ubiquitin in endocytosis and signaling. Science. 2007;315(5809):201–5. pmid:17218518
24. van Wijk SJ, Timmers HM. The family of ubiquitin-conjugating enzymes (E2s): deciding between life and death of proteins. The FASEB Journal. 2010;24(4):981–93. pmid:19940261
25. Michelle C, Vourc'h P, Mignon L, Andres CR. What was the set of ubiquitin and ubiquitin-like conjugating enzymes in the eukaryote common ancestor? Journal of molecular evolution. 2009;68(6):616–28. pmid:19452197
26. Jones D, Crowe E, Stevens TA, Candido EPM. Functional and phylogenetic analysis of the ubiquitylation system in Caenorhabditis elegans: ubiquitin-conjugating enzymes, ubiquitin-activating enzymes, and ubiquitin-like proteins. Genome biology. 2001;3(1):1.
27. Kraft E, Stone SL, Ma L, Su N, Gao Y, Lau O-S, et al. Genome analysis and functional characterization of the E2 and RING-type E3 ligase ubiquitination enzymes of Arabidopsis. Plant physiology. 2005;139(4):1597–611. pmid:16339806
28. Bae H, Kim WT. Classification and interaction modes of 40 rice E2 ubiquitin-conjugating enzymes with 17 rice ARM-U-box E3 ubiquitin ligases. Biochemical and biophysical research communications. 2014;444(4):575–80. pmid:24486490
29. Dong C, Hu H, Jue D, Zhao Q, Chen H, Xie J, et al. The banana E2 gene family: Genomic identification, characterization, expression profiling analysis. Plant Science. 2016;245:11–24. pmid:26940488
30. Jue D, Sang X, Lu S, Dong C, Zhao Q, Chen H, et al. Genome-wide identification, phylogenetic and expression analyses of the ubiquitin-conjugating enzyme gene family in maize. PloS one. 2015;10(11):e0143488. pmid:26606743
31. Wang Y, Wang W, Cai J, Zhang Y, Qin G, Tian S. Tomato nuclear proteome reveals the involvement of specific E2 ubiquitin-conjugating enzymes in fruit ripening. Genome biology. 2014;15(12):1.
32. Sadanandom A, Bailey M, Ewan R, Lee J, Nelis S. The ubiquitin–proteasome system: central modifier of plant signalling. New Phytologist. 2012;196(1):13–28. pmid:22897362
33. Welchman RL, Gordon C, Mayer RJ. Ubiquitin and ubiquitin-like proteins as multifunctional signals. Nature reviews Molecular cell biology. 2005;6(8):599–609. pmid:16064136
34. Li W, Schmidt W. A lysine-63-linked ubiquitin chain-forming conjugase, UBC13, promotes the developmental responses to iron deficiency in Arabidopsis roots. The Plant Journal. 2010;62(2):330–43. pmid:20113438
35. Wen R, Newton L, Li G, Wang H, Xiao W. Arabidopsis thaliana UBC13: implication of error-free DNA damage tolerance and Lys63-linked polyubiquitylation in plants. Plant molecular biology. 2006;61(1–2):241–53. pmid:16786304
36. Cui F, Liu L, Zhao Q, Zhang Z, Li Q, Lin B, et al. Arabidopsis ubiquitin conjugase UBC32 is an ERAD component that functions in brassinosteroid-mediated salt stress tolerance. The Plant Cell. 2012;24(1):233–44. pmid:22214659
37. Zolman BK, Monroe-Augustus M, Silva ID, Bartel B. Identification and functional characterization of Arabidopsis PEROXIN4 and the interacting protein PEROXIN22. The Plant Cell. 2005;17(12):3422–35. pmid:16272432

38. Zhiguo E, Zhang Y, Li T, Wang L, Zhao H. Characterization of the Ubiquitin-Conjugating Enzyme Gene Family in Rice and Evaluation of Expression Profiles under Abiotic Stresses and Hormone Treatments. PloS one. 2015;10(4):e0122621. pmid:25902049
39. Proost S, Van Bel M, Sterck L, Billiau K, Van Parys T, Van de Peer Y, et al. PLAZA: a comparative genomics resource to study gene and genome evolution in plants. The Plant Cell. 2009;21(12):3718–31. pmid:20040540
40. Letunic I, Copley RR, Schmidt S, Ciccarelli FD, Doerks T, Schultz J, et al. SMART 4.0: towards genomic data integration. Nucleic acids research. 2004;32(suppl 1):D142–D4.
41. Gasteiger E, Gattiker A, Hoogland C, Ivanyi I, Appel RD, Bairoch A. ExPASy: the proteomics server for in-depth protein knowledge and analysis. Nucleic acids research. 2003;31(13):3784–8. pmid:12824418
42. Guo A, Zhu Q, Chen X, Luo J. [GSDS: a gene structure display server]. Yi chuan = Hereditas/Zhongguo yi chuan xue hui bian ji. 2007;29(8):1023–6.
43. Edgar RC. MUSCLE: multiple sequence alignment with high accuracy and high throughput. Nucleic acids research. 2004;32(5):1792–7. pmid:15034147
44. Tamura K, Stecher G, Peterson D, Filipski A, Kumar S. MEGA6: molecular evolutionary genetics analysis version 6.0. Molecular biology and evolution. 2013;30(12):2725–9. pmid:24132122
45. Papaleo E, Casiraghi N, Arrigoni A, Vanoni M, Coccetti P, De Gioia L. Loop 7 of E2 enzymes: an ancestral conserved functional motif involved in the E2-mediated steps of the ubiquitination cascade. PloS one. 2012;7(7):e40786. pmid:22815819
46. Schumacher F-R, Wilson G, Day CL. The N-terminal extension of UBE2E ubiquitin-conjugating enzymes limits chain assembly. Journal of molecular biology. 2013;425(22):4099–111. pmid:23871895
47. Winn PJ, Religa TL, Battey JN, Banerjee A, Wade RC. Determinants of functionality in the ubiquitin conjugating enzyme family. Structure. 2004;12(9):1563–74. pmid:15341722
48. Fabi JP, Cordenunsi BR, de Mattos Barreto GP, Mercadante AZ, Lajolo FM, Oliveira do Nascimento JR. Papaya fruit ripening: response to ethylene and 1-methylcyclopropene (1-MCP). Journal of agricultural and food chemistry. 2007;55(15):6118–23. pmid:17602654
49. Yi Sheng JHH, Doherty Ryan, Srikumar Tharan, Shloush Jonathan, Avvakumov George V., Walker John R., Xue Sheng, Neculai Dante, Wan Janet W., Kim Sung K., Arrowsmith Cheryl H., Raught Brian, Dhe-Paganon Sirano. A Human Ubiquitin Conjugating Enzyme (E2)-HECT E3 Ligase Structure-function Screen. Molecular & Cellular Proteomics Mcp. 2012;11(8):329–41.
50. Lin Y-X, Jiang H-Y, Chu Z-X, Tang X-L, Zhu S-W, Cheng B-J. Genome-wide identification, classification and analysis of heat shock transcription factor family in maize. BMC genomics. 2011;12(1):1.
51. Peng X, Zhao Y, Cao J, Zhang W, Jiang H, Li X, et al. CCCH-type zinc finger family in maize: genome-wide identification, classification and expression profiling under abscisic acid and drought treatments. PloS one. 2012;7(7):e40120. pmid:22792223
52. Wang Y, Wang Q, Zhao Y, Han G, Zhu S. Systematic analysis of maize class III peroxidase gene family reveals a conserved subfamily involved in abiotic stress response. Gene. 2015;566(1):95–108. pmid:25895479
53. Zhao Y, Zhou Y, Jiang H, Li X, Gan D, Peng X, et al. Systematic analysis of sequences and expression patterns of drought-responsive members of the HD-Zip gene family in maize. PloS one. 2011;6(12):e28488. pmid:22164299
54. Bachmair A, Novatchkova M, Potuschak T, Eisenhaber F. Ubiquitylation in plants: a post-genomic look at a post-translational modification. Trends in plant science. 2001;6(10):463–70. pmid:11590065
55. Xu L, Ménard R, Berr A, Fuchs J, Cognat V, Meyer D, et al. The E2 ubiquitin-conjugating enzymes, AtUBC1 and AtUBC2, play redundant roles and are

involved in activation of FLC expression and repression of flowering in Arabidopsis thaliana. The Plant Journal. 2009;57(2):279–88. pmid:18798874

56. Jeon EH, Pak JH, Kim MJ, Kim HJ, Shin SH, Lee JH, et al. Ectopic expression of ubiquitin-conjugating enzyme gene from wild rice, OgUBC1, confers resistance against UV-B radiation and Botrytis infection in Arabidopsis thaliana. Biochemical and biophysical research communications. 2012;427(2):309–14. pmid:23000158

CHAPTER 11

Timing of Crop Removal Has Limited Effect on Merlot Grape and Wine Composition

Petra D. King[1], Richard E. Smart[2], Daniel J. McClellan[1]

[1]School of Viticulture and Wine Science, Eastern Institute of Technology, Hawke's Bay, New Zealand

[2]Smart Viticulture, 31 North Corner, Newlyn near Penzance TR185JG, Cornwall, UK

ABSTRACT

Cluster removal during the growing season is a widely utilized vineyard management practice aiming to balance crop load to the capacity of the vine to ripen the fruit. Research was undertaken over two growing seasons (2008-2009) in Hawke's Bay, a cool climate region of New Zealand, to establish the influence of different times of crop removal on Merlot vine growth and fruit and wine composition. The test vineyard was high-yielding, to 23 t/ha, and vigorous. A commercial standard of apical cluster thinning to remove 20 clusters per vine from vines with ca 44 clusters in 2008 and ca 47 in 2009, was carried out on uniform vigour, 7yo grafted Merlot vines at nine times between prebloom and six weeks post veraison. The 2009 season was naturally higher yielding. Timing of crop removal had no significant effect on vine vegetative growth in terms of enhanced shoot growth as measured by cane weights at pruning, or canopy leaf density. Time of thinning also had no effect on overall grape yield, cluster weight, and berry weight. There were limited effects on fruit ripeness in one season (2009) at the veraison time of thinning only, with increased Brix and lower TA levels. Berry anthocyanin concentrations were enhanced by cluster thinning in 2008, and more so when undertaken at or soon after veraison. There was however no influence of removal timing on anthocyanin levels and total phenolics in the wines. Grape ripeness, must and wine composition tended to respond more from crop removal at veraison than the other times evaluated. Data suggest that vine response was modified by excessive leafiness and shading.

Keywords: Crop Removal, Timing, Merlot, Grape Composition, Wine

1. INTRODUCTION

Problems with attaining optimum ripening and minimising variability in maturity within vineyards are due to a range of soil, climatic and cultural factors. An imbalance between vegetative and fruiting phases of vine growth is a major contributor to this variation in ripening and grape composition. Management practices commonly used to compensate for this imbalance involve manipulation of canopy architecture through trellising and training systems, and crop removal to balance the ability of the vine leaf area to ripen fruit. The variety Merlot is different from most other wine grape varieties, with large clusters and large berries, hence a high yield potential. In New Zealand's Hawke's Bay vineyards, crop removal from Merlot vines is undertaken each season as standard management between fruit set and harvest, for perceived benefits of improving grape composition and wine quality. We evaluate the impact of timing of this operation here.

There is a considerable body of literature reporting crop removal responses on grape and wine composition for a wide range of cultivars and growing conditions. A recent example shows advanced maturity, decreased acidity and increased anthocyanins and phenolics in Cabernet Sauvignon in the hot Riverland and Sunraysia regions of Australia with crop thinning [1] . A study with Shiraz in the warm Barossa Valley of Australia found decreasing yield per vine significantly increased wine colour and phenolics, and improved levels of metabolites which contribute to wine aroma and flavour [2] . Reducing Cabernet Sauvignon yield in Napa Valley, Californiaby varying winter bud numbers or cluster showed yield influenced astringency and vegetative flavours, but had little effect on wine aroma [3] . Cluster thinning was noted to increase Brix and enhanced wine colour, aroma and flavour [4] in Pinot noir in cooler Oregon and British Colombia vineyards, and reduced TA, increased pH, Brix, terpenes and increased wine colour, and herbaceous/grassy aroma in Chardonnay Musque cv in Ontario [5] .

Some of this research has involved cluster removal post-bloom (around the modified Eichhorn-Lorenz (E-L) 29 - 31 growth stage as described in [6] - [10] , or at veraison [3] [11] -[13] . Other workers claimed that the best time to cluster thin to affect maturity is prior to veraison [14] . In addition, it was noted that crop levels can be predicted with more certainty later in the season, and that less ripe, later set clusters are more visually evident at veraison [15] .

Research has also examined the influence of the time of thinning relative to the benefits. Early pre-bloom to bunch set thinning enhances berry fresh and dry weights in table grapes [16] , and is commonly used to enhance berry size. Several studies have investigated the influence on vine growth, grape ripening and grape and wine composition of cluster thinning at a range of times from pre-bloom to veraison [5] [15] [17] [18] .

In addition to these mostly positive effects from crop load reduction, there are several studies in which cluster thinning has no effect. Cluster removal in Cabernet Sauvignon vines in the Napa Valley has produced little benefit to either grape ripeness or must composition and wine color [3] [7] [12] , and in British Colombia [18] .

The reported and apparently conflicting results make identification of any optimum time for crop reduction difficult to determine. Some positive effects on grape and wine composition from crop removal at veraison were reported with Merlot in NZ [19] . To investigate the optimum time for removal under Hawke's Bay conditions, the present study was undertaken over two seasons. The objectives were to evaluate a range of crop removal times between prebloom and post-veraison on Merlot grape and wine composition in a high yielding, and vigorous commercial vineyard.

2. MATERIALS AND METHODS

2.1. Study Area and Time

The studies were undertaken over the 2007-2008 and 2008-2009 growing seasons in the premium red wine production area of Hawke's Bay in the North Island of New Zealand. Subsequently growing seasons will be identified by the year of harvest i.e. 2008 or 2009. Study plots were selected on a commercial vineyard within the Bridge Pa Triangle growing area, in a block of Merlot vines, clone 181 on 3309 rootstock, planted in 1999 at 1.6 m vine × 2.4 m row (2604 vines/ha) spacings on a Takapau sandy loam on a gravels soil type [20] . Growth and overall canopy vigour of the vines was even within the block. Vine trunk circumference was measured 30 cm above soil level in winter 2007, and measurement vines selected on the basis of uniformity of trunk circumference (12.6 cm ± 1.2 cm) within a plot. There were nine timing of thinning treatments and an unthinned control.

2.2. Experimental Design

Each treatment was replicated four times in a randomised block design. Within each plot, two vines of the five were permanently tagged and all measurements were confined to these vines. All vines were pruned to two canes with a target of around 25 buds which did not differ by more than 5%, and is the commercial standard. Vines were trained to a vertical shoot positioned (VSP) system. At this site and for this variety, bud break occurs around the beginning of October, 80% capfall mid-December, veraison mid-February and harvest at 23 - 24 Brix at the start of April. Shoot thinning was undertaken at the head of the vine in late October in both seasons to reduce excessive shading. Commercial leaf removal was undertaken on all vines at 6 January 2008 and 29 December 2008 prior to pre-bunch closure, (stage E-L 32) in the two seasons. A Collard air pulsed vine trimmer machine was used to remove some of the leaves in the fruit zone so that fruit clusters were more visible. All other canopy manipulations and viticulture inputs were undertaken in accordance with the commercial specifications and performed uniformly across all treatments. Cluster thinning was undertaken at each of the treatment phenological stages with shoots being thinned to a single cluster per shoot by removal of apical clusters only. There was no attempt to achieve a fixed target of clusters per vine; rather the standard commercial method used on the vineyard was followed.

2.3. Measurement and Data Analysis

Canopy density was quantified using Point Quadrant Analysis (PQA) in mid-February 2008. PQA was undertaken in each plot across 2 measurement vines and at 5 different heights above the fruiting wire (200, 300, 400, 500 and 650 mm). Methodology was the same as described [19] .

Canopy surface area was calculated as per the methodology detailed in [21] based on 2.4 m vine spacing, a 1.3 m high canopy at trimming, and 0.4 m canopy width. Canopy surface areas were similar for each treatment due to the influence of regular canopy trimming, and were 3.0 m^2/m row spacing, 4.8 m^2 per vine, and 12,500 m^2/ha.

Temperatures were continuously measured at one point within the block just below the fruit zone using an "e" Temp Thermocron temperature logger (Rollex Group, Vic. Australia) recording data every 30mins. The loggers were shielded within a well-ventilated screen sited just below the cordon wire.

Harvest was undertaken on 4 April 2008 and 30 March 2009. Grapes were harvested separately for each measurement vine recording yield and cluster number per vine. One measurement vine within each plot was selected at random and the basal cluster was harvested from 3 shoots per vine, randomly selected and spaced along the cordon at the proximal and distal points, and halfway between. These were used for berry weight and anthocyanin measurement using spectrophotometric methods [22] [23] .

All grapes were refrigerated at 2°C for a minimum of 8 hours before further processing for wine. Wines were made from each plot with the grapes from two vines being combined to make 3.5 - 4.5 kg ferments and with four replicates. Grapes were destemmed and crushed using an Enoveneta crusher destemmer (Enoveneta S.p.A., Padova, Italy). Samples of must were taken immediately after crushing and soluble solids (°Brix), must titratable acidity (TA), and pH were measured. Additions of SO_2 at a rate of 30 mg/kg were made using post crush weight. Musts were chaptalized by the addition of sucrose to raise °Brix to the maximum attained within the trial. Fermentations were in food grade 5L plastic containers. Must was warmed to 15°C and inoculated with Lavlin EC1118 yeast (Carter and Associates, Auckland, NZ) at a rate of 250 mg/kg of must. Di-ammonium phosphate (DAP) was added at the rate of 300 mg/L. Superfood (Pacific Rim Oenology, Blenheim, NZ) a yeast nutrient, was added at a rate of 250 mg/L of must.

Fermentations were undertaken in a room maintained at 26°C. Wine was fermented in contact with skins for 18 days, with the cap plunged daily. Wine was pressed in a Pillan 20L, 4 Bar, hydraulic press (Enotecnica PILLAN, Vicenza, Italy). Marc was pressed to give 70% yield of post-crush weight in litres, and 60 mg/l SO_2 was added. Malolactic fermentation was inhibited by the addition of $SO_2$and refrigeration at 0°C. Wines were allowed to settle for two days, racked then returned to 0°C storage for four weeks before a final racking. Wine was sparged with NO_2 gas for 5 minutes before bottling, as were the bottles before filling by siphon.

At bottling, samples were taken from each wine and analysed using a High Performance Liquid Chromatograph (HPLC). TheHPLC was a Shimadzu LC-10AT solvent delivery system equipped with a DGU-14A online de-gassing unit, an SPD M20A photo diode array detector, a CTO-10ASvp column oven, an SIL-10ADvp auto injector set to inject 20 μL of sample, with a 10ADvp sample cooler, connected to an SCL-10Avp system controller and controlled by Shimadzu's LC solutions software.

The column used for the phenolic separations was a Gemini 5μ C18 110A column, 250 × 4.6 mm from Phenomenex, fitted with a guard column of the same material. The gradient used for the separations consisted of 0.2% Tri fluoroacetic acid in water (A) and acetonitrile (B), with a flow rate of 1 mL/minute. Solvents were filtered and de-gassed using 0.45 μm regenerated cellulose filters by Sartorius. Samples were filtered directly into glass sampling vials fitted with Teflon coated liners, using Poly Vinyl Di Fluoride (PVDF) filters from Bonnet equipment with a pore size of 0.45 μm.

Data were collected from the samples at 280, 320, 360, and 520 nm. All retention times for relevant peaks were identified from the relevant standards and calibration curves.

All data were analysed for statistical significance by two-way ANOVA using Minitab 15 statistical software (Minitab Inc., State College, PA).

3. RESULTS

3.1. Canopy Leaf Area

Point Quadrat Analysis was carried out on 14 February 2008 in the first season of the study, measuring only thinning treatments carried out prior to and including veraison plus the control. Canopy leaf area was calculated from this data averaged over the five heights, using the method described in [24] . The results in Table 1 show similar canopy density as expressed as leaf layer number and canopy leaf area over the six treatments.

In the absence of crop removal there was a small (6%) but significant decrease in leaf area of control vines, indicating some compensation of vine growth due to competition with fruit. Data of Table 1 also show the very high Leaf Layer Number (LLN) values relative to the optimum values proposed in [21] which indicate very dense leaf canopies and excessive shade.

3.2. Harvest Data

The yield data for each of the treatments for the 2008 and 2009 harvests are shown in Table 2. The average yield of the unthinned control was high in both years, at 15.9 and 22.7 t/ha respectively. These values are higher than, and around double the normal yield of other wine grape varieties in this region. Yields in 2009, a season of higher cluster number and fruit set, were 43% higher than 2008. In both seasons, cluster removal resulted in significantly reduced yield. There was a trend for yields to be lower for those vines thinned at veraison and later. Mean cluster weights were about 35 g higher in 2009 compared with 2008, which reflected the berry weight being about 20% higher in 2009. Berry weights were not recorded for the three post-veraison treatments due to experimental error. There appeared to be no systematic effect of timing of thinning on neither cluster weight nor berry weight, with treatment differences being relatively small.

3.3. Seasonal Temperatures

The seasonal temperatures for the 2007/08 and 2008/09 seasons are shown in Table 3, as the highest maximum and lowest minimum temperature occurring each month, and Growing Degree Days (GDD) calculated as the sum for each month of the average temperature (maximum plus minimum divided by two, and minus 10°C). Comparison of the GDD from 1 October until harvest confirms the 2009 season as being slightly hotter than 2008 by 8%.

Table 1. Effect of timing of cluster removal on Merlot vine canopies, point quadrat analysis, 14 February 2008.

Time of cluster removal	Leaf Layer No/Vine	Canopy leaf area m^2/vine
Inflorescence swelling	2.8	11.4
80% capfall	2.7	11.0
Fruit set	2.7	11.1
Bunch closure	2.8	11.4
Veraison	2.8	11.3
Control	2.6	10.6
Optimum levels[1]	1.0 - 1.5	-

There was no significant differences between treatments (P = 0.05); Optimum levels [1]see Smart and Robinson (1991).

Table 2. Effect of time of cluster removal on Merlot yield data at harvest 2008 and 2009.

Thinning Treatment	No clusters/vine		Yield kg/vine		Mean Berry wt g		Mean Cluster wt g	
	2008	2009	2008	2009	2008	2009	2008	2009
Control	43.5 b[a]	47.3 b	6.1 b	8.70 a	1.37 a	1.65 bc	126.1 a	181.6 a
Inflorescence swelling	23.3 a	30.6 a	4.46 ab	6.44 ab	1.44 a	1.63 bc	191.6 b	211.2 bc
80% capfall	25.3 a	29.3 a	4.33 ab	7.17 bc	1.37 a	1.67 bc	178.3 b	245.8 d
Fruitset	23.1 a	29.9 a	4.17 ab	6.68 ab	1.34 a	1.60 ab	187.1 b	225.3 cd
Bunch closure	23.6 a	29.1 a	4.38 ab	6.02 ab	1.44 a	1.82 c	182.3 b	205.1 bc
Veraison	21.4 a	25.4 a	3.80 a	4.87 a	1.33 a	1.55 ab	172.2 b	192.6 ab
Veraison + 2 weeks	22.6 b	26.0 a	3.72 a	5.08 a	nm[b]	1.68 bc	169.2 b	194.4 ab
Veraison + 4 weeks	19.2 b	24.1 a	3.31 a	4.86 a	nm	1.48 a	179.8 b	200.8 ab
Veraison + 6 weeks	22.3 a	24.6 a	3.68 a	5.14 a	nm	1.55 ab	177.1 b	210.6 bc

[a]Values with the same letter within the same column do not differ significantly ($P > 0.05$); [b]not measured.

Table 3. Comparison of below vine canopy temperature data of Merlot vines for 2007/08 and 2008/09 season, period 1 October-harvest.

2007/08	October	November	December	January	February	March	Total Oct Harvest	Total Jan Harvest
Monthly max temp °C	24.5	28.4	29.1	31.6	30.9	28.8		
Monthly Min Temp °C	2.5	3.9	7.6	7.7	8.2	7.0		
GDD[b] °C	109	160	252	290	269	238	1318	817
Mean temp °C	13.9	15.4	17.9	19.1	18.6	17.1		
2008/09								
Monthly max temp °C	22.4	27.9	28.6	30.2	29.7	26.3		
Monthly min temp °C	5.2	5.9	10.9	9.9	11.9	8.3		
GDD[b] °C days	57	248	301	311	292	218	1426	831
Mean temp	13.4	16.2	19.1	19.6	19.5	16.4		

Harvest = 4/04/08 and 30/03/09; GDD[b] = Growing Degree Days (base 10°C).

3.4. Must Analysis

Must composition at harvest for both seasons are shown in Table 4. Associated with the lower crop loads of 2008, fruit was riper than in 2009 in terms of higher sugar (Brix) and lower titratable acidity (TA). In 2008 the time of cluster removal had no significant effect on ripeness in terms of Brix, TA and pH. In contrast, both Brix and TA were significantly higher and lower respectively relative to the unthinned control for most times of thinning in 2009, the season of higher yield. Berry anthocyanin levels were significantly increased by all but two timing treatments in 2008, although only the veraison thinning time was significantly higher than the control in 2009.

3.5. Wine Analysis

Levels of total anthocyanins are shown in Table 5. Timing of thinning had no significant effect on wine anthocyanin levels relative to the control in either year. Additionally, there were no significant treatment effects on the proportions of the monomeric anthocyanins in the wines, data not shown. In contrast, thinning effects on wine phenolics were more pronounced. All thinning treatments significantly increased phenolics relative to the control in 2008, apart from the two latest treatments at veraison +4 and +6 wks. The effect on wine phenolics in 2009 was not as evident with only the inflorescence swelling and veraison stages being significantly higher. The higher crop loads in 2009 and the earlier harvest may have influenced the levels of phenolics.

Table 4. Summary of must composition and berry anthocyanin levels in Merlot at havest 2008 and 2009.

Treatment	Brix		TA g/L		pH		Anthocyanins[b] mg/g	
	2008	2009	2008	2009	2008	2009	2008	2009
Control	23.4	22.2 ab[a]	5.10	5.61 c	3.47	3.23	1.4 a	1.29 a
Inflorescence swelling	24.3	23.5 abc	4.30	4.73 ab	3.55	3.32	1.86 b	1.52 ab
80% capfall	24.0	23.0 abc	4.27	4.84 ab	3.54	3.35	1.78 b	1.40 ab
Fruitset	24.0	23.1 abc	4.34	4.95 abc	3.51	3.29	1.73 b	1.50 ab
Bunch closure	23.6	23.3 abc	4.45	5.06 abc	3.53	3.39	1.66 ab	1.39 ab
Veraison	24.2	24.0 c	4.28	4.46 a	3.57	3.35	1.92 b	1.62 b
Veraison + 2 weeks	24.0	23.6 bc	4.39	4.83 ab	3.49	3.34	1.77 b	1.45 ab
Veraison + 4 weeks	23.9	22.0 a	4.55	5.28 bc	3.45	3.27	1.79 b	1.34 a
Veraison + 6 weeks	23.8	22.5 abc	4.52	1.72 ab	3.46	3.23	1.79 b	1.36 ab
	ns[c]		ns		ns	ns		

[a]Values with the same letter within the same column do not differ significantly ($P > 0.05$); [b]level of anthocyanins in whole berryhomogenates; [c]no significant difference.

3.6. Vine Balance

Comparisons of pruning data and vine balance indices between the thinning treatments are shown in Table 6. The thinning treatments had no significant effects on vine vegetative growth in either season based on measurements of mean cane weight and pruning weight. Relative to optimal values cited [21] , the study vines had excessive vegetative growth, in line with excessive leaf area indicated in Table 1. There was no significant effect of crop removal on the yield:pruning weight ratios in 2008, and the low levels for that season indicate vine imbalance with excessive vigour present. The higher crop load carried in 2009 resulted in larger ratios with significant differences between the control and thinning at preflowering, and from veraison and later only. In the 2008 season, the canopy surface area was appropriate to the yield, even for control vines. For 2009, the control vine value was higher than the proposed optimum suggesting that the exposed canopy surface area may be limiting for ripening. However, all thinning treatments especially later in the season were within the optimum range (1.0 to 1.5 kg/m^2).

Data on leaf area was only collected in the 2008 season, and the canopy leaf area: yield index is presented in Table 6. Results ranged from unthinned control vines at 1.7 m^2/kg and for the first five crop thinning treatments the average was 2.7 m^2/kg, both well in excess of the optimum range of 0.8 to 1.2 m^2/kg [21] . Leaf area appeared non limiting for ripening in 2008.

4. DISCUSSION

Timing of thinning had no significant effect on Merlot berry weight in either season of study. This is similar to the results shown with Syrah [15] comparing thinning at bunch closure and veraison where no effect was noted on berry weight. Different trends were noted for white wine grape varieties, for both Vidal blanc [17] and for Chardonnay Musque [5] where cluster removal between bloom and veraison resulted in greater berry weights or cluster weights.

Table 5. Comparison of total phenolics and anthocyanin levels in Merlot wines 2008 and 2009.

Stage of Cluster Thinning	Total Anthocyanins as Malvidin equivalents mg/L		Phenolics as catechin equivalents mg/L	
	2008	2009	2008	2009
Control	356.8 a*	482.3 ab	4233 a	4250 a
Inflorescence swelling	409.8 a	554.5 b	6188 b	5173 cd
80% capfall	423.9 a	515.5 ab	6084 b	4871 abc
Fruitset	403.3 a	532.3 ab	6095 b	4962 abc
Bunch closure	434.1 a	503.8 ab	6038 b	4840 abc
Veraison	441.1 a	532.5 ab	6358 b	5412 c
Veraison + 2 weeks	417.1 a	519.3 ab	5905 b	4798 abc
Veraison + 4 weeks	395.1 a	468.8 a	5520 ab	4633 ab
Veraison + 6 weeks	431.0 a	506.0 ab	5750 ab	4918 abc

a* = values in the same column with the same letter do not differ significantly (P > 0.05).

Table 6. Influence of timing of crop removal on Merlot vine growth and vine balance as measured by winter pruning data 2008 and 2009.

Treatment	Mean cane wt g		Pruning wg kg/m		Yield pruning wt kg/g		Yield: canopy surface area kg/m²		Leaf area: yield m²/kg
	2008	2009	2008	2009	2008	2009	2008	2009	2008
Control	64.1	57.7	0.81	0.60	4.80	8.98 b^a	2.03 b	2.90 c	1.74
Inflorescence swelling	62.5	58.5	0.74	0.63	3.43	5.92 a	1.49 ab	2.17 ab	2.56
80% capfall	62.1	53.0	0.71	0.62	4.06	7.04 ab	1.44 ab	2.39 bc	2.54
Fruitset	64.5	58.2	0.79	0.59	3.21	6.95 ab	1.39 ab	2.22 ab	2.73
Bunch closure	60.2	55.9	0.77	0.58	3.57	6.76 ab	1.46 ab	2.01 ab	2.60
Veraison	53.6	51.5	0.63	0.53	3.49	5.76 a	1.26 a	1.62 a	2.97
Veraison + 2 weeks	56.7	62.7	0.72	0.59	3.21	5.27 a	1.24 a	1.69 a	nm^c
Veraison + 4 weeks	59.1	45.3	0.71	0.50	3.17	5.71 a	1.10 a	1.62 a	nm
Veraison + 6 weeks	59.0	54.6	0.73	0.53	3.15	5.81 a	1.27 a	1.71 a	nm
Significance	ns^b		ns		ns				
Optimum values	20 - 40		0.3 - 0.6		5 - 10		1.0 - 1.5		0.8 - 1.2

^a Values with the same letter do not differ significantly (P > 0.05); ^b no significant differences between the treatment. nm^c = not measured.

The impact of time of thinning on fruit composition varied with the season. In the lower crop season of 2008, fruit was more mature and there were no significant effects of time of thinning on sugar, titratable acidity (TA), or pH. In 2009, the year of higher crop, sugar was increased by thinning at veraison, and TA decreased at five of the eight thinning stages. Berry anthocyanins were higher in 2008 than 2009, and all times of thinning increased values over the control. For the high crop season of 2009, only the thinning treatment at veraison increased the berry anthocyanin. Increased crop load and seasonal effects may have influenced the lower anthocyanin levels in 2009.

The fruit composition results for Merlot in this study are similar to those found for Shiraz in Western Australia [15] . These contrast with another study [8] where no beneficial effect from cluster thinning of Cabernet Sauvignon was found at either one month post-bloom or veraison on grape pH and anthocyanin levels, and significant increases in soluble solids in only two of the five years studied. Differences in fruit composition were influenced more by season than yield or crop load [18] . Seasonal effects on wine composition were also important in this study; wine anthocyanins were highest in 2009, in contrast to berry anthocyanins. There was no significant difference between timing treatments or the control for wine anthocyanin in either season. In 2008, the season of highest phenolics, the first six timing treatments showed an increase in phenolics. For the 2009 season, only the earliest time of thinning, and that at veraison were significantly higher for phenolics than control. In both seasons, thinning at veraison gave the highest or second highest values of anthocyanins and phenolics.

We now attempt to explain these treatment responses using standard viticultural indices, for vigour, balance and shading. The vines were high yielding, 15.9 and 22.7 t/ha for unthinned vines in 2008 and 2009 respectively, a 43% difference. They were also quite severely pruned, to only 25 buds per vine, and the mean pruning weight over both seasons was 1.1 kg per vine. It had been suggested [21] that 30 buds should be retained at winter pruning per kg pruning weight for achieving vine balance; the value in the test vineyard was 23 buds per kg pruning weight, a severe pruning level which will lead to under cropping and excess vigour. The control vines were vigorous, with 0.8 and 0.6 kg pruning weight per m row for 2008 and 2009. The vigour problem was worse for 2008, as published guidelines [21] are 0.3 to 0.6 kg/m. Similar differences were recorded for mean cane weight (64 g in 2008 and 58 g in 2009, with guideline optimum values 20 - 40 g).

There are several indices for vine balance; one is that of yield to pruning weight, called the Ravaz index. Values for unthinned controls are 4.8 and 9.0 for 2008 and 2009 respectively, and the average of all thinned treatments 3.4 and 6.2. Quoted optimum values are 5 - 10, [4] [8] [21] which suggest that vines in 2008 were unbalanced and undercropped, with too much vegetative growth, made worse by thinning, but that in 2009 the balance was better, even for thinned vines. Another index of balance is yield per unit canopy surface area, with optimal values of 1.0 - 1.5 kg/m^2. Unthinned vines had values of 2.0 and 2.9 kg/m^2 for 2008 and 2009 respectively, both indicating "overcropping" relative to the canopy surface area. The situation was improved by thinning, with the average value for thinned vines respectively 1.3 and 1.9 kg/m^2 in 2008 and 2009 respectively.

The ratio of leaf area to fruit weight is another index related to ripening ability. Control unthinned vines had a value of 1.7 m^2 per kg in 2008, and 2.7 m^2/kg the average for the first five thinning treatments. All are substantially in excess of an ideal value of 1.2 m^2 per kg [21] and 0.8 - 1.2 kg/m^2 [25] . Finally, we consider canopy density and shading. This was assessed only in 2008 by point quadrat, the year of highest vigour and lowest crop. The optimum value for leaf layer number is 1.0 - 1.5 [21] ; for control vines the value was 2.6, and the mean of the first five timing treatments 2.8. Excessively dense and shaded canopies are suggested, which is verified by calculating the ratio of canopy leaf area to surface area for 2008. Optimum values are given as less than 1.5 [21] , which is easily exceeded by values of 2.4 for the control, and 2.3 for the average of the first five thinning treatments (data not presented).

In summary then, we can say that in both seasons the control vines were severely pruned, vigorous, excessively leafy, with high leaf to fruit ratio, and that the canopy was very shaded. Vines were better balanced in 2009, but still had a limited canopy surface area for the high yield. Therefore the results obtained are as expected, in that there was little grape and wine composition response to thinning in 2008 when vines were under cropped, and more in 2009 when the vines were in better balance.

Considering this situation one must question whether crop thinning is the most appropriate vineyard management for these unbalanced vines. A more desirable option would be to alter the training system to a divided canopy like the Scott Henry system, allowing more buds to be retained at pruning giving better vine balance, and the canopy surface area would be almost doubled, substantially reducing shading. However, due to the large bunches of Merlot, some thinning may still be required to avoid over cropping in seasons of higher fruitfulness.

Workers in Australia concluded that the Shiraz vines they studied were under cropped based on the low yield:pruning weight ratio [15] , and this may explain why there was a minimal response recorded to the thinning strategies employed. A similar conclusion was drawn from the results with

Chardonnay [5] and with Cabernet Sauvignon [18] . In contrast to these results, other researchers [17] noted that both the Vidal blanc and Chardonnay vines studied had yield:pruning weight indices exceeding 15 and concluded the vines were seriously over- cropped, and this accounted for the beneficial effects on fruit quality from thinning prior to veraison.

5. CONCLUSIONS

Over the two seasons studied, the timing of crop removal by cluster thinning had no significant effect on subsequent vine growth in terms of canopy density as measured by leaf growth, or enhanced shoot growth as measured by cane weights at pruning. Timing also had no effect on overall grape yield, cluster weight, and berry weight, even from thinning at pre-bloom. Grape ripeness at harvest in the 2009 season as measured by increased Brix and lower TA levels was significantly different from the control only at the veraison time of thinning. Higher crop loads in 2009 than 2008 may have influenced this result. Grape anthocyanin concentrations were enhanced by cluster thinning, and significantly more so when undertaken at or soon after veraison. There was no influence of time of thinning on anthocyanin levels and total phenolics in the wines made from these grapes.

Indices of vine balance suggest the vines were excessively leafy and could be considered undercropped, especially in 2008. The vine canopies were limited in exposed surface area, and excessively dense and shaded, which precludes a response to fruit thinning. The limited thinning response noted in the grapes and wine composition may be explained in terms of lowering the yield: canopy surface area ratios, but these values were anyway in the optimal range. Regardless, the time when cluster thinning was undertaken had little significant effect on grape ripeness and must and wine composition, although there was an indication of a greater response from crop removal at veraison.

ACKNOWLEDGEMENTS

Petra D. King, Richard E. Smart, Daniel J. McClellan The authors gratefully acknowledge financial support for this research from the Business Links Fund of the New Zealand Tertiary Education Commission. The authors thank Karen Ball and Melissa Annand for technical assistance, and the owners of Terra Rossa Vineyards for access to the vines and the provision of grapes.

REFERENCES

1. Petrie, P.R. and Clingeleffer, P.R. (2006) Mechanical and/or Hand Thinning of Cabernet Sauvignon. Australian Journal of Grape and Wine Research, 12, 21-29. http://dx.doi.org/10.1111/j.1755-0238.2006.tb00040.x
2. Wolf, T.K., Dry, P.R., Iland, P.G., Botting, D., Dick, J., Kennedy, U. and Ristic, R. (2003) Response of Shiraz Grapevines to Five Different Training Systems in the Borossa Valley, Australia. Australian Journal of Grape and Wine Research, 9, 82-95. http://dx.doi.org/10.1111/j.1755-0238.2003.tb00257.x
3. Chapman, D.M., Matthews, M.A. and Guinard, J.-X. (2004) Sensory Attributes of Cabernet Sauvignon Wines Made From Vines with Different Crop Yields. American Journal of Enology and Viticulture, 55, 325-334.
4. Reynolds, A.G., Yerle, S., Watson, B., Price, S.F. and Wardle, D.A. (1996) Fruit Environment and Crop Level Effects on Pinot Noir. III. Composition and Descriptive Aanalysis of Oregon and British Columbia Wines. American Journal of Enology and Viticulture, 47, 329-339.
5. Reynolds, A.G., Schlosser, J., Power, R., Roberts, R., Willwerth, J. and de Savigny, C. (2007) Magnitude and Interaction of Viticultural and Enological Effects. I. Impact of Canopy Management and Yeast Strain on Sensory and Chemical Composition of Chardonnay Musque. American Journal of Enology and Viticulture, 58, 12-24.
6. Coombe, B.G. (1995) Adoption of a System for Identifying Grapevine Growth Stages. Australian Journal of Grape and Wine Research, 1, 100-110. http://dx.doi.org/10.1111/j.1755-0238.1995.tb00086.x
7. Ough, C.S. and Ngaoka, R. (1984) Effect of Cluster Thinning and Vineyard Yields on Grape and Wine Composition and Wine Quality of Cabernet Sauvignon. American Journal of Enology and Viticulture, 35, 30-34.
8. Bravdo, B., Hepner, Y., Loinger, C., Cohen, S. and Tabacman, H. (1985) Effect of Crop Level and Crop Load on Growth, Yield, Must and Wine Composition, and Quality of Cabernet Sauvignon. American Journal of Enology and Viticulture, 36, 125-131.
9. Prajitna, A., Dami, I.E., Steiner, T.E., Ferree, D.C., Scheerens, J.C. and Schwartz, S.J. (2007) Influence of Cluster Thinning on Phenolic Composition, and Antioxidant Capacity in Chambourcin Wine. American Journal of Enology and Viticulture, 58, 346-350.
10. Guidioni, S.A., Ferrandino, A. and Novello, V. (2008) Effects of Seasonal and Agronomical Practices on Skin Anthocyanin Profile of Nebbiolo Grapes. American Journal of Enology and Viticulture, 59, 22-29.
11. Mattii, G.B. and Ferrini, F. (2005) The Effects of Crop Load on "Sangiovese" Grapevines. In: Williams, L.E., Ed., ISHS Acta Horticulturae 689: VII International Symposium on Grapevine Physiology and Biotechnology, ISHS, Davis, California, USA, 239-242.
12. Nuzzo, V. and Matthews, M.A. (2006) Response of Fruit Growth and Fruit Ripening to Crop Level in Dry-Farmed Cabernet Sauvignon on Four Rootstocks. American Journal of Enology and Viticulture, 57, 314-324.
13. Pena-Neira, A., Caceres, A. and Pastenes, C. (2007) Low Molecular Weight Phenolic and Anthocyanin Composition of Grape Skins from cv. Syrah (Vitis vinifera L.) in the Maipo Valley (Chile): Effect of Clusters Thinning and Vineyard Yield. Food Science and Technology Internationals, 13, 153-158. http://dx.doi.org /10. 1177/1082013207077920
14. Jackson, D.I. and Lombard, P.B. (1993) Environmental and Management Practices Affecting Grape Composition and Wine Quality—A Review. American Journal of Enology and Viticulture, 44, 409-430.
15. McDonald, C., Bell, S.-J., Lantzke, N. and Kennison, K. (2005) Bunch Thinning: When Is the Best Time to Thin, and Is It Worthwhile. Australian & New Zealand Grapegrower & Winemaker, Annual Technical Issue, 40-46.

16. Dokoozlian, N.K. and Hirschfelt, D.J. (1995) The Influence of Cluster Thinning at Various Stages of Fruit Development on Flame Seedless Table Grapes. American Journal of Enology and Viticulture, 46, 429-436.
17. Ferree, D.C., Cahoon, G.A., Scurlock, D.M. and Brown, M.V. (2003) Effect of Time of Cluster Thinning of Grapevines. Small Fruits Review, 2, 3-14. http://dx.doi.org/10.1300/J301v02n01_02
18. Keller, M., Mills, L.J., Wample, R.L. and Spayd, S.E. (2005) Cluster Thinning Effects on Three Deficit-Irrigated Vitis vinifera Cultivars. American Journal of Enology and Viticulture, 56, 91-103.
19. King, P.D., McClellan, D.J. and Smart, R.E. (2012) Effect of Severity of Leaf and Crop Removal on Grape and Wine Composition of Merlot Grapevines in Hawke's Bay, New Zealand. American Journal of Enology and Viticulture, 63, 500-507. http://dx.doi.org/10.5344/ajev.2012.12020
20. Griffiths, E. (2001) Soils of the Heretaunga Plains—A Guide to Their Management. Grifftech and Hawke's Bay Regional Council, Napier, 102 p.
21. Smart, R.E. and Robinson, M. (1991) Sunlight into Wine: A Handbook for Winegrape Canopy Management. Winetitles, Adelaide.
22. Iland, P.G., Cynkar, W., Francis, I.L., Williams, P.J. and Coombe, B.G. (1996) Optimisation of Methods for the Determination of Total and Red-Free Glycosyl Glucose in Black Grape Berries of Vitis vinifera. Australian Journal of Grape and Wine Research, 2, 171-178. http://dx.doi.org/10.1111/j.1755-0238.1996.tb00105.x
23. Iland, P., Ewart, A., Sitters, J., Markides, A. and Bruer, N. (2000) Techniques for Chemical Analysis and Quality Monitoring during Winemaking. Patrick Iland Wine Promotions, Campbelltown, 98-100.
24. Green, S. (2008) Measurement and Modelling the Transpiration of Fruit Trees and Grapevines for Irrigation Scheduling. In: Goodwin, I. and O'Connell, M.G., Eds., Acta Horticulturae 792: Proceedings of the Fifth International Symposium on Horticultural Crops, ISHS, Leuven, 321-327.
25. Kliewer, W.M. and Dokoozlian, N.K. (2005) Leaf Area/Crop Weight Ratios of Grapevines: Influence on Fruit Composition and Wine Quality. American Journal of Enology and Viticulture, 56, 170-181.

CHAPTER 12

Genome Wide Characterization of Short Tandem Repeat Markers in Sweet Orange (*Citrus sinensis*)

Manosh Kumar Biswas[1], Qiang Xu[1], Christoph Mayer[2], Xiuxin Deng[1*]

[1] *Key Laboratory of Horticultural Plant Biology of Ministry of Education (MOE), Huazhong Agricultural University, Wuhan, Hubei, P.R. China,*

[2] *Forschungsmuseum Alexander Koenig, Bonn, Germany*

ABSTRACT

Sweet orange (*Citrus sinensis*) is one of the major cultivated and most-consumed citrus species. With the goal of enhancing the genomic resources in citrus, we surveyed, developed and characterized microsatellite markers in the ≈347 Mb sequence assembly of the sweet orange genome. A total of 50,846 SSRs were identified with a frequency of 146.4 SSRs/Mbp. Dinucleotide repeats are the most frequent repeat class and the highest density of SSRs was found in chromosome 4. SSRs are non-randomly distributed in the genome and most of the SSRs (62.02%) are located in the intergenic regions. We found that AT-rich SSRs are more frequent than GC-rich SSRs. A total number of 21,248 SSR primers were successfully developed, which represents 89 SSR markers per Mb of the genome. A subset of 950 developed SSR primer pairs were synthesized and tested by wet lab experiments on a set of 16 citrus accessions. In total we identified 534 (56.21%) polymorphic SSR markers that will be useful in citrus improvement. The number of amplified alleles ranges from 2 to 12 with an average of 4 alleles per marker and an average PIC value of 0.75. The newly developed sweet orange primer sequences, their *in silico* PCR products, exact position in the genome assembly and putative function are made publicly available. We present the largest number of SSR markers ever developed for a citrus species. Almost two thirds of the markers are transferable to 16 citrus relatives and may be used for constructing a high density linkage map. In addition, they are valuable for marker-assisted selection studies, population structure analyses and comparative genomic studies of *C. sinensis* with other citrus related species. Altogether, these markers provide a significant contribution to the citrus research community

INTRODUCTION

Tandem repeats (TR) are abundant elements in plant genomes. Evidence suggests that TR originate mainly from replication slippage events. Typically, slippage in a TR occurs about once every 1,000 generations, where slippage rates vary with repeat type, unit size as well as among species [1], [2]. TR$_S$ are usually classified according to their unit size into microsatellites, minisatellites, and satellites. There is no consensus regarding the unit size boundary between micro- and minisatellites. Proposed unit size ranges of microsatellites are e.g, 1–6 bp, 2–6 bp, 2–8 bp [3] and 1–10 bp. Microsatellites are also often referred to as simple sequence repeats (SSR$_S$). In this study we consider repeat units of one to ten nucleotides as microsatellites. Their genomic abundance, co-dominant nature, easy assay, non-random distribution, correlation with many phenotypes as well as multi allelic feature have made microsatellites the marker of choice for diverse application in plant genetics. The availability of whole genomic sequences provides the opportunity to investigate the genome wide distribution, density, evolution and putative function of microsatellites. It is well known that the microsatellite frequency differs greatly among species [4] as well as among different genomic regions, *i.e.* introns, exons, CDS, intergeneric regions [5]. Previous studies demonstrated that the SSR distribution in genomic regions has practical implications with regard to their utility as molecular markers. Genic-SSR markers are more transferable to related species than genomic-SSRs. This feature helps to design anchor markers for comparative mapping studies. Since they are often more conserved, genic-SSRs may provide an insufficient degree of polymorphism to discriminate between closely related germplasm. Therefore genomic-SSRs may be valuable complements.

Sweet orange (*Citrus sinensis* [L.] Osbeck) is one of the most widely grown fruit crop in the world with a global production of 69.41 million tons and a total acreage of 4.06 million hectares in 2010. It contributes about 60% of the total world citrus production and consumption [6]. The remaining 40% are distributed among other citrus such as lemon, grape fruit and pummelo. Its value also stems from it being rich in vitamin C and other health beneficial elements. Most of the sweet oranges are diploids with a comparatively small genome size of about 367 Mb [7]having 9 chromosomes. Citrus taxonomy is fairly complex due to the unusual reproduction nature, nucellar embryony, a high frequency of bud mutations and wide cross-compatibility among the species [8], [9]. The high frequency of bud mutation is one of the causes for the narrow genetic basis of the cultivated citrus species. Inter species genetic diversity in sweet orange is relatively low compared to other citrus species as reveled in early studies using different types of molecular markers including RAPD, AFLP, ISSR (Inter simple sequence repeat), IRAP (Inter-retroelement amplified polymorphism) and SSR_s [10]. The development of high density reference linkage maps of citrus is essential for the understanding of genome organization, evolution, tagging important quantitative trait loci (QTL) and map-based cloning of agronomical important traits. Consequently, it is necessary to develop a marker system in citrus that is highly polymorphic and user friendly. Among different marker systems, microsatellite markers are extensively used for genetic mapping and MAS studies in plant breeding. Unfortunately, the number of developed microsatellite markers is still low in citrus. Recently, Biswas et al. [9] and Ollitrault et al. [11] demonstrated the utility of BAC-end derived citrus SSR markers in linkage mapping and phylogenetic studies in citrus; while Chen el al. [12]established the utility of 100 citrus EST-SSR markers for genomic mapping analyses. Despite this progress in the total number of informative, robust and publicly available markers for citrus, their number is still insufficient for many important applications such as the construction of a high density linkage map, closely related cultivar identification, positional cloning, MAS, trait tagging and comparative mapping. The large scale development of these markers was not possible until the whole genomic sequence of citrus was available. In this study we surveyed, developed, and characterized SSR markers from the recently sequenced nuclear genomic sequence of sweet orange *cv* 'Valencia'. Furthermore, we investigated the transferability of these markers to related species.

MATERIALS AND METHODS

Plant materials

Sixteen genotypes were used for the wet lab verification and transferability analysis of the genome wide SSR primers. Plant materials were collected from the National Center of Citrus Breeding, Huazhong Agricultural University, Wuhan, China. In addition, these genotypes represent the major groups of citrus and its close relatives (Table S1 in File S1). Total genomic DNA was extracted from 5 g mature fresh young leaves of each genotype using the CTAB method [13] with subsequent RNase A treatment.

Source of genomic sequences

The genome sequencing of a doubled-haploid callus line of sweet orange cv. 'Valencia' was performed with a whole-genome shotgun approach combined with the DNA-PET technology. The genome was assembled using the Short Oligo-nucleotide Analysis Package (SOAPdenovo). The high quality sequence reads were assembled into 4,811 scaffolds with N50 = 1.7 Mb. The total contig length (320.5 Mb) covers about 87.3% of the sweet orange genome. The frozen version of the sweet orange genome assembly, which consists of scaffolds and BAC end sequences, was used in this study for a genome wide SSR marker characterization.

SSR mining and Primer design

The software SSRLocatorI V1.1 was used for genome wide SSR mining and primer design. To identify SSRs in the citrus genome, we searched for perfect repeats with a unit size of 1 to 10 bp and a length of at least 16 nucleotides. All SSRs were grouped into Class I (≥20 bp total length) and Class II (16–19 bp total length). Primer pairs were design to meet the following restrictions: the amplicon had to be in the range 150–500 bp, the primer annealing temperature was restricted to 55–60°C, the GC content had to be 40–60% and the primer length had to be 19–21 bp.

In silico analysis of SSR polymorphism

In silico polymorphism analysis of SSR markers was perform using the virtual PCR strategy where pairs of primer sequences from Sweet Orange were mapped onto the Clementina genomic sequence scaffolds. (Only di- to hexanucleotide repeat primers where considered.) The specific *in silico*-generated amplicons from Clementina were compared with the expected amplicon size from Sweet Orange and their size differences were recorded. If an amplicon size differed by at least 2 bp, the SSRs was classified as polymorphic, while amplicons of identical size were considered as monomorphic. SSR loci with a 1-bp difference were considered ambiguous and were removed from the analysis.

Functional annotation of genome wide SSR markers

An in house developed perl script was used to isolate flanking sequences of the SSR markers for assigning a putative function to each SSR marker. For primers in protein coding regions, this approach made use of the Blast 2GO tool. The mapping and annotation of the sequences is based on sequence similarity according to gene ontology [14]. Therefore, sequences without BLAST hit have not been annotated. The default settings were used for the annotation parameters (E value filter of 1E-10 and annotation cutoff of 55).

Wet lab verification and utility of the SSR markers

A total of 950 SSR markers with an average distance of 0.5 Mb were selected from the nine chromosomes for the wet lab validation, where a set of 16 citrus genotypes were used as targets species. PCR reactions were conducted as follows: 10 µl of PCR-volume consisted of 25 ng of genomic DNA, 1.5 mmol l^{-1} $MgCl_2$, 0.2 mmol l^{-1} dNTPs, 1.0 U Taq DNA polymerase, 1× PCR buffer and 0.1 µmol l^{-1} of each primer pair. PCR amplification was conducted in a MJ-PTC-200 tm thermal controller (MJ Research, Waltham Mass) using the following program: 94°C for 5 min, 32 cycles at 94°C for 1 min, 55°C for 30 s, 72°C for 1 min, followed by a final step at 72°C for 4 min. After PCR, 8 µl of loading buffer (98% formamide, 2% dextran blue, 0.2 mM EDTA) was added to each sample. Samples were denatured at 90°C for 5 min and then immediately placed on ice. An aliquot (4 µl) of each sample was loaded onto 6% polyacrylamide gel (60 cm×30 cm×0.4 cm), which was run for 2 h and 30 min at 80 V. DNA bands were visualized with silver staining as described by Ruiz et al. [15]. The band size is reported for the most intense amplified band for each SSR or the average of the stutter if the intensity was the same. A 10 bp DNA ladder (Fermentas) was used as the reference point. Polymorphic information content (PIC) of each SSR was calculated using the following formula: $PIC = 1 - \sum p_i^2$, where p_i is the proportion of the *i*th allele.

Gnome wide localization and database of SSR marker

A graphical presentation of the SSR marker distribution in different chromosomes of Sweet Orange was made using the Map Chart 2.2 software. For an easy access and utilization of SSR markers, all markers were stored in a database, which will be made available to the public soon. Markers are named according to the following scheme: E.g. in M2H4Si3025, the M stands for the initial character of the developer name, the 2 depicts the chromosome and H stands for the host institute. 4 depicts the SSR unit size,

S_i the Citrus species (here *Citrus sinensis*) followed by the SSR pattern id, a number that is unique for each unit size.

RESULTS

Frequency and distribution of SSR in the sweet orange genome

In the first step, we analyzed the distribution and frequency of perfect microsatellites with a minimum length of 16-bp and a unit size of 1 to 10-bp. The results are summarized in Table 1. A total of 50,846 SSRs were identified which represents an overall density across the genome of 146.42 SSR/Mbp (i.e., one SSR found in every 6.8 kbp). Based on their length, SSRs are categorized into two classes, namely Class I (≥20 bp) and Class II (16–19 bp). Of the total number of SSR identified in the sweet orange genome, 28,211 (55.48%) were identified as Class I and 22635 (44.52%) as Class II. On average, the estimated frequency of Class I and Class II SSR was 81.24 and 65.18 per Mb, respectively. Altogether, the total lengths of mono- to deca-nucleotide repeats accounted for about 0.461% of the genome. Among the different unit sizes, dinucleotide repeats are the most common. They constitute 32.87% of all SSRs, followed by mono- (27.20%) and trinucleotide repeats (19.99%), while nona- and deca-nucleotide repeats were least abundant. Surprisingly, we found that hepta-nucleotide repeat numbers were two fold higher than hexanucleotides repeat numbers. Furthermore, we estimated the frequency distribution of SSRs in the nine chromosomes of the sweet orange genome and the result reveals that chromosome 4 has the highest frequency of SSRs (237.06 SSRs/Mb), while chromosome 8 contains the lowest (150.46 SSRs/Mb). Overall the frequency of SSRs in the different chromosomes was not significantly different (Table 2). SSR characteristics for different unit sizes are shown in Figure S1 and Table S2 in File S1. The mean number of trinucleotide (16.12) repeats is higher than others, except mononucleotides. The distribution of SSR_s in different genomic region of the sweet orange genome is presented in Figure 1 and as a result it appears that SSR are denser in 5′-UTR region than in other genic regions.

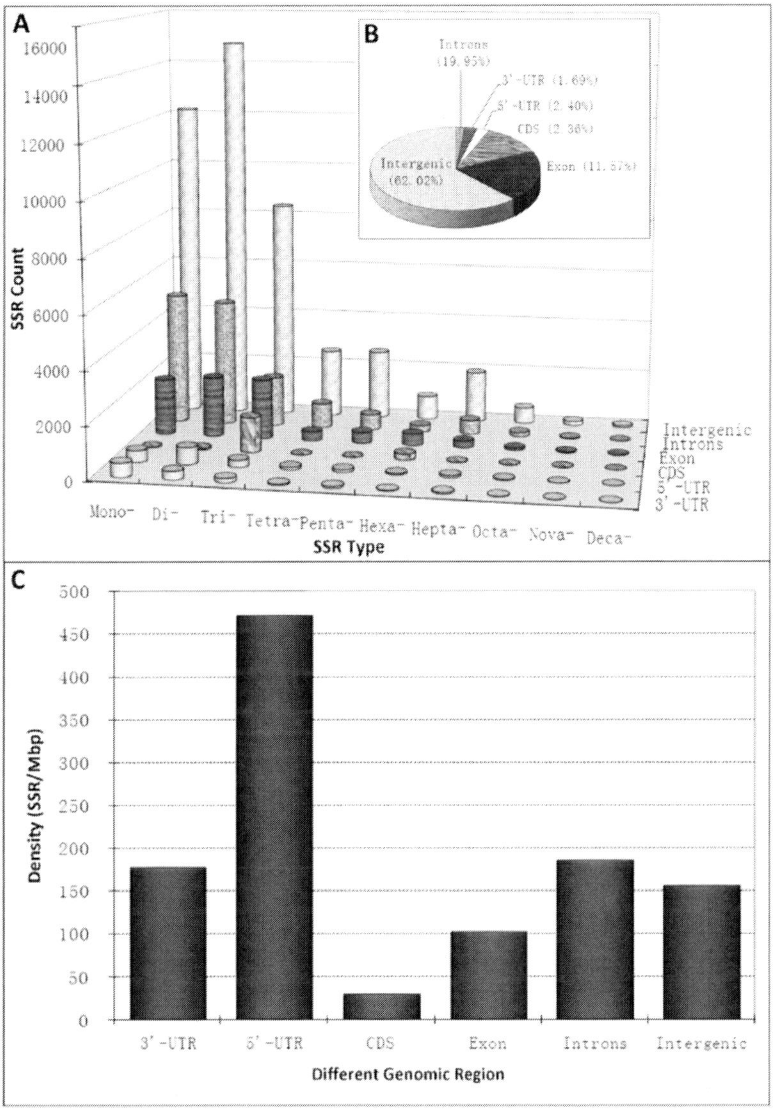

Figure 1. Absolute number count (A), relative number counts (B), and density (C) of SSRs in different genomic regions of the sweet orange genome.

Table 1. SSR characteristics in the genome assembly of *C. sinensis*.

SSR mining			Total	%
Total length of analyzed sequences (bp)			347267366	
Number of identified SSRs			50846	
Number of SSR Loci			46872	
SSR Frequency (Mbp)			146.42	
SSR Density (bp/Mbp)			4605.56	0.461
Distribution of SSRs				
Type	**Class I***	**Class II$**	**Total**	**%**
Mononucleotides	3298	10530	13828	27.20
Dinucleotides	8882	7831	16713	32.87
Trinucleotides	5888	4274	10162	19.99
Tetra-nucleotides	2826	0	2826	5.56
Penta- nucleotides	2960	0	2960	5.82
Hexa- nucleotides	1194	0	1194	2.35
Hepta- nucleotides	2183	0	2183	4.29
Octa- nucleotides	704	0	704	1.38
Nona- nucleotides	193	0	193	0.38
Deca- nucleotides	83	0	83	0.16
Total (%)	28211(55.48)	22635(44.52)		
Frequency (SSR/Mb)	81.24	65.18		

* Class I :SSR loci are greater than 19 nt long;
$ClassII: SSR loci smaller than 20 nt long.
doi:10.1371/journal.pone.0104182.t001

Table 2. Chromosome wide distribution of SSRs in the *C. sinensis* genome.

SSR repeat class	Chromosome										Total
	1	2	3	4	5	6	7	8	9	Un	
Mononucleotides	1266	1395	1746	1333	1843	1031	1345	957	835	2077	13826
Dinucleotides	1578	1665	1914	1666	2015	1071	1834	1032	920	2998	16713
Trinucleotides	958	1005	916	858	1234	664	1075	706	588	2158	10162
Tetranucleotides	273	290	266	251	337	169	303	214	184	539	2826
Pentanucleotides	229	298	319	260	354	207	290	212	185	606	2960
Hexanucleotides	104	140	118	98	143	86	108	96	50	251	1194
Heptanucleotides	198	236	217	186	249	134	215	137	99	512	2183
Octanucleotides	54	70	76	54	81	37	60	48	40	184	704
Nonanucleotides	19	19	15	16	17	9	17	13	10	58	193
Decanucleotides	6	5	9	8	11	3	8	2	6	25	83
Size (Mb)	28.80	30.84	28.71	19.95	36.15	21.18	32.21	22.71	18.45	108.27	347.27
SSR/Mb	162.67	166.78	194.89	237.06	173.85	161.05	163.17	150.46	158.10	86.89	146.42
Total SSR	4685	5143	5596	4730	6284	3411	5255	3417	2917	9408	50846

doi:10.1371/journal.pone.0104182.t002

A more detailed investigation of individual repeat types was performed and presented in Figure 2 and Table S3 in File S1. As expected A/T repeats are more abundant than G/C repeats. The most frequent dinucleotide repeat unit is AT, while GC is very rare. AT dinucleotide repeats account for 16.99% of all SSR_S, which reveals that AT repeats are the overall most common repeat type in sweet orange genome. AAT is the most frequent and CCG the least frequent trinucleotide repeat pattern. The predominant tetranucleotide repeat is AAAT, whereas GC-rich repeats like ACCC, AGGC, AGGG, CCCG are rare. AT-rich motifs are predominant in penta-, hexa- and heptanucleotide repeats, such as AAAAT, AAAAG, AAATT and AATAT, which are the most common pentanucleotide repeats. AAAAAG, AAAAAT and AAAAAC motifs are prevailed among hexanucleotides. Octa-, nona- and decanucleotides are underrepresented repeat types in the sweet orange genome, altogether accounting for only 1.95% of the total SSRs. AT-rich repeat patterns such as AAAAAAAG, AAAAAAAT and AAAAAATT are most frequent among the octanucleotide repeats, whereas AAAAAAAAG, AAAAAAAAT and AAAAAAAAC are most abundant among nonanucleotide repeats. The AACAATTATT and AAAAAAAAAG patterns are most abundant among decanucleotide repeats.

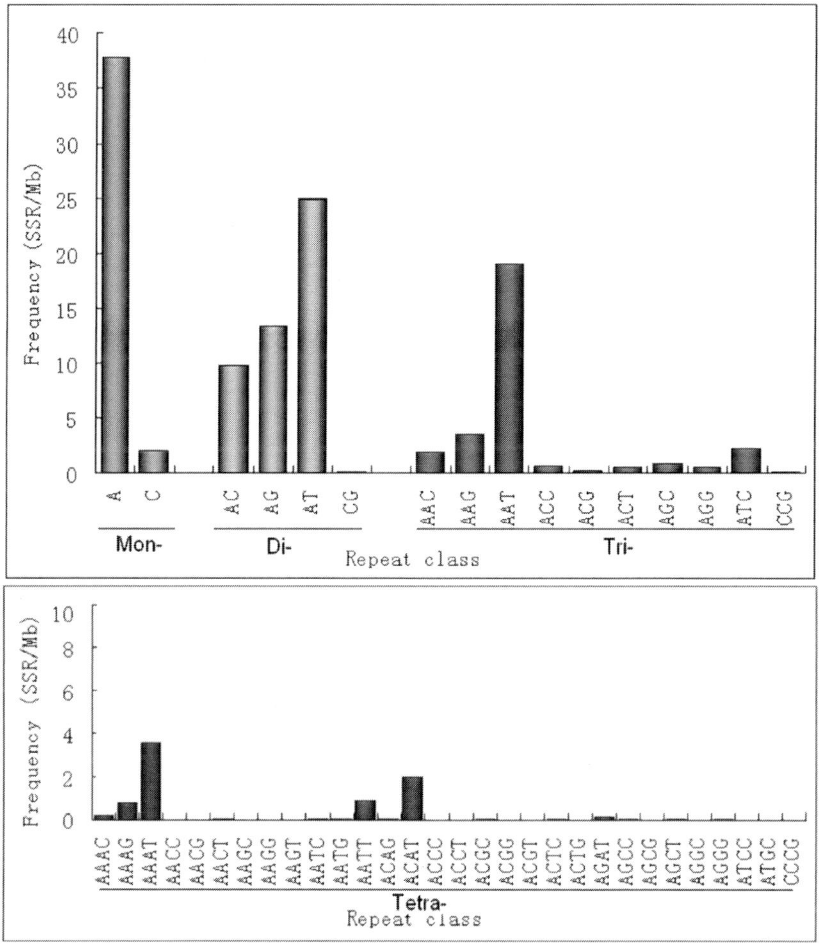

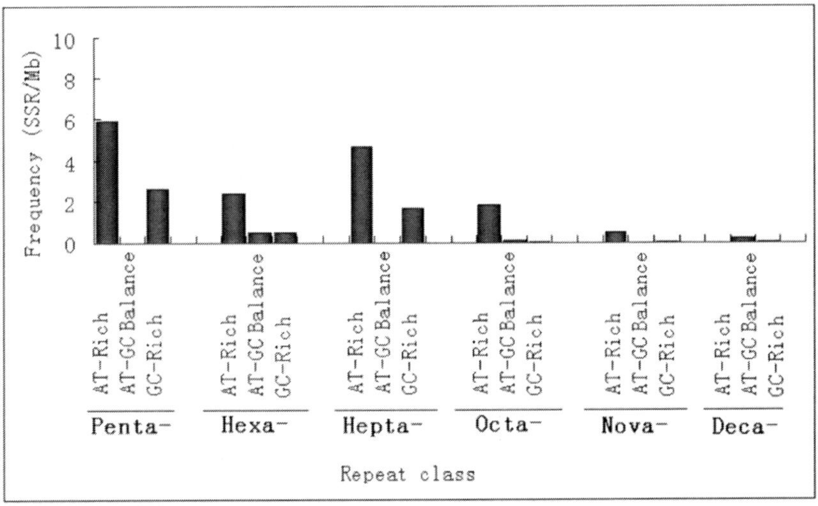

Figure 2. Frequencies of SSRs with certain patterns and base contents.

Genome wide SSR marker development and in silico polymorphism analysis

One of the primary goals of this study was to develop genome-wide SSR markers. Primers were designed for most of the di- to hexanucleotide repeats and results are presented in Table 3. A total of 21,248 SSR markers were successfully designed from the nine chromosomes. The distribution of SSR markers among the nine chromosomes was not significantly different. On average 2360 SSR markers were designed for each chromosome with 89 markers per Mb, covering the whole genome with gaps of less than 12 kb (Figure 3).

Table 3. Chromosome wide SSR marker development.

SSR repeat class	Chr1	Chr2	Chr3	Chr4	Chr5	Chr6	Chr7	Chr8	Chr9	Total
Dinucleotides	882	1403	1669	1546	1258	900	1296	848	765	10567
Trinucleotides	561	807	914	816	717	526	733	519	458	6051
Tetranucleotides	159	252	263	251	231	144	214	180	164	1858
Pentanucleotides	213	256	319	260	210	185	205	190	162	2000
Hexanucleotides	56	120	118	93	100	75	88	81	41	772
Total	1871	2838	3283	2966	2516	1830	2536	1818	1590	21248

doi:10.1371/journal.pone.0104182.t003

Figure 3. Chromosome maps depicting SSR loci of 5824 developed markers along the nine chromosomes of *C. sinensis*.

Virtual PCR (VPCR) was performed for *in silico* polymorphism analysis and results are presented in Figure 4 and Table S4 in File S1. SSR markers are classified either as polymorphic or monomorphic based on the *in silico* amplicon size comparison. A total of 6588 (31%) *C. sinensis* primers amplified specific bands *in silico* from the genome of *C. clementina*. The remaining primers failed to generate specific amplicons or generated no amplicons. Among the 6588 SSR markers with amplicons, 3941 (59%) were polymorphic and 2647 (40%) were monomorphic as indicated in Figure 4. The relationship between degree of polymorphism and repeat length for each SSR type was estimated. We observe that shorter repeats are more monomorphic than long repeats, while polymorphic SSRs include a considerably higher percentage of long repeats. Dinucleotide repeats are more polymorphic than other repeat types and trinucleotide repeats have the highest proportion of monomorphic repeats.

Functional annotation, wet lab validation of SSR markers and utility of newly developed GW-SSR resource for citrus research

Functional annotation of SSR loci was performed by a Blast2Go analysis. A significant GO annotation was found for 1870 loci, whereas 19378 loci had no significant homology to known sequences in the public databases (Figure 5). The majority of SSR loci for which an annotation was found are involved in protein metabolic processes (309), transport (272), and RNA metabolic process (227). When mapped against the molecular function category, 379 (20%), 292 (16%), 246 (13%) and 231 (12%) SSR loci were involved in hydrolase activity, nucleotide binding, protein binding and DNA binding, respectively. When mapped against the cellular component GO terms, 333 SSR loci (18%) were involved in nucleus and 144 (8%) were involved in chloroplast function.

A subset of 950 markers had been selected for wet lab experiments. Of these, 64.11% could be amplified in all 16 citrus accessions with prominent PCR products having the expected size. Among the tested primer pairs, 578 amplified in orange accessions, while 56.11%, 57.58% and 50.42% amplified in *C. grandis*, lemon (*C.limonia*) and trifoliata orange (*Poncirus trifoliata*), respectively. Highest transferability to the relative species was found in lemon followed by *C. grandis* and Kumquat (*Fortunella*) (Table S10 in File S1). A total of 2547 alleles were recorded from 609 SSR loci with an average of 4 alleles per loci. The majority of the primer amplified 4 alleles, followed by 2 and 3 respectively (Figure 6). The PIC value varied from 0.10 to 0.95 with an average of 0.73. The majority of the PIC values were found in the interval from 0.60 to 0.79 (Figure 6).

Database of SSR marker and genome wide localization

A total of 21,248 SSR primer pairs were designed. For an easy access and utilization of these markers, we developed the sweet orange SSR Marker database, which stores the exact positions of these SSR_S in the sweet orange genome assemble (Figure 3), the repeat pattern, expected PCR product length, primer temperature as well as the virtual PCR result are available in Table S11 in File S2.

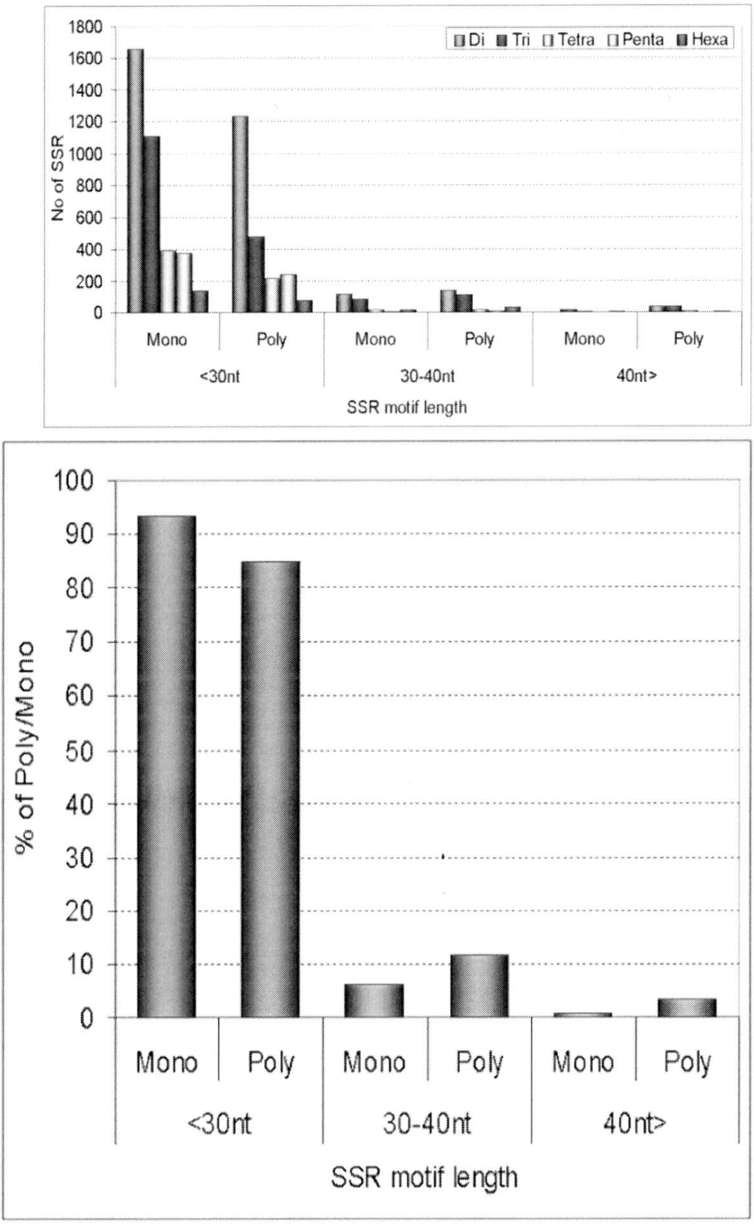

Figure 4. Frequency distribution of *C. sinensis* SSRs by repeat length, in monomorphic and polymorphic SSRs obtained from *in silico* PCR analyses.

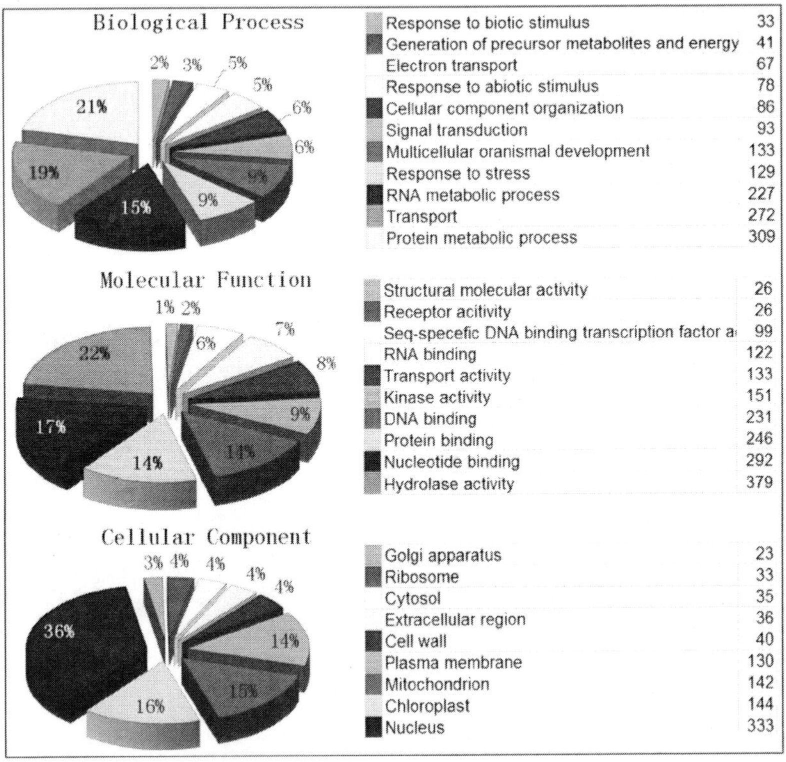

Figure 5. Gene Ontology annotation of genome wide developed sweet orange SSR marker flanking regions (A), the GO biological process, (B) molecular function and (C) cellular component.

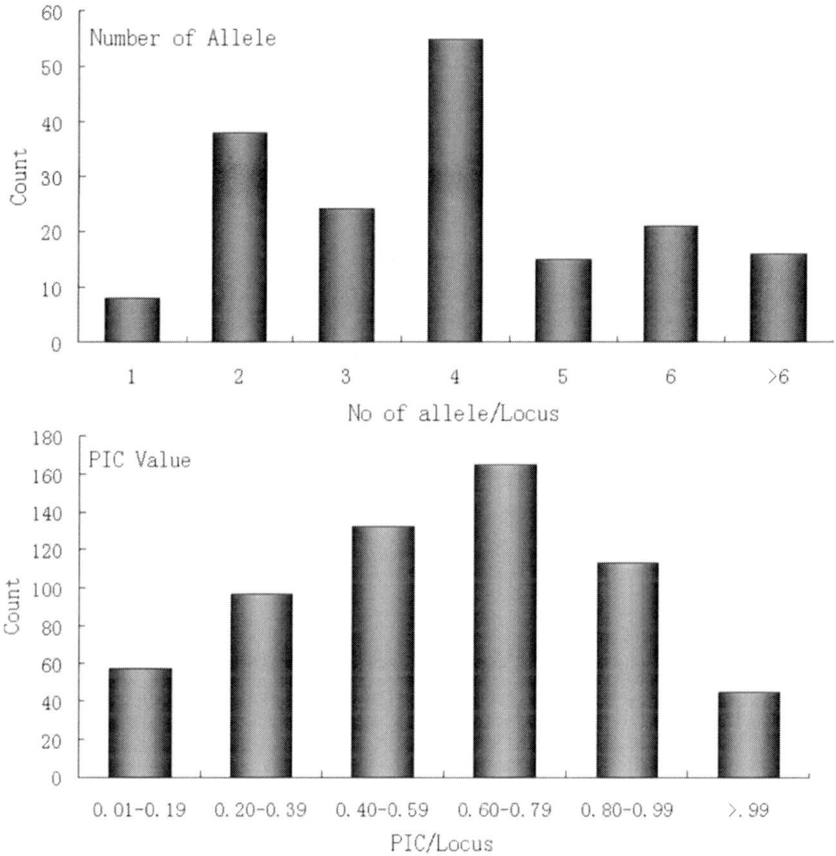

Figure 6. Number counts of allele frequencies and PIC values for the 609 SSRs analyzed in 16 Citrus germplasms.

DISCUSSION

Frequency and distribution of SSRs in the sweet orange genome

We analyzed the genome-wide SSR distribution, frequency and density in the unit size range of 1 to 10 bp in *C. sinensis*. SSR_s contributed 0.5% to the 347 Mbp genome assembly of *C. sinensis* and we found 1 SSR per 6.8 Kbp. This result is comparable with the SSR densities reported for other plant species [16], [17], [18]. In general, the frequency of SSR_s is considerably higher in dicot species compare to monocots [4]. Our comparison of the SSR frequency in the *C. sinensis* and the *S. bicolor* genome shows a 2 fold higher frequency in the sweet orange genome (see Table S5, S6 in File S1). The majority of the SSRs were identified in chromosome 5, although the density of SSRs is the highest in chromosome 4. Many reports have demonstrated that SSRs are non-randomly distributed in genomes of various species [5], [19]. We performed a chi-square test to test for differences in the SSR distribution among different genomic regions and also the distribution among nine pseudo chromosomes (Table S7 in File S1). The chi square test supports a non-random distribution of SSRs in the sweet orange genome. We observed that the number of SSRs is negatively correlated with the genome size ($r=-0.70$). Similar observations have also been reported by Cavagnaro et al. [20]. Different SSR characteristics have been documented for different genomic regions in plant genomes. For example, Mun et al. [18] reported that SSR frequencies are higher in intergenic regions than in transcribed regions. In agreement with this observation, our results show a 1.5 fold higher SSR frequency in intergenic regions compared to transcribed regions of the sweet orange genome. Furthermore, we observed that among untranscribed regions, SSR_s are densest in 5′-UTR, followed by introns and 3′-UTR regions. Similar observations for SSR densities in different genomic regions have been made for fungi [21], [22] and plants [4]. The presence of certain polymorphic SSRs in coding regions could modify the coding protein. SSRs in UTRs or introns could affect the level of gene expression, which could even lead to phenotypic changes. Li et al. [23] and Zhang et al. [24] demonstrated that variations in 3′-UTR or 5′-UTR SSRs could be responsible for regulating the translation of proteins and for mRNA stabilization. The SSR elements in 5′-UTR region are essential for some gene regulation adaptation as well as phenotypic changes on short time scales. In addition, evidence for SSRs in coding region which affect phenotypes in human MMR genes has been reported by Duval and Hamelin [25] as well as Vassileva et al. [26]. In this light, the high densities of SSRs in 5′-UTRs of the sweet orange genome should be seen as an opportunity to study the influence of SSRs on citrus gene regulation.

In sweet orange, most SSR_s are mono-, di- and trinucleotide repeats, which together account for about 80% of all of the SSR_s. We found that dinucleotide repeats are most frequent in the sweet orange genome, where trinucleotide repeats are most frequent in the cucumber [20]genome. This difference could arise from different search parameters used [20]. Cavagnaro et al. [20] explained how different SSR frequencies can be obtained in different studies due to differences in SSR search parameters and search algorithms. Therefore, in order to compare the frequency of SSRs in different plant species, the same program should be used with exactly the same search parameters. For this reason, we calculate the SSR frequencies of 11 plant genomes including citrus (see Table S5, S6 in File S1). Our data reveal that dinucleotide repeats are predominant in both monocot and dicot genomes and that the occurrence of different types of SSRs greatly vary with the search parameters. Surprisingly, heptanucleotide repeats show a higher density than hexanucleotide repeats in the sweet orange genome. Similar trends are also found in dicot species but not in monocot plant species. So it has been postulated that the decrease in SSR frequency with unit length is higher in monocot species as compared to dicot species. The distribution of microsatellites across the different genomic regions show that all repeat types except tri- and hexanucleotide repeats were comparatively less frequent in CDS regions compared to the other genomic region of the sweet orange genome. The high frequency of trinucleotide repeats in protein coding regions (CDS) has previously been reported for several plant species [16], [17], [18], [27], [28] and other eukaryotes including insects and human [29], [30], [31]. The predominance of tri- and hexanucleotide repeats in CDS regions can be explained by the fact that their mutations won't disrupt the reading frame. We also found evidence for positive selection for specific repeats in the sweet orange genome by comparing the repeat density of a specific patter with its reverse complement on the sense strand in transcribed regions. This effect is called a standedness [5]. For example, the CT pattern is over-represent in 5′-UTR but not 3′-UTR compared to the AG pattern (Table S8, S9 in File S1). It is believed that oligo pyrimidine tracts in the 5′-terminal may be involved in the regulation of translation in vertebrate mRNA and also known as the plant translational apparatus [32]. Evidences suggested that CT microsatellites in 5′UTR of *Arabidopsis thaliana* are involved in their antisense transcription [33].

As shown in Figure 1, there is a remarkable variation in the frequency of individual repeat patterns in the sweet orange genome. The base composition of sweet orange SSR patterns is strongly biased toward A and T for all unit sizes. To give an example, the density of mono repeats A/T was 20 fold higher than for G/C patterns. This result is consistent with the previous studies where AT rich repeats have been found to be characteristic for dicot plants but not for monocots. The prevalence of AT over GC rich repeats seems to correlate with the overall genome composition. Indeed, the GC content in dicot genomes is comparatively lower than in monocot genomes. High differences among repeat pattern densities have also been reported for example for the Brassicaceae where the GAA/TTC and AAG/CTT trinucleotide patterns were the most frequent [34], while Hong et al. [35] found AAG/CTT to be the most abundant pattern in *Arabidopsis* and *B. rapa*. In the Solanaceae, the GAA/TCC and AGA/TCT patterns are most frequent, while CCG/CGG are the most frequent patterns found in the Solanaceae family [35] and in the coccolithophore *Emiliania huxleyi* [36].

Genome wide SSR marker development, in silico polymorphism analysis, Functional annotation and wet lab validation of SSR markers

As noted for example by Cavagnaro et al. [20] mononucleotide repeats are not suitable for marker development. Therefore, we only considered di- to hexanucleotides repeats for primer modeling. In this size range, 33855 SSRs were identified in the sweet orange genome, but only 21248 (62.76%) loci are suitable for SSR marker development. SSR loci can be unsuitable for marker development due to insufficiently flanking regions. *In silico* polymorphisms of the developed markers were estimated by using the virtual PCR strategy. That allowed us to understand the possible relationships between the degree of polymorphism and particular features of sweet orange microsatellites. It is known that long repeats are more prone to mutations, which shrink the repeat, both in plants and animals [20], [37], [38]. Similar trends are found in our study. Comparing *C. sinensis* and *C. clementina*, we found that most SSR alleles from *C. sinensis* showed a length reduction in the corresponding *C. clementina* alleles. This result could be biased by the selection of the SSR loci analyzed, or the differences in SSR mutation rates between the two genotypes. Usually larger repeats are selected during SSR marker development; increasing the chances for a biased selection.

As expected, the majority of the SSR markers had no GO assignment since most SSRs are located in the intergenic regions of the sweet orange genome. However, a total 9% of the SSR markers had significant Gene Ontology hits. SSR loci with GO terms are good candidates as molecular markers for association studies.

In total, 950 SSRs that had been evaluated *in silico* were confirmed by wet lab experiments (i.e. PCR and gel electrophoresis) and for most of them, the results are consistent with the virtual PCR result. We obtained a high PCR amplification efficiency in this study (609 primer yielded scorable amplicons) which is consistent with earlier studies of marker development in plant species [39]. As expected, we found that the degree of marker transferability is higher in intra-specific populations than in inter-specific populations. The high degree of intra and inter specific transferability of markers will have a broad utilization in taxonomic, population conservation as well as mapping studies. Finally, it will assist breeding programs of citrus relatives, especially for species for which only few markers have been developed to data (such as lemon, lime, citron and kumquat etc.). The transferability of sweet orange SSR markers across *Citrus*species is higher than reported for other plant species [40], [41]. In general, the transferability rate within genus and among the genus greatly varies with the phylogenetic distances of the examined species and the genomic region used for marker development. EST-derived SSRs are more conserved than genomic-SSRs, therefore EST-SSRs are more transferable to related genera than genomics SSRs.

CONCLUSION

The present study contributes to a detailed characterization and utilization of genome wide SSR markers in sweet orange. The sweet orange genome has a prevalence for AT-rich SSRs and SSRs are non-randomly distributed. A large number of markers have been developed and almost two thirds of these are transferable as well as polymorphic among citrus relatives. The knowledge of these markers significantly contributes to enhance the genomic resources in citrus species and will facilitate a number of genetic and genomic studies in citrus, including genetic diversity evaluation, population genetics, high density linkage map, positional cloning, and comparative genomics in other citrus species.

REFERENCES

1. Weber JL, Wong C (1993) Mutation of human short tandem repeats. Hum Mol Genet 2: 1123–1128
2. Kruglyak S, Durrett RT, Schug MD, Aquadro CF (1998) Equilibrium distributions of microsatellite repeat length resulting from a balance between slippage events and point mutations. Proc Natl Acad Sci U S A 95: 10774–10778
3. Richard GF, Kerrest A, Dujon B (2008) Comparative genomics and molecular dynamics of DNA repeats in eukaryotes. Microbiol Mol Biol Rev 72: 686–727
4. Sonah H, Deshmukh RK, Sharma A, Singh VP, Gupta DK, et al. (2011) Genome-wide distribution and organization of microsatellites in plants: an insight into marker development in Brachypodium. PLoS One 6: e21298
5. Mayer C, Leese F, Tollrian R (2010) Genome-wide analysis of tandem repeats in Daphnia pulex - a comparative approach. BMC Genomics 11: 277
6. FAOSTAT (2012) Food and Agriculture Organization of the United Nations, Rome: FAO.
7. Arumuganathan K, Earle E (1991) Nuclear DNA content of some important plant species. Plant Molecular Biology Reporter 9: 208–218
8. Biswas MK, Chai L, Amar MH, Zhang X, Deng XX (2011) Comparative analysis of genetic diversity in Citrus germplasm collection using AFLP, SSAP, SAMPL and SSR markers. Scientia Horticulturae 129: 798–803
9. Biswas MK, Chai L, Mayer C, Xu Q, Guo W, et al. (2012) Exploiting BAC-end sequences for the mining, characterization and utility of new short sequences repeat (SSR) markers in Citrus. Molecular Biology Reports 1–14
10. Biswas MK, Xu Q, Deng XX (2010) Utility of RAPD, ISSR, IRAP and REMAP markers for the genetic analysis of Citrus spp. Scientia Horticulturae 124: 254–261

11. Ollitrault F, Terol J, Pina JA, Navarro L, Talon M, et al. (2010) Development of SSR markers from Citrus clementina (Rutaceae) BAC end sequences and interspecific transferability in Citrus. Am J Bot 97: e124–129
12. Chen C, Zhou P, Choi YA, Huang S, Gmitter FG Jr (2006) Mining and characterizing microsatellites from citrus ESTs. Theor Appl Genet 112: 1248–1257
13. Cheng YJ, Guo WW, Yi HL, Pang XM, Deng X (2003) An efficient protocol for genomic DNA extraction fromCitrus species. Plant Molecular Biology Reporter 21: 177–178
14. Bohra A, Dubey A, Saxena RK, Penmetsa RV, Poornima KN, et al. (2011) Analysis of BAC-end sequences (BESs) and development of BES-SSR markers for genetic mapping and hybrid purity assessment in pigeonpea (Cajanus spp.). BMC Plant Biol 11: 56
15. Ruiz C, Paz Breto M, Asíns MJ (2000) A quick methodology to identify sexual seedlings in citrus breeding programs using SSR markers. Euphytica 112: 89–94
16. Cardle L, Ramsay L, Milbourne D, Macaulay M, Marshall D, et al. (2000) Computational and experimental characterization of physically clustered simple sequence repeats in plants. Genetics 156: 847–854
17. Temnykh S, DeClerck G, Lukashova A, Lipovich L, Cartinhour S, et al. (2001) Computational and experimental analysis of microsatellites in rice (Oryza sativa L.): frequency, length variation, transposon associations, and genetic marker potential. Genome Res 11: 1441–1452
18. Mun JH, Kim DJ, Choi HK, Gish J, Debelle F, et al. (2006) Distribution of microsatellites in the genome of Medicago truncatula: a resource of genetic markers that integrate genetic and physical maps. Genetics 172: 2541–2555
19. Kuntal H, Sharma V, Daniell H (2012) Microsatellite analysis in organelle genomes of Chlorophyta. Bioinformation 8: 255–259
20. Cavagnaro PF, Senalik DA, Yang L, Simon PW, Harkins TT, et al. (2010) Genome-wide characterization of simple sequence repeats in cucumber (Cucumis sativus L.). BMC Genomics 11: 569
21. Labbé J, Murat C, Morin E, Tacon F, Martin F (2011) Survey and analysis of simple sequence repeats in the Laccaria bicolor genome, with development of microsatellite markers. Current Genetics 57: 75–88
22. Murat C, Riccioni C, Belfiori B, Cichocki N, Labbe J, et al. (2011) Distribution and localization of microsatellites in the Perigord black truffle genome and identification of new molecular markers. Fungal Genet Biol 48: 592–601
23. Li B, Xia Q, Lu C, Zhou Z, Xiang Z (2004) Analysis on frequency and density of microsatellites in coding sequences of several eukaryotic genomes. Genomics Proteomics Bioinformatics 2: 24–31
24. Zhang L, Zuo K, Zhang F, Cao Y, Wang J, et al. (2006) Conservation of noncoding microsatellites in plants: implication for gene regulation. BMC Genomics 7: 323
25. Duval A, Hamelin R (2002) Genetic instability in human mismatch repair deficient cancers. Ann Genet 45: 71–75
26. 26.Vassileva V, Millar A, Briollais L, Chapman W, Bapat B (2002) Genes involved in DNA repair are mutational targets in endometrial cancers with microsatellite instability. Cancer Res 62: 4095–4099
27. Morgante M, Hanafey M, Powell W (2002) Microsatellites are preferentially associated with nonrepetitive DNA in plant genomes. Nat Genet 30: 194–200
28. Peng JH, Lapitan NL (2005) Characterization of EST-derived microsatellites in the wheat genome and development of eSSR markers. Funct Integr Genomics 5: 80–96
29. Toth G, Gaspari Z, Jurka J (2000) Microsatellites in different eukaryotic genomes: survey and analysis. Genome Research 10: 967–981
30. Subramanian S, Mishra RK, Singh L (2003) Genome-wide analysis of microsatellite repeats in humans: their abundance and density in specific genomic regions. Genome Biol 4: R13
31. Weng Y, Azhaguvel P, Michels GJ Jr, Rudd JC (2007) Cross-species transferability of microsatellite markers from six aphid (Hemiptera: Aphididae) species and their use for evaluating biotypic diversity in two cereal aphids. Insect Mol Biol 16: 613–622

32. Shama S, Meyuhas O (1996) The translational cis-regulatory element of mammalian ribosomal protein mRNAs is recognized by the plant translational apparatus. Eur J Biochem 236: 383–388
33. Martienssen RA, Colot V (2001) DNA methylation and epigenetic inheritance in plants and filamentous fungi. Science 293: 1070–1074
34. da Maia LC, de Souza VQ, Kopp MM, de Carvalho FI, de Oliveira AC (2009) Tandem repeat distribution of gene transcripts in three plant families. Genet Mol Biol 32: 822–833
35. Hong CP, Piao ZY, Kang TW, Batley J, Yang TJ, et al. (2007) Genomic distribution of simple sequence repeats in Brassica rapa. Mol Cells 23: 349–356
36. Read BA, Kegel J, Klute MJ, Kuo A, Lefebvre SC, et al. (2013) Pan genome of the phytoplankton Emiliania underpins its global distribution. Nature 499: 209–213
37. Ellegren H (2000) Heterogeneous mutation processes in human microsatellite DNA sequences. Nat Genet 24: 400–402
38. Wierdl M, Dominska M, Petes TD (1997) Microsatellite instability in yeast: dependence on the length of the microsatellite. Genetics 146: 769–779
39. Cavagnaro PF, Chung SM, Manin S, Yildiz M, Ali A, et al. (2011) Microsatellite isolation and marker development in carrot - genomic distribution, linkage mapping, genetic diversity analysis and marker transferability across Apiaceae. BMC Genomics 12: 386
40. Zhang LY, Bernard M, Leroy P, Feuillet C, Sourdille P (2005) High transferability of bread wheat EST-derived SSRs to other cereals. Theor Appl Genet 111: 677–687
41. Zhao H, Yu J, You FM, Luo M, Peng J (2011) Transferability of microsatellite markers from Brachypodium distachyon to Miscanthus sinensis, a potential biomass crop. J Integr Plant Biol 53: 232–245

CHAPTER 13

Antioxidant and Antiproliferative Activities of Twenty-Four *Vitisvinifera* Grapes

Zhenchang Liang[1,2], Lailiang Cheng[3], Gan-Yuan Zhong[4]*, RuiHai Liu[2]*

[1] *Institute of Botany, the Chinese Academy of Sciences, Beijing, China,*

[2] *Department of Food Science, Cornell University, Ithaca, New York, United States of America,*

[3] *Department of Horticulture, Cornell University, Ithaca, New York, United States of America,*

[4] *USDA -ARS Grape Genetics Research Unit, Geneva, New York, United States of America*

ABSTRACT

Grapes are rich in phytochemicals with many proven health benefits. Phenolic profiles, antioxidant and antiproliferative activities of twenty-four selected *Vitisvinifera* grape cultivars were investigated in this study. Large ranges of variation were found in these cultivars for the contents of total phenolics (95.3 to 686.5 mg/100 g) and flavonoids (94.7 to 1055 mg/100 g) and antioxidant activities (oxygen radical absorbance capacity 378.7 to 3386.0 mg of Trolox equivalents/100 g and peroxylradical scavenging capacity14.2 to 557 mg of vitamin C equivalents/100 g), cellular antioxidant activities (3.9 to 139.9 μmol of quercetin equivalents/100 g without PBS wash and 1.4 to 95.8 μmol of quercetin equivalents /100 g with PBS wash) and antiproliferative activities (25 to 82% at the concentrations of 100 mg/mL extracts).The total antioxidant activities were significantly correlated with the total phenolics and flavonoids. However, no significant correlations were found between antiproliferative activities and total phenolics or total flavonoids content. Wine grapes and color grapes showed much higher levels of phytochemicals and antioxidant activities than table grapes and green/yellow grapes. Several germplasm accessions with much high contents of phenolics and flavonoids, and total antioxidant activity were identified. These germplasm can be valuable sources of genes for breeding grape cultivars with better nutritional qualities of wine and table grapes in the future.

INTRODUCTION

Fruits, as an important part of diet, are beneficial for human health.The Dietary Guidelines for Americans provide three reasons to eat more fruits and vegetables [1]. First, most fruits and vegetables are major contributors of a number of nutrients that are under-consumed in the United States, including folate, magnesium, potassium, dietary fiber, and vitamins A and C. Second, consumption of fruits and vegetables is associated with reduced risk of many chronic diseases. Specifically, moderate evidence indicates that intake of at least 2.5 cups of vegetables and fruits perday is associated with a reduced risk of cardiovascular disease, including heart attack and stroke. Some fruits and vegetables may be protective against certain types of cancer. Third, most fruits and vegetables, when prepared without added fats or sugars, are relatively low in calories. Consumption of fruits and vegetables instead of higher calorie foods can help adults and children achieve and maintain healthy weight. Currently, very few Americans consume diets that meet Dietary Guideline recommendations of fruits and vegetables. In fact, most Americans only consume about 42% of goal intake for fruits. For improving public health and reducing the risk of developing chronic diseases, the Dietary Guidelines for Americans recommend that most people should eat at least 9 servings of fruits and vegetables per day based on a 2000 kcal diet [1].

Grapes (*Vitis sp.*) are one of the most important fruit crops worldwide. They are consumed as fresh fruits as well as wine, juice and other processed products. There are about 60 grape species in the genus of *Vitis*, and the species *V. vinifera*, or European grapes, is most widely cultivated. There are thousands of *V. vinifera* cultivars extant in the world. To obtain an assessment of the full range of phytochemicals in grapes, we recently analyzed 36 phenolic compounds in the berry samples of 344 representative *V. vinifera* cultivars [2] and supported previous findings that grapes are rich in phytochemicals [3], [4]. Demonstration of health benefits of these phytochemicals in grapes has been actively pursued. It has been reported that grape extracts exhibited antioxidant activities, including scavenging of free radicals, inhibition of lipid oxidation, and reduction of hydroperoxide formation [4], [5], and inhibit cardiovascular diseases and certain types of cancers, reducing plasma oxidative stress, and slowing aging [6], [7], [8].While there were many reports on the assessment of phytochemical compounds and their antioxidant activities of grapes, these studies were limited to only a few cultivars [9]–[15].

In this study, we determined the phytochemical profiles of 24 *V. vinifera* grape cultivars and analyzed their total antioxidant and antiproliferative activities. These 24 cultivars were selected on the basis of a previous study in which we demonstrated that these cultivars contained a wide range of the composition and content of phenolic compounds [2]. The contribution of flavonoids to total phenolics and correlative relationships among phenolics, flavonoids, antioxidant activities and antiproliferative activities were also investigated.

MATERIALS AND METHODS

Chemicals and reagents

Methanol (MeOH), ethanol (EtOH), acetone, hexane, ethyl acetate, hydrochloric acid (HCl), acetic acid (HAC), potassium chloride (KCl), sodium acetate (NaAC), sodium carbonate (NaCO3), sodium hydroxide (NaOH), potassium phosphate monobasic (KH_2PO4), and potassium phosphate dibasic (K_2HPO4) were of analytical grade and purchased from Mallinckrodt Chemicals (Phillipsburg, NJ). Ascorbic acid, 2′, 7′-dichlorofluorescin diacetate (DCFH-DA), fluorescein disodium salt, sodium borohydride ($NaBH_4$, reagent grade), chloranil (analytical grade), vanillin (analytical grade), quercetin dehydrated, catechin hydrated, Folin-Ciocalteu reagent, 6-hydroxy-2, 5, 7, 8-tetramethylchroman-2-carboxylic acid (Trolox), trifluoroacetic acid (TFA, chromatographic grade), and acetonitrile (chromatographic grade) were purchased from Sigma-Aldrich, Inc. (St. Louis, MO). Tetrahydrofuran (THF, analytical grade), dimethyl sulfoxideand aluminum chloride ($AlCl_3 \bullet 6H_2O$, analytical grade) were purchased from Fisher Scientific (Fair Lawn, NJ). Gallic acid was obtained from ICN Biomedicals, Inc. (Aurora, OH). 2, 2′-Azobis (2-amidinopropane) dihydrochloride (ABAP) was purchased from Wako Chemicals USA, Inc. (Richmond, VA). HepG2 liver cancer cells were obtained from the American Type Culture Collection (ATCC) (Rockville, MD). Williams' Medium E (WME) and Hanks' Balanced Salt Solution (HBSS) were purchased from Gibco Life Technologies (Grand Island, NY). Fetal bovine serum (FBS) was obtained from Atlanta Biologicals (Lawrenceville, GA).

Plant materials

Berry samples of 24 *V. vinifera* grape cultivars were collected from the United States Department of Agriculture-Agricultural Research Service (USDA-ARS) *Vitis* clonal repository in Davis, California (Table 1). The grape berries were harvested upon ripening from the 2010 vintages. All the vines received standard fertilization, irrigation, pruning, and insect and disease control. About 100 of representative berries were collected from each individual cultivar. Berry color was scored as green-yellow, pink, red, dark-red violet, blue black or red-black and compared with those recorded in the Germplasm Resources Information Network [16] when a cultivar was introduced. The number of berries was counted and the berry weight was recorded for each sample before being frozen and stored at −80°C for further processing. The frozen berries were then crushed using a mortar and pestle. After removing all the seeds, flesh and peel tissues were ground in an IKA A11 mill (IKA Works, Inc, NC, USA) into powder while frozen using liquid nitrogen. All data collected for each grape cultivar were reported as mean ± SD with at least three replications.

Table 1. Description of V. vinifera grape samples used in the study and total phenolics (μmol of Gallic acid equiv /100 g FW) and flavonoids (μmol of Catechinequiv /100 g FW) content, ratio of total flavonoids to total phenolics (F/P, %), ORAC (mg of Trolox equivalents/100 g FW) and PSC (mg of vitamin C equivalents/100 g FW) values, CAA with PBS washing and no PBS washing (μmol of QE/100 g of grape) values.

Accession	Cultivar	Primary use	Color	Phenolics	Flavonoids	F/P	ORAC	PSC	CAA with wash	CAA with no wash
DVIT2630	Agwam	Wine grape	Red	102.7±3.0	128.6±8.6	73.5±5.2	547.3±50.5	73.5±7.5	1.4±0.2	5.3±0.3
DVIT0632	Araclinos	Wine grape	Blue-black	347.9±7.1	406.1±11.2	68.9±2.7	1171.3±58.3	232.1±19.9	24.6±1.0	33.9±2.1
DVIT0677	Cabernet Sauvignon	Wine grape	Red-black	448.3±5.2	603.1±34.5	79.0±5.1	1394.7±54.3	226.0±18.0	22.0±2.6	43.6±5.3
DVIT0696	Corbeau	Wine grape	Red-black	582.8±11.7	687.1±43.5	69.0±3.5	2320.0±129.0	420.5±30.8	30.7±1.2	46.1±3.3
DVIT0364	Coudsi	Table grape	Green-yellow	159.5±4.8	201.7±14.4	74.0±3.2	710.3±63.6	31.2±1.4	9.8±1.2	14.7±1.5
DVIT0314	Demir Kara	Table grape	Red-black	345.0±13.2	341.0±13.3	58.0±1.0	1400.7±130.0	167.3±6.6	25.5±2.5	59.7±3.0
DVIT2705	Dornfelder	Table grape	Blue-black	391.6±9.5	530.0±36.4	79.3±4.7	1479.7±149.0	404.5±18.1	49.6±5.7	54.6±3.2
DVIT2642	Madam Matijas	Table grape	Green-yellow	201.5±5.4	146.4±7.6	42.6±2.4	566.0±67.0	73.0±6.5	10.2±0.9	12.2±0.1
DVIT0099	Mathilde	Table grape	Red	201.9±9.1	170.8±11.3	49.6±5.1	630.7±14.0	109.3±10.8	3.4±0.3	8.0±1.0
DVIT0825	Melon	Wine grape	Green-yellow	466.8±9.9	587.5±16.0	73.1±2.9	1651.7±138.0	103.8±10.9	11.9±0.8	42.5±0.6
DVIT1042	Mermark	Wine grape	Red-black	664.4±18.2	568.0±21.7	50.1±0.8	2422.7±221.0	244.0±15.6	14.1±1.3	47.5±1.5
DVIT0465	Muscat of Alexandria	Table grape	Green-yellow	95.3±1.8	94.7±3.0	58.3±1.2	392.0±34.9	14.2±0.6	1.8±0.2	3.9±0.1
DVIT2646	Pearl of Zola	Table grape	Green-yellow	291.2±1.5	230.0±4.1	46.3±0.7	733.0±65.2	58.8±3.5	17.0±1.2	26.4±0.9
DVIT2700	Pirobelle	Table grape	Blue-black	207.6±10.5	210.6±15.2	59.4±2.5	571.0±37.7	117.6±3.7	21.2±0.3	23.1±0.7
DVIT1119	Plavac Mali	Wine grape	Blue-black	351.1±11.0	502.6±31.8	84.2±6.5	1186.0±122.0	138.2±8.8	33.0±1.8	34.3±1.5
DVIT0503	Ribier	Table grape	Blue-black	294.3±7.9	251.9±10.1	50.2±2.6	909.3±93.5	46.9±6.0	19.9±2.5	21.5±2.6
DVIT0934	Robola	Wine grape	Green-yellow	384.4±13.6	425.2±27.4	65.1±5.2	1298.3±135.0	91.5±7.3	14.5±0.5	47.4±0.2
DVIT0937	Royalty	Wine grape	Red-black	686.5±1.9	1054.6±78.1	90.0±6.8	3386.3±221.0	557.0±50.0	95.8±5.0	139.9±9.2
DVIT0940	Ruby Cabernet	Wine grape	Red-black	486.7±13.2	758.2±48.1	91.3±3.9	1752.3±174.0	183.7±16.3	73.3±7.2	79.8±4.7
DVIT1060	Saint Macaire	Wine grape	Red-black	368.2±17.6	301.2±19.1	47.9±5.1	1402.7±158.0	285.0±26.6	16.9±1.0	50.0±2.9
DVIT0944	Salvador	Wine grape	Red-black	407.5±8.9	485.6±30.8	69.8±3.4	2504.7±271.0	376.2±36.7	81.1±7.7	91.0±8.8
DVIT0535	Thompson Seedless	Table grape	Green-yellow	151.1±6.6	201.4±7.3	78.3±3.2	378.7±24.8	24.1±2.2	7.0±0.6	10.5±0.9
DVIT1061	Touriga	Wine grape	Red-black	482.5±20.8	331.5±19.8	40.3±3.6	1960.3±85.6	302.0±11.0	21.1±0.1	54.4±2.3
DVIT0605	Yourutico	Table grape	Blue-black	419.1±6.0	534.6±29.3	74.7±3.8	1365.0±74.7	134.9±7.3	11.4±0.9	45.9±3.6

Extraction of total phenolic compounds

Total phenolics were extracted from ground berry samples using the method previously reported from our laboratory [17]. Briefly, 10 g of grape powder was homogenized with 30 mL of 80% chilled acetone/water using a Virtis High Speed Homogenizer for 3 min on the ice. After centrifugation at 3000 g for10 min at 4°C, the supernatant was collected, and 30 mL of fresh extraction solution were added and the extraction was repeated twice. All extracts were transferred into a round bottom flask, and were evaporated using a rotary evaporator at 45°C until the weight of the evaporated filtrate was less than10% of the weight of the original filtrate. The extracts were brought to a final volume of 10 mL with 70% methanol. All extracts were stored at −40°C until use. All extractions were analyzed in triplicates.

Determination of total phenolic contents

Total phenolic content was determined using the Folin-Ciocalteu colorimetric method [18], which was modified by our laboratory [19]. Briefly, all extracts were diluted 1:20 with distilled water to obtain readings within the standard curve ranges of 0.0–600.0 μg of gallic acid/mL. The grape extracts were oxidized by the Folin-Ciocalteu reagent and the reaction was acted in sodium carbonate. The absorbance was measured at 760 nm, after 90 min at room temperature by a MRX II Dynex plate reader (Dynex Technologies Inc., Chanilly, VA). The absorbance values were then compared with those of standards with known gallic acid concentrations. All values were expressed as the mean (milligrams of gallic acid equivalents per 100 g of fresh sample) ± SD of three replications.

Determination of total flavonoid contents

The total flavonoid content was determined using the sodium borohydride /chloranil-based (SBC) assay developed by our laboratory [20] with modifications [21]. Briefly, 0.2 mL extracts of tested samples were added into test tubes (15×150 mm), dried under nitrogen gas, and reconstituted in 1 mL of terahydrofuran/ethanol (THF/EtOH, 1:1, v/v). Catechin standards (0.1–10.0 mM) were prepared fresh before use in 1 mL of THF/EtOH (1:1, v/v). Then 1 mL of 50 mM $NaBH_4$ solution and 0.5 mL of 74.6 mM $AlCl_3$ solution were added into each test tube with 1 mL of sample solution or 1 mL of catechin standard solution. The test tubes were shaken in an orbital shaker at room temperature for 30 min. An additional 0.5 mL of 50.0 mM $NaBH_4$ solution was added into each test tube with shaking continued for another 30 min at room temperature. Two millilitersof cold 0.8 M acetic acid solution was added into each test tube, and the solution was kept in the dark for 15 min after a thorough mix. One milliliter of 20.0 mMchloranil was added into each tube, which was heated at 95 °C with shaking for 60 min. The reaction solution was cooled using tap water, and the final volume was brought to 4 mL using methanol. One milliliter of 1052 mM vanillin was added into each tube and mixed. Then 2 mL of 12 M HCl was added to each tube, and the reaction solution was kept in the dark for 15 min after a thorough mix. Aliquots of final reaction solutions (200 μL) were added into each well of a 96-well plate, and the absorbance was measured at 490 nm using a MRX Microplate Reader with Revelation workstation (Dynex Technologies, Inc., Chantilly, VA). Total flavonoid content was expressed as milligrams of catechin equivalents per 100 g of fresh weight of sample. Data were reported as mean ± SD with at least triplicates.

Oxygen radical absorbance capacity (ORAC) assay

The antioxidant activity was determined using oxygen radical absorbance capacity (ORAC) assay [22] and modified in our laboratory [23]. Phenolic extract dilutions were prepared with 75 mM phosphate buffer (pH 7.4). The assay was performed in black-walled 96-well plates (Corning Scientific, Corning, NY). The outside wells of the plate were not used as there was much more variation from them than from the inner wells. Each well contained 20 μL extracts or 20 μLTrolox standard (range 6.25–50 μM), and 200 μL fluoroscein (final concentration 0.96 μM), which were incubated at 37°C for 20 min. After incubation, 20 μL of 119 mM ABAP was added to each well. Fluorescence intensity was measured using Fluoroskan Ascent FL plate-reader (Thermo Labsystems, Franklin, MA) at excitation of 485 nm and emission of 520 nm for 35 cycles every 5 min. ORAC values were expressed as milligrams of TE/100 g FW. Data were reported as mean ± SD with triplicates.

Peroxylradicalscavenging capacity (PSC) assay

The antioxidant activity of extracts was determined by using a PSC assay developed in our lab [24] with modifications [25]. Just prior to use in the reaction, 107 μL of 2.48 mMdichlorofluoresceindiacetate was hydrolyzed to dichlorofluorescein with 893 μL of 1.0 mM KOH for 5 min in a vial to remove the diacetate moiety and then diluted with 7 mL of 75 mM phosphate buffer (pH 7.4). ABAP (200 mM) was prepared fresh in the buffer and was kept at 4°C between runs. In an assay, 100 μL of extracts was diluted in 75 mM phosphate buffer (pH 7.4) and then transferred into reaction cells on a 96-well plate, and 100 μL of dichlorofluorescein was added. The 96-well plate was loaded into a Fluoroskan Ascent fluorescence spectrophotometer (Thermo Labsystems, Franklin, MA), and the solution in each cell was mixed by shaking at 1200 rpm for 20 s. The reaction was then initiated by adding 50 μL of ABAP from the autodispenser on the equipment. Each set of dilutions for a replicate and control was analyzed three times in adjacent columns. The reaction was carried out at 37°C, and fluorescence was monitored at 485 nm excitation and 538 nm emission with the fluorescence spectrophotometer. The buffer was used for control reactions. Data were acquired with Ascent software, version 2.6 (Thermo Labsystems). Data were reported as mean ± SD with three replicates.

Cell Culture

HepG2 cells (The American Type Culture Collection, ATCC, Rockville,MD) were grown in growth medium (WME supplemented with 5% FBS, 10 mMHepes, 2 mM L-glutamine, 5 μg/mL insulin, 0.05 μg/mL hydrocortisone, 50 units/mL penicillin, 50 μg/mL streptomycin, and 100 μg/mL gentamicin) and were maintained at 37°C and 5% CO_2 as described previously [26], [27]. Cells used in this study were between passages or generations18 and 28 at which the cells were at the most stable physiological status.

Cytotoxicity

Cytotoxicity was measured using the methylene blue assay developed by our laboratory (Felice et al., 2009) with modifications [28]. Briefly, HepG2 cells were seeded at 4×10^4/well on a 96-well plate in 100 μL of growth medium and incubated for 24 h at 37°C. The medium was removed, and the cells were washed with PBS. Treatments of fruit extracts or antioxidant compounds in 100 μL of treatment medium (WME supplemented with 2 mM L-glutamine and 10 mMHepes) were applied to the cells, and the plates were incubated at 37°C for 24 h. The treatment medium was removed, and the cells were washed with PBS. A volume of 50 μL/well methylene blue staining solution (98% HBSS, 0.67% glutaraldehyde, 0.6% methylene blue) was applied to each well, and the plate was incubated at 37°C for 1 h. The dye was removed, and the plate was immersed in fresh deionized water six times, or until the water became clear. The water was tapped out of the wells, and the plate was allowed to air-dry briefly before 100 μL of elution solution (49% PBS, 50% ethanol, 1% acetic acid) was added to each well. The microplate was placed on a bench-top shaker for 20 min to allow uniform elution. The absorbance was read at 570 nm with blank subtraction using the MRX II DYNEX spectrophotometer (DynexInc., Chantilly, VA). Concentrations of fruit extracts that decreased the absorbance by >10% when compared to the control were considered to be cytotoxic (Felice et al., 2009).

Cellular Antioxidant Activity (CAA) of grape extracts

The CAA was determined using the method previously developed in our laboratory [29]. Briefly, HepG2 cells were seeded at a density of 6×10^4/well on a 96-well microplate in 100 μL of growth medium/well. Twenty-four hours after seeding, the growth medium was removed, and the wells were washed with PBS. Wells were then treated in triplicate for 1 h with 100 μL of treatment medium containing tested fruit extracts plus 25 μM DCFH-DA. When a PBS wash was utilized, wells were washed with 100 μL of PBS. When no PBS wash was done between treatments, the activity may have been higher, but the CV also tended to be higher. This was likely due to the interaction of the samples and oxidants with other factors in the residual medium on the cells. Washing the cells with PBS removed most of the interfering medium components and increased the consistency of the results. Then 600 μM ABAP was applied to the cells in 100 μL of HBSS, and the 96- well microplate was placed into a Fluoroskan Ascent FL plate reader (ThermoLabsystems, Franklin, MA) at 37°C. Emission at 538 nm was measured after excitation at 485 nm every 5 min for 1 h.

After blank subtraction and subtraction of initial fluorescence values, the area under the curve for fluorescence versus time was integrated to calculate the CAA value at each concentration of fruit as [30]

$$CAA(unit) = 1 - \left(\int SA / \int CA\right)$$

Where ∫SA is the integrated area under the sample fluorescence versus time curve and ∫CA is the integrated area from the control curve. The median effective dose (EC_{50}) was determined for the fruits from the median effect plot of log(fa/*f*u) versus log(dose), where *f*a is the fraction affected (CAA unit) and *f*u is the fraction unaffected (1 - CAA unit) by the treatment. The EC_{50}values were stated as mean ± SD for triplicate sets of data obtained from the same experiment. EC_{50} values were converted to CAA values, expressed as micromoles of quercetin equivalents (QE) per 100 g of grape, using the mean EC_{50} value for quercetin from five separate experiments.

Measurement of inhibition of HepG2 cell proliferation

The antiproliferative activity of different grape extracts was assessed by measurement of the inhibition of $HepG_2$ (The American Type Culture Collection, ATCC, Rockville, MD) human cancer cell proliferation. Antiproliferative activities were determined by the colorimetric methylene blue method developed by our laboratory [31] with modifications [28], [32]. Human hepatocellular carcinoma HepG2 cells were seeded at a density of 2.5×10^4/well on a 96-well microplate in 100 μL of growth medium/well. The outside wells of the plate were not used as there was much more variation from them than those from the inner wells. After 4 h of incubation, the growth medium was removed and media containing various concentrations (10, 20, 40, 60, 80 and 100 mg/mL) of grape extracts were added to the cells. Control cultures received the extraction solution minus the grape extracts, and blank wells contained 100 μL of growth medium without cells. Cell proliferation (percent) was determined at 96 h from the absorbance reading at 570 nm for each concentration when compared to the control, using at least three replications for each sample. The effective median dose (EC_{25} and EC_{50}) was determined and expressed as milligrams of grape extracts per milliliter ± SD.

Statistical analysis

Data from this study were reported as mean ± SD with at least three replicates for each sample extract.All graphical representations were performed using Sigmaplot 10.0 for Windows (SPSS, USA), and all data were analyzed using SPSS (Statistics for Social Science) 13.0 for Windows. Results were subjected to ANOVA, and differences between means were located using Tukey B multiple comparison test. Correlations between various parameters were also investigated. Significance was determined at $P<0.05$.

RESULTS

Total phenolic content

Table 1shows the total phenolic contents of 24 *V. vinifera* grape cultivars. Among all the grape cultivars analyzed, Royalty (red flesh cultivar) and Mermark had the highest total phenolic content (686.5±3.2 and 664.4±31.5 mg of gallic acid equivalents/100 g FW, respectively), and were significantly higher than other grape cultivars ($P<0.05$). Agwam and Muscat of Alexandria had the lowest total phenolic contents ($P<0.05$). There were 7-fold differences in the total phenolic contents between the highest (Royalty) and lowest (Muscat of Alexandria) ranked cultivars. Most green-yellow grape cultivars had lower total phenolic contents with no more than 300 mg of gallic acid equivalents/100 g FW except for Melon and Robola. On the contrary, colored grape cultivars had higher total phenolic contents and were more than 300 mg of gallic acid equivalents/100 g FW except for Ribier and two red cultivars of Mathilde and Agwam. All wine grape cultivars had higher total phenolic contents with more than 347.9 mg of gallic acid equivalents/100 g FW, but Agwam was an exception. All table grape cultivars had low phenolic contents except for Yourutico and Dornfelder.

Total flavonoid content

Total flavonoid contents of the 24 grape cultivars are presented in Table 1. The Royalty grape presented the highest flavonoid content and was significantly higher than other cultivars ($P<0.05$). Compared to most other cultivars, Pearl of Zola, Pirobelle, Coudsi, Thompson Seedless, Mathilde, Madam Matijas, and Agwam had relatively lower levels of total flavonoid contents. Muscat of Alexandria had the lowest flavonoid content. There were almost 11-fold differences in total flavonoid contents between the highest and lowest ranked cultivars. Melon and Robola had higher total flavonoid contents than other green-yellow cultivars in which no significant differences were detected. In general, colored cultivars had higher total flavonoid contents than green-yellow cultivars ($P<0.05$). Compared to wine grape cultivars, all table grape cultivars had lower flavonoid contents (less than 250 mg of catechin equivalents/100 g FW) except for Yourutico, Dornfelder and Demir Kara. Total flavonoid contents of wine grape cultivars were higher than 300 mg of catechin equivalents/100 g FW except for Agwam.

Contribution of total flavonoids to total phenolics

The contributions of total flavonoids to total phenolics, calculated on the basis of micromoles of the total flavonoids and total phenolics, are presented in Table 1. The contributions of total flavonoids to the total phenolicsin the 24 grape cultivars ranged from 40.3 to 91.3% with a mean 65.5%. The contributionsof flavonoids to total phenolics in Ruby Cabernet and Royalty (91.3 and 90% of the total phenolics, respectively) were the highest among all cultivars. In contrast, the total flavonoid contents were no more than 50% of the total phenolic contents in Mathilde, Saint Macaire, Pearl of Zola, Madam Matijas and Touriga.

Oxygen radical absorbance capacity (ORAC)

The ORAC values for the 24 cultivars, expressed as mg of Trolox equivalents per 100 g of fresh grape, are shown in Table 1.Variation of ORAC values ranged from 378.7 (Thompson Seedless) to 3386.3 (Royalty) mg of Trolox equivalents/100 g FW, and the mean was 1339 mg. Royalty had the highest ORAC values, significantly higher than other cultivars ($P<0.05$). The other red flesh cultivars Salvador, Mermark and Corbeau also had relatively high ORAC values ($P<0.05$). Most green yellow cultivars had ORAC value no more than 733.0 mg of TE/100 g FW except for Melon and Robola. Most colored cultivars had ORAC values more than 909.0 mg of TE/100 g FW except for one blue-black cultivar (Pirobelle) and two red cultivars (Mathilde and Agwam). All table grape cultivars had lower ORAC values (lower than 1000 mg of TE/100 g FW) except for Dornfelder, Demir Kara and Yourutico. All ORAC values of wine grape cultivars were higher than 1170.0 mg of TE/100 g FW except for Agwam.

Peroxyl radical scavenging capacity (PSC)

Total antioxidant activities measured by peroxyl radical scavenging capacity (PSC) for the 24 cultivars are presented in Table 1with the PSC values expressed as mg of vitamin C equivalents/100 g FW. Royalty had the highest PSC values, and was significantly higher than other cultivars (P<0.05). Corbeau, Dornfelder and the other red flesh cultivar Salvador also had high PSC values (P<0.05). PSC values of all green yellow cultivars were lower than 100 mg of vitamin C equivalents/100 g FW except for Melon. In addition, most PSC values of colored cultivars were higher than 109 mg of vitamin C equivalents/100 g FW except for a red cultivar (Agwam) and a blue black cultivar (Ribier). Dornfelder had the highest PSC value in all table grapes, followed by Demir Kara, Yourutico, Pirobelle and Mathidle. The other PSC values of table grapes were no more than 73 mg of vitamin C equivalents/100 g FW. Most wine grapes had high PSC values except for Agwam, Robola and Melon.

Cellular antioxidant activities

CAA values for 24 grape cultivars using the no PBS wash protocol are shown in Table 1. CAA values ranged from 3.9 to 139.9 μmol of QE/100 g of grape, and the mean was 41.5 μmol of QE/100 g of grape. Royalty had the highest CAA value (P<0.05), followed by Salvador and Ruby Cabernet (P<0.05). Demir Kara, Dornfelder, Touriga, Saint Macaire, Mermark, Robola, Corbeau and Yourutico, Cabernet Sauvignon, Melon had intermediate high CAA activities, but were not significantly different from each other (P>0.05). Then the rankings of CAA activities were followed by Plavac Mali, Araclinos, Pearl of Zola, Pirobelle, Ribier and Coudsi. Madam Matijas, Thompson Seedless, Mathilde, Agwam and Muscat Alexandria had the lowest CAA values among the 24 grape cultivars tested.

In using the PBS wash protocol, CAA values ranged from 1.4to 95.8 μmol of QE/100 g of grape, and the mean was 25.7 μmol of QE/100 g of grape (Table 2). Royalty had the greatest cellular antioxidant activity among the cultivars tested (P<0.05). Salvador ranked second, and Ruby Cabernet and Dornfelder were third and fourth in CAA activities, respectively (P<0.05).

Plavac Mali and Corbeau and other cultivars were in descending order in their CAA activities. Muscat of Alexandria and Agwam had the lowest CAA values. In both PBS wash and no PBS wash protocols, the CAA values were significantly higher in wine grapes than in table grapes. This was also true in colored grapes when compared with green-yellow grapes ($P<0.05$, Table 2).

Table 2. Mean contents of total phenolics, total flavonoids, ORAC and PSC values among different types of grapes (mean±SD).

	Primary use		Color type	
	Table grape	Wine grape	Colored grape	Green yellow grape
Total phenolics (GAE/100 g)	$250.8\pm31.6^{b*}$	444.6 ± 42.0^{a}	399.3 ± 37.7^{a}	250.0 ± 51.4^{b}
Total flavonoids (CE/100 g)	264.8 ± 44.0^{b}	526.3 ± 64.2^{a}	462.8 ± 57.4^{a}	269.6 ± 65.8^{b}
ORAC values (TE/100 g)	830.6 ± 122.0^{b}	1769.0 ± 206.1^{a}	1553.0 ± 186.4^{a}	818.6 ± 181.4^{b}
PSC values (VE/100 g)	107.4 ± 33.2^{b}	248.7 ± 39.1^{a}	236.4 ± 33.8^{a}	56.5 ± 13.1^{b}
CAA with PBS wash (QE/100 g)	17.5 ± 12.9^{b}	31.6 ± 29.2^{a}	32.0 ± 26.8^{a}	10.3 ± 4.8^{b}
CAA with no PBS wash (QE/100 g)	27.7 ± 18.4^{b}	51.4 ± 33.8^{a}	49.3 ± 31.8^{a}	22.5 ± 16.0^{b}

Note: GAE, CE, TE and VE represent mg of gallic acid, catechin, Trolox and vitamin C equivalent respectively; QE represents μmol of quercetin equivalent.
*Values with no letters in common in each row are significantly different ($P<0.05$).

Correlationsamong total phenolics, total flavonoids and antioxidant activities

Correlations among the total phenolics, total flavonoids, ORAC values and PSC values were analyzed (Table 3). Significantly positive correlations were found between total phenolics and total flavonoids (r^2=0.888), and between ORAC and PSC values (r^2=0.863). The ORAC values had significantly positive correlations with total phenolics (r^2=0.924) and total flavonoids (r^2=0.862). In comparison, the PSC values also had significantly positive correlations with total phenolics (r^2=0.757) and total flavonoids (r^2=0.737), but the correlation coefficients were lower than those observed for the ORAC values.

Table 3. The relationship among total phenolics, total flavonoids, ORAC, PSC and CAA values.

	Total phenolics	Total flavonoids	ORAC values	PSC values
Total phenolics	1			
Total flavonoids	0.888*			
ORAC values	0.924	0.862		
PSC values	0.757	0.737	0.863	
CAA values(no PBS wash)	0.788	0.848	0.899	0.809
CAA values(PBS wash)	0.574	0.741	0.736	0.739

*All correlation coefficients were significant at $P<0.01$.

The relationships of total phenolics content, total flavonoids content, ORAC values, and PSC values with CAA values were also determined. CAA values were significantly positively correlated with total phenolics (r^2=0.574), total flavonoids (r^2=0.741), ORAC values (r^2=0.736) and PSC values (r^2=0.739) when no PBS wash protocol was used (P<0.05). While significantly positive correlation relationships were also found between CAA values and total phenolics, total flavonoids, ORAC values and PSC values using the PBS wash protocol (P<0.05),the coefficients of correlation were lower than those with no PBS wash.

Inhibition of human cancer cell proliferation

The effects of extracts from the 24 grape cultivars against the growth of HepG2 human liver cancer cells *in vitro* are presented in Figure 1. Overall, grape extracts had potent antiproliferative activities against HepG2 human liver cancer cell growth in dose-dependent manners. The inhibition of cancer cell proliferation by the 24 grape extracts ranged from 25 to 82% at the concentration of 100 mg/mL extract applied (Figure 1). Robola, Plavac Mali, Royalty, Saint Macaire, Touriga, Coudsi, Ruby Cabernet,andDemir Kara exhibited relatively higher antiproliferative activities towards $HepG_2$ cells (Figure 1and Table 4). The EC_{25} and $EC_{50}$of the antiproliferative activities of different grape cultivars are presented in Table 4. Lower EC_{25} and EC_{50}values represent higher antiproliferative activities. There were no significant differences in antiproliferative activities between wine and table grapes, but in general, the antiproliferative activities of red grapes were higher than those of yellow-green grapes. There were no detected cytotoxicities towards HepG2 cells *in vitro* at the concentration of 150 mg/mL of grape extracts in all 24 grape cultivars tested, indicating that antiproliferative activities were not caused by cytoxicity [32].

DISCUSSION

Regular consumption of fruits and vegetables has been linked to reduced risk of developing chronic diseases [33], [34]. Grapes and their products are rich in phytochemicals, which are proven to possess many health benefits. Grape phytochemicals include anthocyanins, flavanols, flavonols, stilbenes (resveratrol) and phenolic acids and their contents and profiles can vary significantly among different *V. vinifera* cultivars [2], [35].There have been studies attempting to link phytochemical profiles with their total antioxidant activities in *V. vinifera*, but these studies were limited to just a few cultivars [4]. Here we reported phytochemical profiles of 24 *V. vinifera* cultivars and studied the relationships among total phenolics, total flavonoids, total antioxidant activities, cellular antioxidant activities, and antiproliferative activities in these cultivars. These 24 *V. vinifera* cultivars were a part of the core collection of *V. vinifera*germplasm preserved at the USDA-ARS *Vitis* clonal repository in Davis, California and they were selected for this study on the basis of their compositions and contents of polyphenolic compounds reported in a previous study [2].

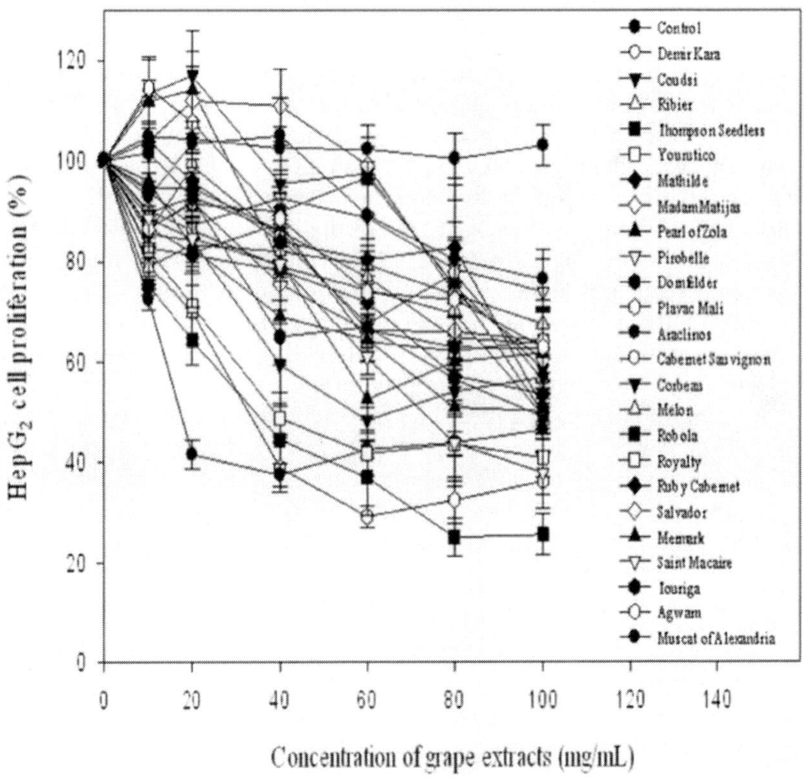

Figure 1. Inhibition of HepG2 cell proliferation by 24 grape extracts.

Table 4. Antiproliferative activities (EC25 and EC50) and cytotoxicities (CC50) of grape extracts towards human HepG2 liver cancer cells.

Cultivar	Antiproliferative Activity	
	EC_{25} (mg/mL)	EC_{50} (mg/mL)
Araclinos	1.53 ± 0.25	ND*
Robola	10.40 ± 1.41	33.63 ± 4.04
Plavac Mali	11.57 ± 2.43	38.70 ± 11.73
Royalty	14.50 ± 1.57	54.97 ± 2.40
Corbeau	25.43 ± 4.40	ND
Mathilde	30.83 ± 11.17	ND
Saint Macaire	31.13 ± 2.81	79.90 ± 13.14
Mermark	39.43 ± 2.82	ND
Pearl of Zola	39.97 ± 2.17	ND
Demir Kara	42.40 ± 3.76	138.23 ± 10.17
Touriga	43.90 ± 1.40	97.00 ± 6.53
Salvador	48.30 ± 8.02	ND
Yourutico	49.50 ± 6.15	ND
Cabemet Sauvignon	55.80 ± 4.90	ND
Ribier	66.07 ± 6.54	ND
Dornfelder	66.50 ± 6.99	ND
Ruby Cabernet	68.93 ± 3.16	132.80 ± 3.03
Agwam	71.47 ± 2.34	ND
Melon	89.10 ± 1.14	ND
Coudsi	90.80 ± 4.74	129.40 ± 2.91
Madam Matijas	95.67 ± 2.30	ND
Thompson Seedless	101.27 ± 3.30	ND
Pirobelle	108.00 ± 16.81	ND
Muscat of Alexandria	112.77 ± 6.52	ND

*ND, not detected, CC50>150 mg/mL.

The present study, using a different, more accurate detection method for the total phenolic compounds, confirmed our previous result that the 24 cultivars selected for this study had a wide range of variation in their total phenolic contents. The mean total phenolic content ranged from 95.3 (Muscat of Alexandria) to 686.5 (Royalty) mg of gallic acid equivalents/100 g FW. In contrast, Yang et al. [4] reported that the total phenolic contents of 14 wine grapes ranged from 201.1 to 424.6 mg of gallic acid equivalent/100 g of fruits. The difference in these observations, to a large extent, was certainly attributed to different cultivars used in the studies. It should be pointed out that there could be some other factors which may contribute to these differences. For example, samples collected from different vineyards or vintages could have significant influence on the total phenolic content as reported by Arnous et al [36] and Kallithraka et al. [37]. The wide range of variation of total phenolic contents among the grape cultivars in this study offered us a unique opportunity to investigate the correlative relationships of phenolic compounds with some health benefit measurements such as antioxidant and antiproliferative activities. In consistence with our earlier study [2], the total phenolic contents of wine grapes were significantly higher than those of table grapes ($P<0.05$); and total phenolics of colored grape cultivars were higher than those of green yellow grape cultivars ($P<0.05$, Table 2).

This study was the first to determine total flavonoid content in grapes using a new method developed [38]. Determination of total flavonoids in foods is challenging. The most common methods for determining the content of total flavonoids include Aluminum Chloride ($AlCl_3$) colorimetric assay [4], [38] and high performance liquid chromatography (HPLC). However, these two methods each have certain limitations. $AlCl_3$ colorimetric assay only measures partial flavonoids and therefore cannot be used accurately to determine total flavonoids [38], [39]. HPLC is an excellent method to determine individual flavonoids, but cannot be used to determine the total flavonoids because the method is limited by the availability of flavonoid standards and many un-identified flavonoids are present in foods [33], [38]. The SBC assay developed by our lab can detect all types of flavonoids, including flavones, flavonols, flavonones, flavononols, isoflavonoids and anthocyanins [38]. The total flavonoid contents in the 24 grape cultivars, measured by SBC assay varied from 94.7 to 1055 mg of catechin equivalents/100 g FW. Anthocyanins were the dominant in total flavonoids in colored grape cultivars according to our previous study [2]. The total flavonoid contents of wine grapes were significantly higher than those of table grapes ($P<0.05$); and total flavonoids of colored grape cultivars were higher than those of green yellow grape cultivars ($P<0.05$, Table 2) because colored grape cultivars had high anthocyanins content and most wine grape cultivars were colored grapes. These results were consistent with those reported as the sum of individual flavonoids using HPLC [2].

The contribution of total flavonoids to total phenolics, calculated on the micromole basis, ranged from 40.3 to 91.3%. Further analysis found no significant differences in the contributions of total flavonoids to total phenolics between table and wine grapes, or between colored and green yellow grapes. These results suggested that flavonoids are one of the major phytochemicals in grapes and their relative contributions to the total phenolics are not dependent on the types of grape cultivars (i.e. wine vs. table grapes).

The mean ORAC values of wine grapes were 1769±206.1 mg of Trolox equivalents/100 g FW and significantly higher than those in table grapes ($P<0.05$, Table 2). In addition, the ORAC values of colored grapes were significantly higher than those of green yellow grapes ($P<0.05$, Table 2).

The PSC values for the 24 grape cultivars ranged from14.2 (Muscat of Alexandria) to 557 (Royalty) mg of vitamin C equivalents/100 g FW, indicating that there was a wide-range of variation across all 24 grape cultivars. Adom and Liu [24] reported the PSC value of red grapes was 371 mg of vitamin C equivalents/100 g FW, which was in the range of the results we reported here. The mean of PSC values of wine grapes was significantly higher than that of table grapes ($P<0.05$, Table 2). Similarly, the PSC values of colored grape cultivars were higher than those of green yellow grape cultivars (Table 2).

The 24 cultivars were also evaluated for their antioxidant activity in the CAA assay. The CAA values of Royalty, Salvador, Ruby Cabernet, Demir Kara, Dornfelder and Corbeauwere higher than other cultivars (Figure 1). These cultivars also had higher total phenolics and total flavonoids, and higher ORAC and PSC values. Among these cultivars, Royalty, Salvador and Corbeau are red flesh grape cultivars and tend to be rich in anthocyanins. The mean CAA values were 32.0 μmol of QE/100 g of grape for colored grape cultivars and 10.3 μmol for green yellow grape cultivars using the PBS wash protocol. But higher CAA values were observed when no PBS wash protocol was used (49.3 μmol of QE/100 g of grape for colored grape cultivars and 22.5 μmol for green yellow grape cultivars). We also observed that the CAA values in wine grapes were significantly higher than these in table grapes. The general trends of these results were in agreement with those of Wolfe et al. [29], [30], but this study covered a much larger range of cultivar variation in terms of both color and end-product types.

The inhibition of HepG2 liver cancer cell proliferation by extracts from the 24 grape cultivars ranged from 25 to 82% at the concentration of 100 mg/mL extracts applied (Figure 1). Yang et al. [4] reported that the phytochemical extracts of Pinot Noir and Chardonnay varieties exhibited strong antiproliferative activity towards HepG2 cells with the lowest EC_{50}values of 17.0±0.8 mg/mL and 18.1±0.1 mg/mL ($p<0.05$), respectively. On the other hand, the phytochemical extracts of Vidal Blanc and DeChaunac showed a relatively weak antiproliferative activity with higher EC_{50}values of 52.1±2.1 and 52.2±3.0 mg/mL, respectively [4]. Our results in this study were similar to the results of Vidal Blanc and DeChaunacas reported by Yang et al. [4]. In consistent with other results, the colored grape cultivars had higher antiproliferative activities than those of green yellow grape cultivars.

Correlations among total phenolics, total flavonoids, total antioxidant activity, cell proliferation, and cellular antiocxdant activities were analyzed. There were significantly correlative relationship among total phenolics, total flavonoids and antioxidant activities, which were in agreement with what was reported by Arnous et al [36] and Kallithraka et al. [37]. No significantly correlative relationships were found between total antioxidant and antiproliferative activities against HepG2 liver cancer cells ($P>0.05$). Additionally, the total phenolic and flavonoid contents of grapes did not correlate with the antiproliferative activities. These results were consistent with what were previously reported in literature [17], [39], suggesting that specific phytochemicals, rather than the total phytochemicals, might be responsible for the antiproliferative activities. Different phytochemicals might act additively and/or synergistically to contribute to the total antiproliferative activities in grapes. The CAA values for fruits were significantly positively related to total phenolic content and total flavonoids when log-transformed data were analyzed ($P<0.05$)in this study. The correlation coefficients for CAA values and total phenolics, total flavonoids, ORAC and PSC values were higher for the no PBS wash protocol than these for the PBS wash protocol in this study. These results were consistent with what Wolfe et al. reported previously [23]. This is in contrast to a study involving vegetable extracts, in which prevention of DCFH oxidation in HepG2 cells by vegetable extracts was not correlated to ORAC or total phenolics [40]. Our current study suggested that total phenolic content might be a better predictor for the cellular antioxidant activity of grapes than the ORAC value, despite the commonality of measuring peroxyl radical scavenging abilities in both of the antioxidant activity assays.

Among the 24 cultivars evaluated, Royalty grape showed highest total phenolic content, total flavonoid content, ORAC, PSC and CAA values. It was the one which had consistent and high correlations among these measurements. Several other cultivars such as Corbeau, Salvador and Mermark also had high levels of phytochemicals, and antioxidant activities. On the other hand, Cabernet Sauvignon, a well-known wine grape cultivar, had relative high levels of total phenolic and flavonoid contents, but its ORAC and PSC values were about average. Some of the differences are certainly due to varietal genetic background. For example, Royalty is a red flesh cultivar with abundant anthocyanins produced in whole fruits, while many other cultivars including Cabernet Sauvignon are not.

In summary, grapes are rich in phenolics and flavonoids. Although viticulture practices, environmental conditions, and post-harvest processing can affect the content of total phenolics, total flavonoids, or individual bioactive compounds in grapes and grape products, varietal or genetic difference is one of the most important factors. Here we observed tremendous variation in total phenolic and total flavonoid contents, and ORAC, PSC and CAA values among the24 grape cultivars tested. The total antioxidant activities of grapes were significantly correlated with the total phenolics and flavonoids. However, no significant correlations were found between total phenolic content, total flavonoid content, ORAC, PSC values, or CAA values with the antiproliferative activities in this study. Lack of such correlations provides both challenges and opportunities in future studies. Total phenolics, total flavonoids, ORAC, PSC and CAA values of wine grapes and colored grapes were significantly higher than those of table grapes and green-yellow grapes. Another important finding from this study was the identification of several germplasm accessions with much high contents of phenolics and flavonoids, and total antioxidant activity. This information is important for breeding grape cultivars with better nutritional qualities of wine and table grapes in the future.

ACKNOWLEDGMENTS

We are grateful to Mr. Bernard Prins of the USDA-ARS National Clonal Germplasm Repository for Fruit and Nut Crops in Davis, California and Dr. Christopher Owens of the USDA-ARS Grape Genetics Research Unit in Geneva, New York for providing *V. vinifera* berry samples. USDA-ARS is an equal opportunity provider and employer.

REFERENCES

1. USDA, Center for Nutrition Policy and Promotion. Dietary Guidelines for Americans (2010) Available: http://www.cnpp.usda.gov/DGAs2010-PolicyDocument.htm. Accessed 2014 July 25.
2. Liang ZC, Christopher OL, Zhong GY, Cheng LL (2011) Polyphenolic profiles detected in the ripe berries of *Vitisvinifera* germplasm. *Food Chem* 129: 940–950.
3. Xia EQ, Deng GF, GuoYJ, Li HB (2010) Biological activities of polyphenols from grapes. *Int J MolSci* 11: 622–646.
4. Yang J, Martinson TE, Liu RH (2009) Phytochemical profiles and antioxidant activities of wine grapes. *Food Chem* 116: 332–339.
5. Bitsch R, Netzel M, Frank T, Strass G, Bitsch I (2004) Bioavailability and biokinetics of anthocyanins from red grape juice and red wine. *J Biomed Biotechnol* 2004: 293–298.
6. Bub A, Watzl B, Blockhaus M, Briviba K, Liegibel U, et al. (2003) Fruit juice consumption modulates antioxidative status, immune status and DNA damage. *J NutrBiochem* 14: 90–98.
7. Garcia-Alonso M, Minihane AM, Rimbach G, Rivas-Gonzalo JC, de Pascual-Teresa S (2009) Red wine anthocyanins are rapidly absorbed in humans and affect monocyte chemoattractant protein 1 levels and antioxidant capacity of plasma. *J Nutr Biochem*20: 521–529.
8. Jang MS, Cai EN, Udeani GO, Slowing KV, Thomas CF, et al. (1997) Cancer chemopreventive activity of resveratrol, a natural product derived from grapes. *Science*275: 218–220.
9. Del Caro A, Cacciorro AF, Fenu PAM, Piga A (2010) Polyphenols, colour and antioxidant activity changes in four Italian red wines during storage. *Acta Aliment Hung*39: 192–210.
10. Liang ZC, Wu BH, Fan PG, Yang CX, Duan W, et al. (2008) Anthocyanin composition and content in grape berry skin in *Vitis* germplasm. *Food Chem* 111: 837–844.
11. Nikfardjam MSP, Mark L, Avar P, Figler M, Ohmacht R (2006) Polyphenols, anthocyanins, and trans-resveratrol in red wines from the Hungarian Villany region. *Food Chem* 98: 453–462.
12. Pirniyazov AZ, Abdulladzhanova NG, Mavlyanov SM, Kamaev FG, Dalimov DN (2003) Polyphenols from *Vitisvinifera* seeds. *Chem Nat Comp* 39: 349–354.
13. Tagliazucchi D, Verzelloni E, Bertolini D, Conte A (2010) In vitro bio-accessibility and antioxidant activity of grape polyphenols. *Food Chem* 120: 599–606.

14. Valdez LB, Alvarez S, Zaobornyj T, Boveris A (2004) Polyphenols and red wine as antioxidants against peroxynitrite and other oxidants. *Bio Res* 37: 279–286.
15. Vita JA (2005) Polyphenols and cardiovascular disease: effects on endothelial and platelet function. *Am J ClinNutr* 81: 292s–297s.
16. USDA-ARS, GRIN. Available: http://www.ars-grin.gov. Accessed 2014 July 25.
17. Sun J, Chu YF, Wu XZ, Liu RH (2002) Antioxidant and anti proliferative activities of common fruits. *J Agr Food Chem* 50: 7449–7454.
18. Singleton VL, Orthofer R, Lamuela-Raventos RM (1999) Analysis of total phenols and other oxidation substrates and antioxidants by means of Folin-Ciocalteu reagent. In *Oxidants and Antioxidants, PtA* 299: 152–178.
19. Okarter N, Liu CS, Sorrells ME, Liu RH (2010) Phytochemical content and antioxidant activity of six diverse varieties of whole wheat. *Food Chem* 119: 249–257.
20. He XJ, Liu D, Liu RH (2008) Sodium borohydride/chloranil-based assay for quantifying total flavonoids. *J Agr Food Chem* 56: 9337–9344.
21. Zhang MW, Zhang RF, Zhang FX, Liu RH (2010) Phenolic profiles and antioxidant activity of black rice bran of different commercially available varieties. *J Agr Food Chem*58: 7580–7587.
22. Ou BX, Hampsch-Woodill M, Prior RL (2001) Development and validation of an improved oxygen radical absorbance capacity assay using fluorescein as the fluorescent probe. *J Agricul Food Chem* 49: 4619–4626.
23. Wolfe KL, Liu RH (2008) Structure-activity relationships of flavonoids in the cellular antioxidant activity assay. *J Agr Food Chem* 56: 8404–8411.
24. Adom KK, Liu RH (2005) Rapid peroxyl radical scavenging capacity (PSC) assay for assessing both hydrophilic and lipophilic antioxidants. *J Agr Food Chem* 53: 6572–6580.
25. Shin Y, Ryu JA, Liu RH, Nock JF, Watkins CB (2008) Harvest maturity, storage temperature and relative humidity affect fruit quality, antioxidant contents and activity, and inhibition of cell proliferation of strawberry fruit. *Postharvest BiolTechnol* 49: 201–209.
26. Liu RH, Jacob JR, Tennant BC, Hotchkiss JH (1992) Nitrite and nitrosamine synthesis by hepatocytes isolated from normal woodchucks (*Marmotamonax*) and woodchucks chronically infected with woodchuck hepatitis virus. *Cancer Res* 52: 4139–4143.
27. Liu RH, Jacob JR, Hotchkiss JH, Cote PJ, GerinJL, et al (1994) Woodchuck hepatitis virus surface antigen induces nitric oxide synthesis in hepatocytes: possible role in hepatocarcinogenesis. *Carcinogenesis* 15: 2875–2877.
28. Yoon H, Liu RH (2008) Effect of 2 alpha-hydroxyursolic acid on NF-kappa B activation induced by TNF-alpha in human breast cancer MCF-7 cells. *J Agr Food Chem* 56: 8412–8417.
29. Wolfe KL, Liu RH (2007) Cellular antioxidant activity (CAA) assay for assessing antioxidants, foods, and dietary supplements. *J Agr Food Chem* 55: 8896–8907.
30. Wolfe KL, Kang XM, He XJ, Dong M, Zhang QY, et al. (2008) Cellular antioxidant activity of common fruits. *J Agr Food Chem* 56: 8418–8426.
31. Felice DL, Sun J, Liu RH (2009) A modified methylene blue assay for accurate cell counting. *J Funct Foods* 1: 109–118.
32. Yoon H, Liu RH (2007) Effect of selected phytochemicals and apple extracts on NF-kappa B activation in human breast cancer MCF-7 cells. *J Agr Food Chem* 55: 3167–3173.
33. Liu RH (2004) Potential synergy of phytochemicals in cancer prevention: Mechanism of action. *J Nutr* 134: 3479S–3485S.
34. Liu RH (2003) Health benefits of fruit and vegetables are from additive and synergistic combinations of phytochemicals. *Am J ClinNutr* 78: 517S–520S.
35. Jia ZS, Tang MC, WU JM (1999) The determination of flavonoid contents in mulberry and their scavenging effects on superoxide radicals. *Food Chem* 64: 555–559.

36. Arnous A, Makris DP, Kefalas P (2002) Correlation of pigment and flavanol content with antioxidant properties in selected aged regional wines from Greece. *J Food Comp Analysis* 15: 655–665.
37. Kallithraka S, Aliaj L, Makris DP, Kefalas P (2009) Anthocyanin profiles of major red grape (*Vitisvinifera* L.) varieties cultivated in Greece and their relationship with in vitro antioxidant characteristics. *Inter J Food SciTechnol* 44: 2385–2393.
38. He X, Liu RH (2008) Phytochemicals of apple peels: isolation, structure elucidation, and their antiproliferative and antioxidant activities. *J Agr Food Chem* 56: 9905–9910.
39. Chu YF, Sun J, Wu XZ, Liu RH (2002) Antioxidant and anti proliferative activities of common vegetables. *J Agr Food Chem* 50: 6910–6916.
40. Eberhardt MV, Kobira K, Keck AS, Juvik JA, Jeffery EH (2005) Correlation analyses of phytochemical composition, chemical, and cellular measures of antioxidant activity of broccoli (*Brassica oleracea* L. var. italica). *J Agr Food Chem* 53: 7421–7431.

CHAPTER 14

Use of Nondestructive Devices to Support Pre- and Postharvest Fruit Management

Guglielmo Costa*, Lorenzo Rocchi, Brian Farneti, Nicola Busatto, Francesco Spinelli and Serena Vidoni

Department of Agricultural Science, Alma Mater Studiorum, University of Bologna, Bologna 40127, Italy

ABSTRACT

Fruit quality is greatly affected by the ripening stage at harvest. In order to preserve quality traits, increase product marketability, and extend both the storage time and the shelf life, it is crucial to tailor postharvest strategies to ripening and physiological stages, and these need to be determined precisely. Surveying instruments working with modern technologies such as visible spectrometry can be precise and effective in assessing ripening stage and in grouping fruit in homogeneous classes. This paper reviews results using original nondestructive devices developed at the University of Bologna to define the fruit ripening of several fruit species through a new index (Index of Absorbance Difference (I_{AD})) to compare relationships among fruit ripening stage, fruit quality, and postharvest life. The devices defining the I_{AD} can be used in the field (by the Difference Absorbance (DA)-Meter), at the packinghouse (with the DA-head, a stationary device), and at the cold storage level (with the DA Fruit Logger (DAFL)).

Keywords: fruit ripening; index of absorbance difference (I_{AD}); DA-Meter; DA-Head; DAFL

1. INTRODUCTION

Fruit quality is a combination of several features that depend on the stage of maturity reached at harvest, postharvest fruit management and disease susceptibility, and storage duration, each influencing general consumer appreciation. In fact, it has been demonstrated that, when fruit are harvested before they reach the proper ripening stage on the tree, fruit quality is poor, as they do not reach their characteristic aroma and flavor, causing, in some situations, disaffection of the consumers for the fruit that do not fulfill their expectations. In this context, the proper choice of the best harvest time plays a crucial role, representing a compromise between the achievement of sufficient quality and the mechanical resistance to harvest and postharvest practices.

Unfortunately, the most common techniques used to determine the ripening stage are often based on destructive methods (°Brix, texture analysis, starch degradation evolution, etc.), and thus they are not applicable to the entire harvest, but only to a small subset of fruits used as a statistical sample. Because the ripening dynamics on-tree can significantly diverge based on the fruit position within the canopy or on the position of the tree in the orchard, even in case of multiple harvest dates performed over a wide harvest window, the fruit batch should be considered heterogeneous [1,2,3,4,5]. A deeper and more precise knowledge of the ripening degree, obtained using a strict protocol and proper devices, could help by segregating the fruit into more uniform groups that each could then be managed with the best storage and marketing strategies. In addition, the possibility of monitoring fruit ripening changes occurring during cold storage could provide critical information for deciding the post-storage temperature that needs to be re-established to guarantee the most appropriate marketing strategy [6]. As a consequence, the ripening stage of fruit at harvest has to be determined with great accuracy following a strict protocol, using proper techniques and devices. However, although fruit quality has always been recognized as a crucial aspect [7,8], only a few quality traits and characteristics are commonly determined at harvest by simple and outdated analyses.

To overcome the problems arising from an imprecise determination of fruit quality, the University of Bologna has focused its efforts in recent years on developing nondestructive devices based on visible/near infrared (VIS/NIR) properties, allowing the definition of fruit ripening and fruit quality attributes by assessing the level of chlorophyll degradation, defined as Index of Absorbance Difference (I_{AD}), and providing a general indication about the fruit ripening stage. The devices have been tested at field and packinghouse levels and in the cold storage room to monitor fruit maturation and ripening evolution from the "field to the fork" level.

In this review, the main results obtained with these nondestructive devices are reported, focusing on the most significant results achieved with different fruit species as far as the relationships among fruit ripening definition, fruit quality enhancement, and disease susceptibility are concerned.

2. THE VIS/NIR DEVICES USED IN THE RESEARCH

The 3 different devices developed for defining fruit ripening as I_{AD} can be used along the supply chain at different levels: (a) in the field with the DA-Meter; (b) in the packinghouse with the DA-Head, a stationary device; and (c) in cold storage with the Difference Absorbance Fruit Logger (DAFL).

2.1. The DA-Meter

The University of Bologna developed and patented three different devices, the DA-Meter, Kiwi-Meter and Cherry-Meter, that have been tested on different fruit species. All three meters do not require any complex calibration and can be used along the productive chain [9]. The first device patented was the DA-Meter (University of Bologna patent n° MO 2005000211) (Figure 1A). The DA-Meter is a portable, user-friendly device for measuring the I_{AD}. The DA-Meter has been tested on several pome and stone fruit cultivars. The I_{AD} is calculated as the difference between the absorbance values at 670 nm and 720 nm, near the chlorophyll *a* absorbance peak.

The Kiwi-Meter (University of Bologna patent n° PD2009A000081) has been tested on *Actinidia deliciosa* and *Actinidia chinensis* kiwi fruit varieties. The device was developed specifically for kiwifruit, and it differs from the DA-Meter by the wavelengths used: 540 and 640 nm with 800 nm as a reference point. The difference between absorbance at 540 nm versus 800 nm is used for the *Actinidia deliciosa* fruit, and the difference between 640 nm versus 800 nm is used for the *Actinidia chinensis* fruit [10]. In *Actinidia deliciosa*, a higher I_{AD} implies more ripe fruits with a brilliant green colored flesh, while with *Actinidia chinensis*, I_{AD} gradually decreases, staying in a constant range for some time followed by a rapid decrease with the onset of the flesh color break from green to yellow.

Figure 1. The three devices assessing fruit ripening (I_{AD}): (**A**) the DA-Meter at field level; (**B**) the DA-Head at the packinghouse level; and (**C**) the DAFL in cold storage.

The Cherry-Meter is used on cherry (*Prunus avium*, *Prunus cerasus*) fruit, allowing the determination of I_{AD} with a value range from 0 to 2.6 positively correlated with ripening. The device is made up of six diode LEDs positioned around a photodiode detector. A set of two diode LEDs emit at 560 nm, 640 nm, and 750 nm. The fruit is illuminated alternately with the two monochromatic sources, and for each the amount of light re-emitted by the fruit is measured. Light is detected by a photodiode positioned centrally to a diode crown, and is converted to a digital signal through an analogue to digital converter. I_{AD} is obtained from the differences of the absorbance between two wavelengths for anthocyanin (560 and 640 nm) and the reference value at 750 nm [3].

2.2. The DA-Head

The DA-Head (Figure 1B) is a stationary device that measures the ripening stage as I_{AD} in a similar way to the DA-Meter. This device has been tested on several fruit species (kiwifruit, apricot (*Prunus armeniaca*), pear (*Pyrus communis*), apple (*Malus × domestica*), and peach (*Prunus persica*) [5,10,11], although the first and more complete trials have been carried out with the apricot cultivar "La vallée", with the collaboration of the Office Cantonal de Arboriculture of Sion, Switzerland. The device can be considered a pre-commercial prototype [12]. The DA-Head, besides grading fruit on the basis of their size, contemporaneously groups the fruit according to their ripening stage using 2 or 3 remote sensors (DA-heads) which are able to read the I_{AD} of fruit passing on a moving belt. The DA-Heads can be adapted to standard commercial grading machines and operate and select fruit at the same speed of the sorting machine itself, approximately 15 fruit/s.

From the practical point of view, it is a very important achievement that the I_{AD} groups the fruit in homogeneous classes according to their ripening stage, starting in field conditions with the portable DA-Meter and at the packinghouse with the DA-Head.

2.3. The Difference Absorbance Fruit Logger (DAFL)

The last device along the productive chain is the DAFL (Figure 1C). It is a small device positioned over a batch of fruit stored in a cold room that continuously monitors the I_{AD} and the temperature data of each fruit remotely at programmed times (minutes, hours, days, etc.). The data are transmitted via radio signal to a receiver unit connected to a computer, and the information is available through Internet connection [6].

3. MAJOR ACCOMPLISHMENTS WITH THE USE OF THE DEVICES

3.1. Assessing Proper Harvest Time

The portable DA-Meters (DA-Meter, Kiwi-Meter and Cherry-Meter) are mainly used to monitor the ripening process of fruit still attached to the tree and to assess the proper time to harvest. In particular, in some fruit species (i.e., peach and nectarine) [11,13], the ripening stage can be precisely defined since the I_{AD} and chlorophyll *a* amount in the outer mesocarp are highly correlated. In addition, in stone fruit, and in particular in peach, the I_{AD} correlates with ethylene evolution, the ripening hormone responsible for flesh softening, color development, and sugar accumulation [14,15]. It is worth noting that a trial carried out for several years with different cultivars of peach and nectarine showed that each cultivar has a typical I_{AD} value that remains constant year after year, coincident with the highest ethylene production peak. In addition, the I_{AD} value is strongly correlated with the transcript levels of ripening related genes [13,16]. As a consequence, the determination of the I_{AD} in nectarines and peach can precisely predict the optimal harvest time [17]. The ripening stage assessed at harvest by the I_{AD} was also recognized and appreciated by consumers who were also able to distinguish fruit of different ripening stages. In a consumer test trial, the most desirable "Stark Red Gold" nectarine fruit were those that reached an I_{AD} value of 0.4–0.5 [18], while "Plus Plus" peach fruit had to reach an I_{AD} value <0.7 to be as desirable (Table 1).

Table 1. "Plus Plus" peach fruit segregated with the DA-stationary machine into 4 ripening classes, the % of the fruit in each class, and consumer likelihood of purchasing each class.

Class	I_{AD}	Fruits (%)	Likelihood of Purchasing (%)		
			Yes	Maybe	No
0	0–0.4	5.5	34.0	57.4	19.4
1	0.4–0.7	22.2	31.9	59.5	23.4
2	0.7–1	61.1	19.1	46.8	23.4
3	>1	11.2	8.5	40.4	21.3

For pome fruit, the I_{AD} was also used to define the proper harvest time of several apple varieties, including "Gala", "Golden Delicious", "Red Delicious", "Fuji", "Granny Smith", and "Pink Lady" [18,19,20] and of "Abbé Fétel"pear [5]. A consumer test trial performed with "Gala" apples indicated that the most desirable fruit were harvested at an I_{AD} value of 0.6–0.9 [19]. Using "Abbé Fètel" pear, a consumer test trial performed with more than 100 people at the "AGER-Innovapero" Conference (Ferrara, Italy, 18 October 2013) [21] at the end of the storage period showed that the most desirable fruit were those harvested when the I_{AD} had reached value of 1.8–1.9, but when the I_{AD} value was >2 the fruit were not as desirable to consumers.

For cherry, the ripening stage expressed as I_{AD} correlates with anthocyanin content (Figure 2) as well as with other traits normally used to assess fruit quality [3].

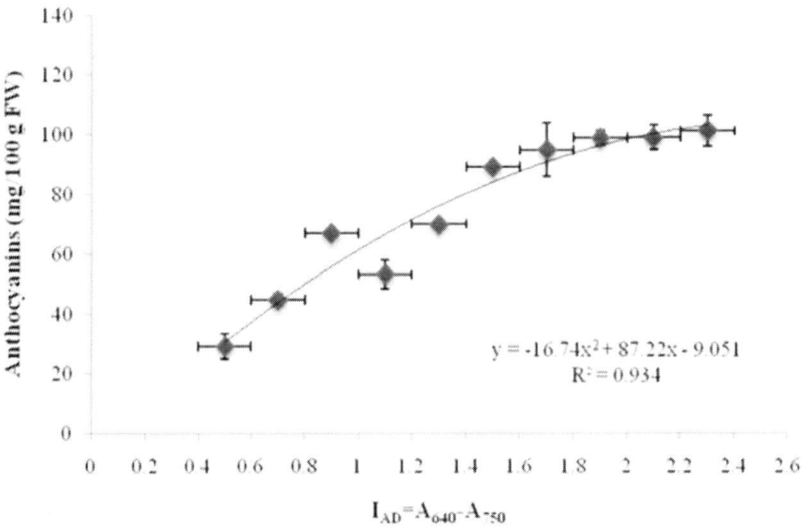

Figure 2. Correlation of skin anthocyanin content and I_{AD} classes at harvest of "Lala Star" cherry.

With *Actinidia deliciosa* kiwifruit, the robustness of I_{AD} use is represented by the fact that the I_{AD} value significantly correlates with the standard fruit quality trait values (Figure 2) normally used for assessing fruit maturity [10,22]. "Hayward" kiwifruit must be harvested when the soluble solids concentration (SSC) reaches a minimum value of 6.2 °Brix [23]. In some seasons, ripening of *Actinidia deliciosa* fruit, expressed as SSC values, evolves slowly in the last period of fruit growth, complicating the definition of the evolution of fruit ripening. When the Kiwi-meter is used in such circumstances, the I_{AD} might not be fully able to detect small changes in ripening, and readings must be taken with a high accuracy to have robust and suitable data. In contrast, with *Actinidia chinensis* fruit, harvest must be performed when the flesh color changes from green to yellow, a change normally occurring at a hue angle of 103°–105° determined with a colorimeter (Figure 3) [10,24]. However, this determination requires fruit destruction and, as a consequence, can be only performed on a limited number of fruit. The Kiwi-Meter allows a much larger fruit sample, and the I_{AD} results are significantly correlated with hue angle (R^2 = 0.819). As a result the ripening stage expressed as I_{AD} is more reliable and repeatable [12,23] on most of the yellow-fleshed cultivars. In addition, the I_{AD} robustness is also underlined by the fact that the value is constant across years as shown in pome and stone fruit [12,13,25].

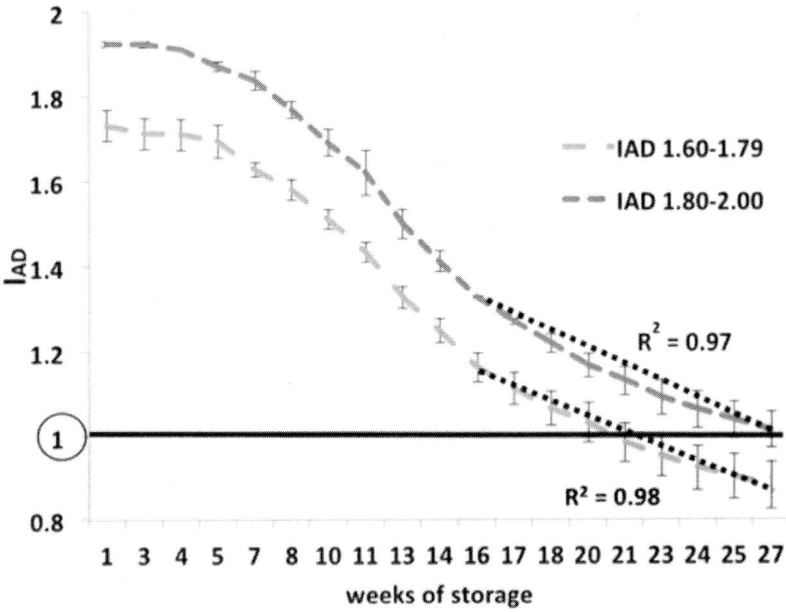

Figure 3. Ripening trends expressed as I_{AD} class during storage of "Abbé Fétel" pear fruit. An I_{AD} value = 1 shown in the figure represents the beginning of skin color change from green to yellow. In the last phase of storage, the I_{AD} change tends to be linear.

3.2. Grouping Fruit in Ripening Homogeneous Classes

The possibility of grouping fruit in homogeneous classes according to their ripening stage is an important use for the I_{AD}, especially for fruit undergoing longer storage periods. Such information can be easily obtained by the I_{AD} value determined at harvest with the DA-Meter on a limited number of fruit, or at the packinghouse with the DA-Head, so that all of the harvested fruit can be grouped into homogeneously-ripening classes. Such trials have been carried out on peach, apricot, apple, pear and kiwifruit. For instance, the DA stationary machine was able to detect within commercial classes of fruit differences that were not previously noticed using standard methods of evaluation, as in the case of classes of "Plus Plus" peaches that were further grouped into 4 subclasses of fruit according to the more precise I_{AD} ripening measure.

Moreover, the differences detected by these new devices, and represented by the I_{AD} index, are economically useful since the consumers are able to recognize the differences in ripening determined by the DA-Meter and appreciate more ripened fruit (Table 1). The same information was also obtained with several cultivars of apricot where the fruit were grouped into classes of uniform ripening with the DA-head device. Notably, consumer appreciation was related to fruit quality/ripening and but was influenced very little by appearance for "Swigold" apricot. In fact, although fruit ripening was quite different among the classes (4 classes characterized by the following I_{AD} values, 0–0.5; 0.5–0.7; 0.7–0.9, and 0.9–2.0, and by the soluble solids content, 14.9; 14.2; 13.2 and 12.5 °Brix, respectively, fruit appearance (skin color and fruit size) in the 4 classes was practically identical (Figure 4) [12], demonstrating that the traditional methodologies, such as colorimeter or caliper measurements, are not suitable for effectively determining fruit physiological stage.

Figure 4. "Swigold" apricot grouped by the DA-head into 4 classes of ripening.

3.3. Fruit Ripening Monitoring in Cold Storage

The Difference Absorbance Fruit Logger (DAFL) (Figure 1C) monitors ripening stage of the fruit maintained in cold room during storage. Trials were carried out with apple, pear and kiwifruit and indicated that fruit collected at higher I_{AD} values at harvest maintained higher values for the entire storage period as compared to fruit harvested at a lower I_{AD}. In pear, the I_{AD} value precisely followed ripening evolution in a sigmoid pattern (Figure 3). In the last phase of storage, the I_{AD} value was linear for both lower and higher initial I_{AD} values, allowing a tentative prediction of the span of time needed to reach the optimal I_{AD} value before removing fruit from cold storage (arbitrarily defined as 1) and driving marketing decisions [6].

3.4. Reducing Postharvest Losses

Several studies have pointed out that fruit disease susceptibility is related to the ripening stage reached by the fruit at harvest [26,27,28]. In apple, superficial scald susceptibility trials were performed with 3 varieties, "Red Delicious", "Granny Smith", and "Pink Lady", and revealed that when the harvest was performed too early, the fruit were significantly affected by scald. For the susceptible variety "Granny Smith", fruit were harvested at an I_{AD} value of 1.8–2.0 (early ripening, fruit not fully ripe). Scald incidence reached 100% after just two months of storage. In contrast, when fruit were harvested later at a more advanced ripening stage (I_{AD} of 1.6–1.8), fruit were not affected by superficial scald within the first two months of storage (Figure 5) [20].

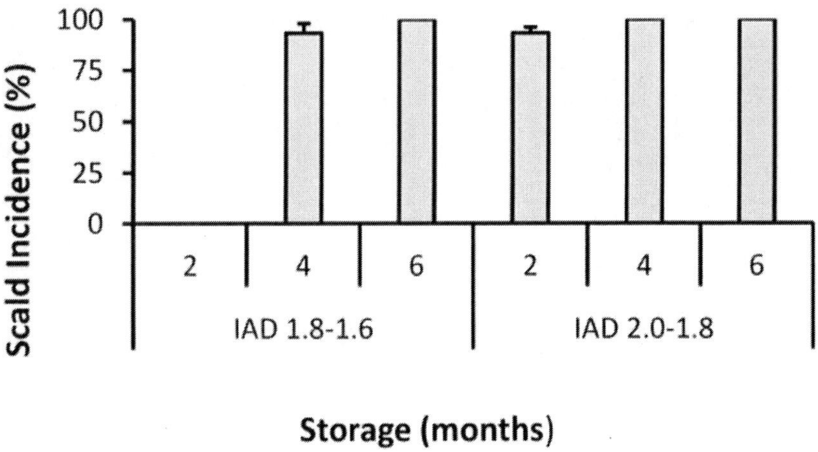

Figure 5. Scald incidence (%) in Granny Smith apples harvested at two different ripening stages determined by I_{AD}. The scald incidence was evaluated on fruit left at room at temperature for 1 week after 2, 4 and 6 months of commercial storage.

Also, in peach, the appearance of brown rot (*Monilia fruticola*) was related to the ripening stage. The damage caused by the fungus was more severe on fruit characterized by an advanced ripening stage [29].

4. FINAL REMARKS

The results of research carried out over the last few years on some of the main temperate fruit species indicated that the I_{AD} represents a powerful and reliable tool for assessing fruit ripening. The I_{AD} devices allow monitoring the evolution of ripening from the field until removal from cold storage. Fruit ripening measured in the field with the DA-Meter establishes the proper harvest time and gives preliminary information about the homogeneity of fruit ripening. Furthermore, at the packinghouse, use of the DA-Head allows grouping of the fruit into homogeneously-ripening classes driving decisions on the best storage strategies and subsequent marketing. Finally, in cold storage, the DAFL allows continuous monitoring of the evolution of fruit ripening.

The I_{AD} can find useful applications in the fruit production chain and represents an essential decision support system tool that can drive pre- and postharvest management of ripening fruit.

ACKNOWLEDGMENTS

Research funded by AGER (Project "Agroalimentare e Ricerca"), grant No. 2010–2119.

REFERENCES

1. Smith, G.S.; Gravett, I.M.; Edwards, C.M.; Curtis, J.P.; Buwalda, J.G. Spatial analysis of the canopy of kiwifruit vines as it relates to the physical, chemical and postharvest attributes of the fruit. *Ann. Bot.* **1994**, *73*, 99–111.
2. Dallabetta, N.; Costa, F.; Pasqualini, J.; Wehrens, R.; Noferini, M.; Costa, G. The influence of training system on apple fruit quality. *Acta Hortic.* **2013**, *1058*, 55–62.
3. Nagpala, E.G.L.; Noferini, M.; Farneti, B.; Piccinini, L.; Costa, G. Cherry-Meter: An innovative nondestructive (vis/NIR) device for cherry fruit ripening and quality assessment. In Proceedings of the 7th ISHS Internation Symposium on Cherry, Plasencia, Spain, 23–27 June 2013; ISHS: Leuven, Belgium in press.
4. Bonora, E.; Noferini, M.; Vidoni, S.; Costa, G. Modeling fruit ripening for improving peach homogeneity in planta. *Sci. Hortic.* **2013**, *159*, 166–171.
5. Vidoni, S.; Rocchi, L.; Donati, I.; Spinelli, F.; Costa, G. Combined use of Planttoon® and IAD to characterize fruit ripening homogeneity in "Abbé Fétel" pears. *Acta Hortic.* **2015**.
6. Vidoni, S.; Fiori, G.; Rocchi, L.; Spinelli, F.; Musacchi, S.; Costa, G. DAFL: New innovative device to monitor fruit ripening in storage. *Acta Hortic.* **2015**, *1094*, 549–554.
7. Harman, J.E. Kiwifruit maturity. *Orchard. N. Z.* **1981**, *54*, 126–127.
8. Shewfelt, R.L. What is quality? *Postharvest Biol. Technol.* **1998**, *15*, 197–200.
9. Costa, G.; Fiori, G.; Torrigiani, P.; Noferini, M. Use of Vis/NIR spectroscopy to assess fruit ripening stage and improve management in post-harvest chain. In *New Trends in Postharvest Management of Fresh Produce I*, Fresh Produce 3 (Special Issue 1) ed.; Sivakumar, D., Ed.; Global Science Book: Ikenobe, Japan, 2009; pp. 1–6.
10. Costa, G.; Bonora, E.; Fiori, G.; Noferini, M. Innovative nondestructive device for fruit quality assessment. *Acta Hortic.* **2011**, *913*, 575–581.

11. Costa, G.; Bonora, E.; Noferini, M. Modeling system and vis-NIR devices to improve pre- and post-harvest management. *Acta Hortic.* **2015**, *1084*, 809–821.
12. Costa, G.; Fiori, G.; Rocchi, L.; Vidoni, S.; Berthod, N.; Besse, S.; Knieling, S.; Rossier, J. A prototype of a stationary DA-device to group the apricots in classes of homogeneous ripening. In Proceedings of the ISHS XVI International Symposium on Apricot Breeding and Culture, Shenyang, China, 29 June–3 July 2015.
13. Ziosi, V.; Noferini, M.; Fiori, G.; Tadiello, A.; Trainotti, L.; Casadoro, G.; Costa, G. A new index based on Vis spectroscopy to characterize the progression of ripening in peach fruit. *Postharvest Biol. Technol.* **2008**, *49*, 319–329.
14. Ruperti, B.; Cattivelli, L.; Pagni, S.; Ramina, A. Ethylene-responsive genes are differentially regulated during abscission, organ senescence and wounding in peach (*Prunuspersica*). *J. Exp. Bot.* **2002**, *53*, 429–437.
15. Trainotti, L.; Zanin, D.; Casadoro, G. A cell-oriented genomic approach reveals a new and unexpected complexity of the softening in peaches. *J. Exp. Bot.* **2003**, *54*, 1821–1832.
16. Tadiello, A.; Ziosi, V.; Negri, A.S.; Noferini, M.; Fiori, G.; Busatto, N.; Espen, L.; Costa, G.; Trainotti, L. On the role of ethylene, auxin and a GOLVEN-like peptide hormone in the regulation of peach ripening. *BMC Plant Biol.* **2016**.
17. Bonora, E.; Noferini, M.; Stefanelli, D.; Costa, G. A new simple modeling approach for the early prediction of harvest date and yield in nectarines. *Sci. Hortic.* **2014**, *172*, 1–9.
18. Costa, G.; Noferini, M. Use of Nondestructive devices as a decision support system for fruit quality enhancement. *Acta Hortic.* **2013**, *998*, 103–115.
19. Costamagna, F.; Giordani, L.; Costa, G.; Noferini, M. Use of AD Index to define harvest time and characterize ripening variability at harvest in "Gala"apple. *Acta Hortic.* **2013**, *998*, 117–123.
20. Farneti, B.; Gutierrez, M.S.; Novak, B.; Busatto, N.; Ravaglia, D.; Spinelli, F.; Costa, G. Use of the index of absorbance difference (IAD) as a tool for tailoring post-harvest 1-MCP application to control apple superficial scald. *Sci. Hortic.***2015**, *16*, 110–116.
21. Costa, G.; Rossi, D.; Tamburini, E.; Donegà, V.; Loberti, R. Innovazioni di processo per unapericoltura di qualità. *L'Inf. Agrar.* **2013**, *36*, 44–46. (In Italian)
22. Pellegrino, S.; Costamagna, F.; Noferini, M.; Costa, G. Monitoring of "Hayward" (*Actinidiadeliciosa*) fruit ripening in North-West Italy. *Acta Hortic.* **2011**, *913*, 665–669.
23. Rocchi, L.; Vidoni, S.; Ceccarelli, A.; Fiori, G.; Costa, G. Use of the DA-index™ for monitoring fruit ripening evolution in *A. chinensis* to precisely assess harvesting time "in planta". *J. Berry Sci.* **2016**.
24. Costa, G.; Vidoni, S.; Rocchi, L.; Cellini, A.; Buriani, G.; Donati, I.; Spinelli, F. Innovative nondestructive device for fruit quality assessment and early disease diagnosis. *Acta Hortic.* **2015**, *1096*, 69–78.
25. Costa, G.; Noferini, M.; Fiori, G.; Ziosi, V.; Berthod, N.; Rossier, J. Establishment of the optimal harvest time in Apricot ("Orangered" and "Bergarouge") by means of a new Index based on Vis spectroscopy. *Acta Hortic.* **2010**, *862*, 533–539.
26. Salzman, R.A.; Tikhonova, I.; Bordelon, B.P.; Hasegawa, P.M.; Bressan, R.A. Coordinate accumulation of antifungal proteins and hexoses constitutes a developmentally controlled defense response during fruit ripening in grape. *Plant Physiol.* **1998**, *117*, 465–472.
27. Guillén, F.; Castillo, S.; Zapata, P.J.; Martínez-Romero, D.; Valero, D.; Serrano, M. Efficacy of 1-MCP treatment in tomato fruit: 2. Effect of cultivar and ripening stage at harvest. *Postharvest Biol. Technol.* **2006**, *42*, 235–242.
28. Cantu, D.; Vicente, A.R.; Greve, L.C.; Dewey, F.M.; Bennett, A.B.; Labavitch, J.M.; Powell, A.L.T. The intersection between cell wall disassembly, ripening, and fruit susceptibility to *Botrytis cinerea*. *Proc. Natl. Acad. Sci. USA* **2008**, *105*, 859–864.

29. Spadoni, A.; Cameldi, I.; Noferini, M.; Bonora, E.; Costa, G.; Mari, M. An innovative use of DA-meter for peach fruit postharvest management. *Sci. Hortic.* **2016**, *201*, 140–144.

CHAPTER 15

The Nematicidal Effect of *Camellia* Seed Cake on Root-Knot Nematode *Meloidogyne javanica* of Banana

Xiujuan Yang[1], Xuan Wang[2], Kang Wang[1], Lanxi Su[1], Hongmei Li[2], Rong Li[1]

[1] *Jiangsu Key Lab and Engineering Center for Solid Organic Waste Utilization, National Engineering Research Center for Organic-based Fertilizers, Jiangsu Collaborative Innovation Center for Solid OrganicWaste Resource Utilization, Nanjing Agricultural University, Nanjing, China,*

[2] *Key Laboratory of Integrated Management of Crop Diseases and Pests, Ministry of Education, Department of Plant Pathology, Nanjing Agricultural*

ABSTRACT

Suppression of root-knot nematodes is crucially important for maintaining the worldwide development of the banana industry. Growing concerns about human and environmental safety have led to the withdrawal of commonly used nematicides and soil fumigants, thus motivating the development of alternative nematode management strategies. In this study, *Meloidogyne javanica* was isolated, and the nematicidal effect of *Camellia* seed cake on this pest was investigated. The results showed that in dish experiments, *Camellia* seed cake extracts under low concentration (2 g/L) showed a strong nematicidal effect. After treatment for 72 h, the eggs of *M. javanica* were gradually dissolved, and the intestine of the juveniles gradually became indistinct. Nematicidal compounds, including saponins identified by HPLC-ESI-MS and 8 types of volatile compounds identified by GC-MS, exhibited effective nematicidal activities, especially 4-methylphenol. The pot experiments demonstrated that the application of *Camellia* seed cake suppressed *M. javanica*, and promoted the banana plant growth. This study explored an effective nematicidal agent for application in soil and revealed its potential mechanism of nematode suppression.

INTRODUCTION

Bananas (*Musa* spp.) are among the most important crops in the world as a staple food, and they are also the main source of income for local farmers in many developing countries [1]. However, the production of bananas is hampered by many diseases and pests [2]. Among the most damaging banana pests are the widespread plant-parasitic nematodes. In particular, the root-knot nematodes, *Meloidogyne* spp., are economically important soil-borne pathogens that are reported to infect almost all of the world's major crop plants [3]. Infection with root-knot nematodes causes root damage that not only leads to severe crop losses in commercial banana plantations for export but also seriously limits the production and viability of other banana types [4]. *M. javanica,* one species of the root-knot nematodes, attacks banana plants at an early stage in the field [5] and induces the root cells into giant cells. Through these giant cells, the root-knot nematode sucks up nutrients and prevents the plant growth, thereby results the symptom of stunting and yellowing. Thus, it is both necessary and urgent to find a better way to suppress this harmful nematode and maintain worldwide development of the banana industry.

Nematode management is traditionally achieved by application of nematicides. However, growing concerns about human and environmental safety have led to the withdrawal of several commonly used nematicides and soil fumigants [6]. In a previous study, 55 banana accessions were evaluated for resistance to nematode species [7], but no source of resistance was found. Soil fumigation as a pre-plant treatment is reported to be effective in suppressing the nematode population in some plants [8]; however, the effect is short-lived compared to the life of a banana crop. Moreover, plant-parasitic nematodes tend to repopulate an area fairly quickly after fumigation, and some fumigants are potential environmental contaminants [9]. These findings motivate the development of alternative nematode management strategies.

Effective control strategies for *M. javanica* may include the use of plant-origin nematicidal agents [10]. *Camellia* seed cake is an organic substance made of the residual from *Camellia oleifera* Abel seeds after oil extraction. The saponin released from this by-product can kill muciferous mollusks

[11] and could therefore be used to prevent the damage caused by root-knot nematodes. Oil cakes in combination with *Bradyrhizobium* sp. and *Paecilomyces lilacinus*have already been studied for the control of mungbean root knot nematode [12]. Tea-oil *Camellia* seed cake extracted with methanol or water was tested for its effects on controlling *Bursaphelenchus xylophilus* [13]. Because of the high nutritional value and essential elements together with mammalians safety, the tea-oil *Camellia* seed cake using as a non-conventional fertilizer is increasingly gaining priority [14–15]. However, to our knowledge, the nematicidal effects of *Camellia* seed cake and its potential mechanisms have not yet been studied. In particular, no reports have described the effects of using *Camellia* seed cake to amend banana-planting soil for the suppression of *M. javanica*.

In this study, the nematicidal ability of *Camellia* seed cake was investigated in detail to explore a green strategy for managing *M. javanica* in banana plantings. In addition, the nematicidal substance extracted from tea-oil *Camellia* seed cake by ultrapure water was identified using HPLC-ESI-MS, and the nematicidal ability of 8 VOCs (volatile organic compounds) detected using GC-MS were confirmed. This paper highlights how tea-oil *Camellia* seed cake can be used to manage *M. javanica* to reduce its harmful effects on banana plants and examines some of the potential mechanisms involved in the control of *M. javanica* by tea-oil *Camellia* seed cake.

MATERIALS AND METHODS

Ethics statement

Our study was carried out on the farmers' land (18°23′ N, 108°44′ E) with property rights in China (1996–2035) and farmer Yusheng Li should be contacted for future permissions. No specific permits were required for the described field studies and the locations are not protected. The field studied did not involve endangered or protected species.

Isolation and identification of *M. javanica*

Individual seedlings of *Musa* AAA *Cavendish* cv. Brazil were taken from the field in Ledong, Hainan province, China. Tweezers were used to collect all visible suspect *Meloidogyne* spp. eggmasses after cleaning the roots with deionized water. Then, all the eggmasses (18 in total) were put into plates (one in each plate, numbered C1 to C18) with 5 mL water and hatched at

25°C for 72 h to obtain second-stage juveniles (J2s). Then, the J2s (10 from each plate) were taken for molecular identification *via* species-specific primer pairs [16]. DNA samples of J2s were prepared according to Li et al. 2008 [17]. Primers Far/Rar, Finc/Rinc, JMV1/JMV2 and Fjav/Rjav used for PCR amplification were synthesized by Invitrogen (Shanghai, China).

The J2s of other suspect *Meloidogyne* spp. from each plate were inoculated onto banana roots to produce disease. Banana seedlings were cultured in a seedling matrix that was sterilized at 121°C for 30 min prior to use so that no nematodes were observed in the plant roots. Pot experiments were carried out in a greenhouse located at Hainan Wan Zhong Co., Ltd., Hainan, China. The temperature and relative humidity were in accordance with the environment, and the season was suitable for banana planting. Each pot containing 500 g sterilized soil and banana roots were inoculated with 100 J2s. Banana plants grown in soil with no inoculation were treated as the control (CK).

After 6 months, disease symptoms of plants were observed, the root knots were sampled, and the nematode eggmasses and females were harvested for identification. Nematodes inside banana roots were stained using the sodium hypochlorite acid fuchsin method [18] and observed using an Olympus ZX10 stereo microscope. The selected females of *Meloidogyne*spp. were determined by observation of perineal patterns using an Olympus BX51 microscope (Japan, CCD DP72). Meanwhile, the J2 from one of *Meloidogyne* spp. in plate C11 was selected for further molecular identification. PCR products were cloned into the pMD19-T vector (TaKaRa) and transformed into competent *Escherichia coli* DH5α cells. The 18S rRNA gene sequence was sequenced, and the molecular phylogenetic analysis was performed using MEGA version 4.0 [17].

Nematicidal effect of Camellia seed cake extracts

Camellia seed cake was soaked in sterile ultrapure water in different concentrations (1, 1.5, 2, 2.5, 3, 4, 5, 6, 7, 8, 9, 10, 15, 20, 30, 50, 100 g/L) for 72 h. After filtration, 10 mL of extracts with different concentrations were transferred to 6 cm petri dishes (300 individual freshly hatched J2s of *M. javanica* per dish) in order to evaluate their nematicidal effects. An equal volume of deionized water was used as a control. The dishes were allowed to incubate for 36 h, and then the nematodes were observed with an Olympus ZX10 stereo microscope in order to calculate the corrected mortality. Nematodes were considered alive if they moved or appeared as a winding shape and were considered dead if they did not move when probed with a fine needle [19]. Then, the nematodes in each treatment were transferred to distilled water for 48 h to ascertain whether the dead nematodes regained mobility or not. The corrected mortality was calculated according to the following formula: mortality (%) = (mortality of treatment-mortality of CK) / (1-mortality of CK) × 100. Morphological variations of *M. javanica* juveniles during 72 h treatment with 5 g/L extract were observed using an Olympus BX51 microscope.

To evaluate the effects of *Camellia* seed cake extract on the relative juveniles hatching ratio, eggmasses contained total about 500 eggs were inoculated in dishes with 5 g/L, 2 g/L and 1 g/L extract and hatched at 25°C for 120 h. Each extract has 5 replicates. Hatching J2s were observed with an Olympus ZX10 stereo microscope. The hatching ratio was calculated at 24 h intervals as (%) = (incubation of treatment-incubation of CK) / (1-incubation of CK) × 100. Morphological variations of eggs treated with 5 g/L extract for 72 h were observed using an Olympus BX51 microscope.

Nematicidal effect of the extracted camellia saponin

The camellia saponin was extracted according to the method of Zhong et al. [20]. Briefly, 13.25 g of *Camellia* seed cake was ground, wrapped in a filter paper package and placed in a drying oven maintained at 70°C for 2 h to remove water. Then, extraction proceeded using 200 mL methanol (HPLC grade) in a soxhlet extractor. Reflux extraction was performed at 5 min per cycle for 4 h. The extract was then evaporated under reduced pressure using a centrifugal evaporator at room temperature. Residual organic material was re-dissolved in a total of 2 mL of methanol. The re-dissolved extract was diluted to different concentrations (6, 7, 8, 9, 10, 13, 15, 20, 30, 50, 70, 100 g/L). Effects of the extracts on the *M. javanica* J2s were tested as previously described.

The molecular weight of the saponin was detected by HPLC-ESI-MS (HPLC: 1200 series; ESI-MS: 6410 Triple Quad LC/MS, Agilent, USA) using a C18 column (250 mm × 4.6 mm, 5 μm) as a chromatographic column at a flow-rate of 0.5 mL/min. The column temperature was 20°C, injection volume was 10 μL and wave length for all detections was 280 nm. Mobile phase A was water with 0.1% acetic acid, and mobile phase B was acetonitrile. An elution gradient from 20% to 15% mobile phase A volume fraction, over a gradient time of 0 to 20 min, was used. MS analysis was performed by electrospray ionization in the positive ion mode.

Nematicidal effect of the volatile compounds of Camellia seed cake

The split plates (Petri dish with vents, Greiner company, Germany) were used to test the nematicidal activity of the volatiles produced by the *Camellia* seed cake. One side of the split plate was tiled with 5 mL sterile ultrapure water with different concentrations of *Camellia* seed cake (10, 12, 15, 20, 30, 50, 100, 150, 200 g/L), and the other side held 5 mL of ultrapure water with 300 individuals of freshly hatched juveniles of *M. javanica* (S1 Fig.). An equal volume of ultrapure water was used as a control. Corrected mortality was determined after the plates stood for 72 h.

The volatiles produced by the *Camellia* seed cake extract were collected using the solid-phase micro-extraction (SPME) technique [21], which detects volatiles produced at the headspace of the serum bottle. The SPME syringe, equipped with fiber material (50/30 DVB/Carboxen on PDMS stable flex fiber) [22], was then inserted into the center of the parafilm covering the serum bottle. Fiber material was exposed to the volatiles in the headspace to entrap the volatiles. Three days later, the syringe containing the volatiles was inserted into a Gas Chromatography-Mass Spectrometer (GC-MS). GC-MS analysis was performed in electron ionization (EI) mode (70 eV) using a Finnigan gas chromatograph equipped with an MS detector. A Finnigan capillary column (15 m length × 0.5 mm id × 0.25 mm film thickness) was first used, with the following temperature program: the temperature was held at 35°C for 3 min, then increased to 180°C at 10°C/min; this temperature was held for 1 min, and finally increased to 240°C at 40°C min^{-1} and held at 240°C for 10 min. Helium was used as the carrier gas at a constant flow rate of 1.0 mL/min. The samples were analyzed in split mode (1: 20) with an injection and EI source temperature of 220°C and then scanned in the mass range from 30 m/z to 650 m/z. The compounds produced by *Camellia* seed cake were identified using the NIST (National Institute of Standards and Technology) database on the mass spectrometer.

The nematicidal activity of the various VOCs identified by GC-MS was examined using pure standard substances and split plates. All the selected substances were dissolved with ethanol and later diluted with the sterile ultrapure water. The concentration of diluted solvents was 3000, 300, 30 and 3 mg/L, respectively. Nematicidal activity was tested by placing sterile filter paper discs with 100 μL of diluted solvent in one side of the split plates and 5 mL water with 300 freshly hatched J2s was in another side. The 100 μL of water diluted ethanol was as the control. The corrected mortality of each treatment was determined after standing for 72 h.

Pot experiments

Pot experiments were performed in a greenhouse located at Hainan Wan Zhong Co., Ltd., Hainan, China, during three banana growing seasons from May to July 2012, September to October 2012 and May to July 2013. Banana seedlings (*Musa* AAA *Cavendish* cv. Brazil) without nematodes in their roots were used in this experiment. The soil for the pot experiments was collected from a field with serious root-knot nematode-wilt disease in Hainan province, China. The soil had a pH value of 6.52, an organic matter content of 10.12 g/kg, and available N, P, K contents of 38.39, 202.19, and 174.22 mg/kg, respectively.

In each season of pot experiments, four treatments were designed as follows. In the control, CK, no *Camellia* seed cake was added to the soil. In treatment A, 5 g/kg of *Camellia* seed cake was added to the soil. In treatment B, 2 g/kg of *Camellia* seed cake was added to the soil. In treatment C, 1 g/kg of *Camellia* seed cake was added to the soil. Each treatment was supplemented with identical nutrient content. Each treatment had three blocks; each block contained nine pots.

In the first-season of pot experiments, the soil for all pots (3 kg soil for each pot) was first sterilized. Then, juveniles of *M. javanica* were added to each pot (600 individuals per 1 kg soil). Then, *Camellia* seed cake was added to the pot. Seven days later, one banana seedling free of nematodes was transplanted to the pot. All pots were kept in greenhouse under room temperature for 60 days and the seedlings were removed for the measurement of agronomic characteristics and biomasses. The number of eggmasses on roots was also calculated.

In the second- and third-season of pot experiments, the crude soil with nematodes was put into pot and *Camellia* seed cake was added. Again seven days later, one banana seedling free of nematodes was transplanted to the pot. The conditions for keeping seedlings were the same as described above. Sixty days later, the seedlings were removed from the pots and the agronomic characteristics and biomasses were measured. The roots of each seedling were macerated in a blender and root-knot nematodes were extracted by the Baermann funnel and calculated using an Olympus ZX10 stereo microscope. The total nematodes in soil of each pot were extracted by the Baermann funnel and calculated. The nematodes species were determined using morphological and feeding-habit based classifications. The number of culturable microorganisms including bacteria, fungi and actinomycetes in soil of each pot was determined [23].

Statistical analysis

Differences among treatments were assessed using a one-way ANOVA analysis. All means were calculated from the values of five replicates and subjected to Duncan's multiple range tests at $P = 0.05$ and the statistical analysis was carried out using software SPSS version 17.0 (SPSS Inc., Chicago, IL). Different letters indicated in figures means the statistically significant differences at the 0.05 probability level according to the Duncan test.

RESULTS

Isolation and identification of *M. javanica*

To isolate *M. javanica*, 18 individual suspect eggmasses with or without females were first isolated and numbered C1 to C18. All the eggmasses were then hatched into *Meloidogyne* spp. J2s for molecular identification. No PCR products were amplified from all DNA templates using the primers Far/Rar and JMV1/JMV2 (data no shown). Two bands at approximately 1200 bp were observed using the primer Finc/Rinc for C9 and C10; these nematodes were therefore preliminarily identified as *M. incognita* [16]. Approximately 670 bp band was detected in the rest of other J2s using the primer Fjav/Rjav (Fig. 1A); these nematodes were therefore preliminarily identified as *M. javanica* [16], which confirmed to be the dominant species on banana.

After identification, all the hatched juveniles of *M. javanica* were inoculated into the banana plants for disease production. Six months after transplanting, the plants were dwarfed and the leaves were pale, and significant root knots were observed in the treatment plants. The root knots were sampled, and white females were harvested. Two obvious lateral lines were observed in the perineal patterns (Fig. 1B) of the females, and the hyaline tail of juvenile was shown in Fig. 1C. The disease symptoms increased with increasing numbers of inoculated juveniles, and no root knots were observed in the control. The infection of C11 juvenile to banana root is shown in Fig. 1D. Due to C11 has the highest activities for juveniles hatching and root infection than others, therefore, it was further selected for molecular identification. An approximately 670 bp band from primer Fjav/Rjav was amplified and sequenced. The phylogenetic tree of the sequence showed that nematode C11 was related to the *M. javanica*lineage and closely clustered with similar species (Fig. 1E). Based on the above phenotypic characteristics and phylogenetic analysis, nematode C11 was confirmed to be *M. javanica*. The combination of these results verified that the isolated *M. javanica* C11 is from the previous inoculation; and was selected for further study.

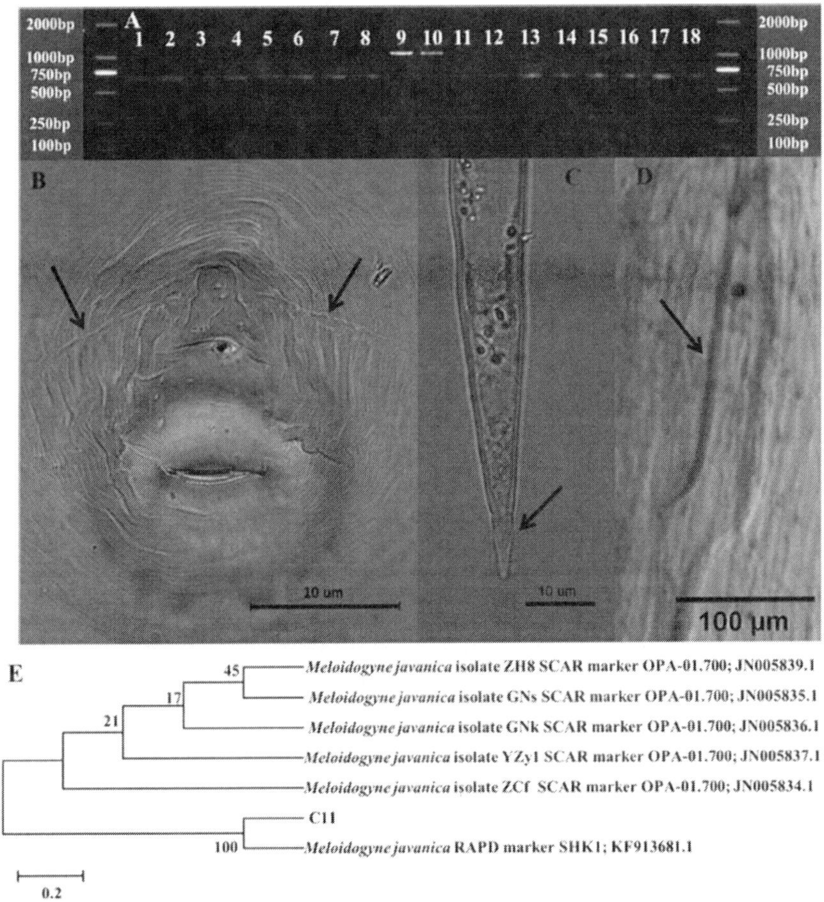

Figure 1. Molecular identification of *M. javanica* detected by primers Fjav/Rjav and *M. incognita* by Finc/Rinc (A); the female perineal pattern of *M. javanica* showing two obvious lateral lines (B); the hyaline tail observed in J2 of *M. javanica* (C); the infection of J2s from C11 of *M. javanica* in banana roots (D); phylogenetic tree of the sequence of nematode C11 (E).

The nematicidal activity of the Camellia seed cake extracts

A juvenile mortality of approximately 99% was observed for 100 g/L to 5 g/L of *Camellia* seed cake extracts. The rate showed a continual decline from 5 g/L to 1 g/L and tended to be lowest for 1 g/L, at approximately 2.4%. Moreover, the nematicidal rate was 57.11% for 2 g/L of *Camellia* seed cake extract (Fig. 2A). Therefore, concentrations of 5 g/L, 2 g/L and 1 g/L were selected for further study.

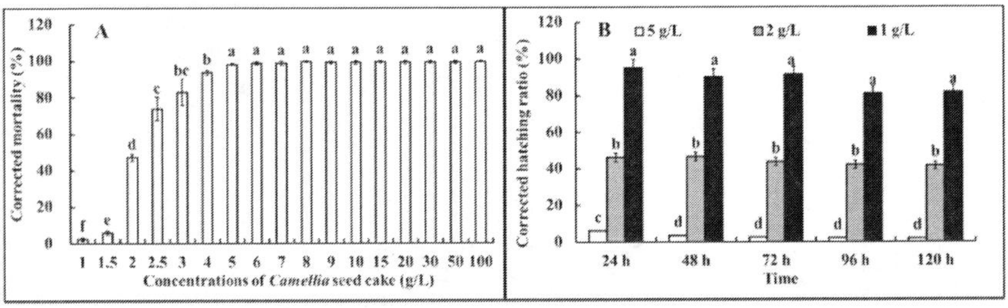

Figure 2. Effects of different concentrations of *Camellia* seed cake extract on the corrected mortality of *M. javanica* J2s (A); and the corrected hatching ratio of eggs (B).

Approximately no juveniles were hatched from 0 to 72 h at 5 g/L. At 2 g/L, the hatching ratio was nearly 50% from 24 to 120 h, and at 1 g/L, no inhibition effect was observed compared to the control (Fig. 2B).

The juvenile had a smooth cuticle and brighter content at 0 h, while the cuticle was crimpled and the intestine was obviously dissolved at 72 h (Fig. 3A). No apparent changes were observed on the anterior part of the body (Fig. 3B) and tail region (Fig. 3C) of J2s. However, the junction region of esophagus (Fig. 3D) and intestine part (Fig. 3E) of J2s was gradually became destroyed as the treatment time increasing. No obvious morphological changes were observed on J2s in water control. The eggs treated with the extracts for 72 h showed the content apparently dissolved and the eggshell layers obviously reduced (Fig. 4A), however, the eggs in water control had a smooth and bright surface and a clear content (Fig. 4B).

Nematicidal effect of camellia saponin and its molecular weight

In Fig. 5A, nearly 100% nematicidal rates were observed from 100 g/L to 50 g/L. From 50 g/L to 6 g/L, the nematicidal rate showed a continual decline to approximately 3.88%.

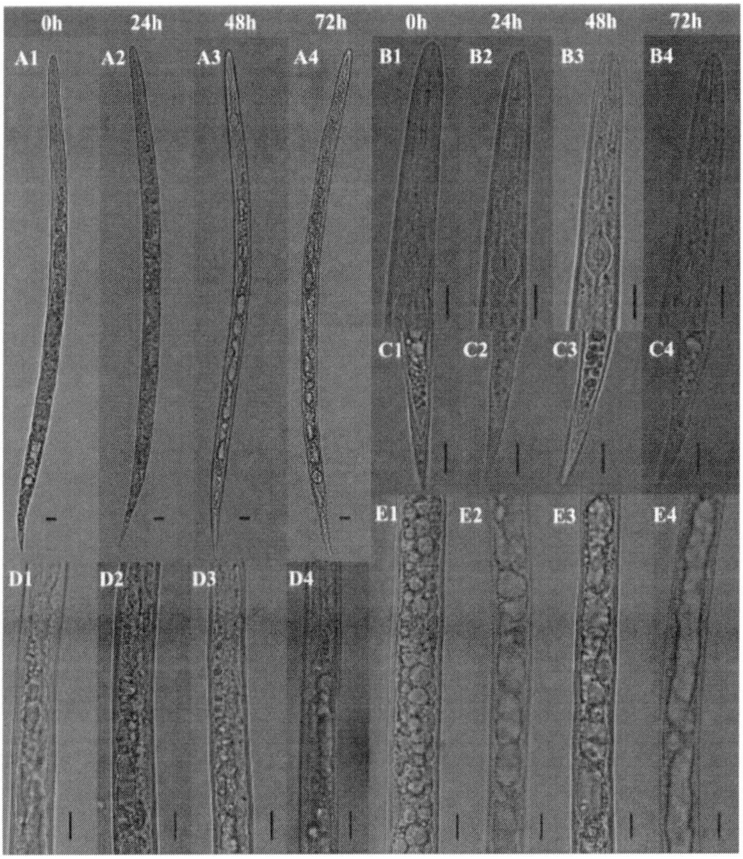

Figure 3. Morphological variations of *M. javanica* J2s (the whole body (A), anterior part of the body (B), tail region (C), junction region of esophagus and intestine (D) and intestine (E)) after treatment with 5 g/L *Camellia* seed cake extract for different times. Bars: 10 μm.

One chromatographic peak with retention time of 2.0 min was obtained after HPLC analysis (Fig. 5B). Mass spectrometry analysis detected three major compounds [Rt (min) = 1.85, 2.0, 2.22] in the total ion chromatogram of the mixture ions peak (Fig. 5C). The first and third compound has the same molecular mass as 1001.6 Da $[M+H]^+$, however, no evidences were found to support them as saponin homologues (S2 and S3 Figs.). The second compound has a molecular mass of 1241.50 Da $[M+H]^+$ (Fig. 5D) in the positive-ion mode, showing that the molecular weight of the compound was 1240.50, which was conjectured as saponin homologues [24–26].

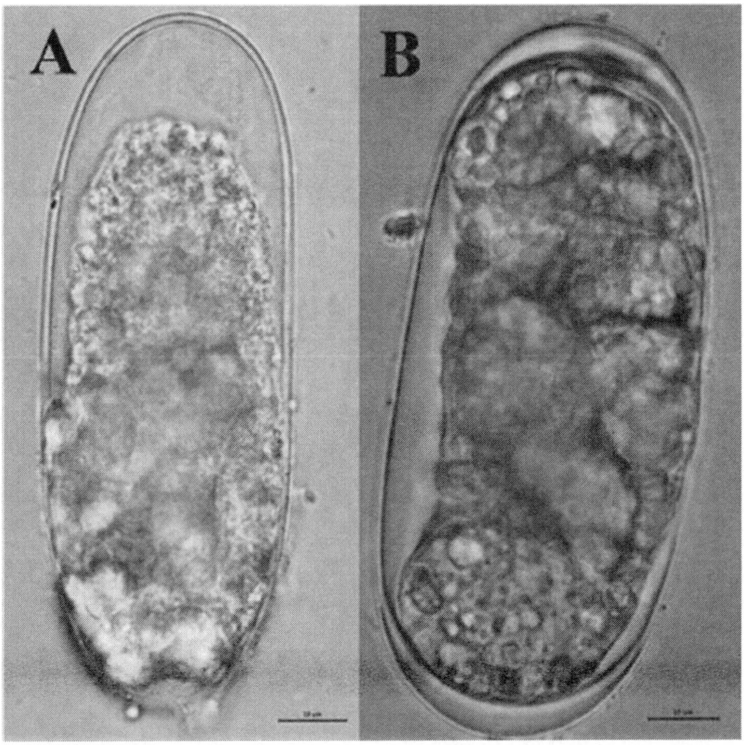

Figure 4. Morphological variations of *M. javanica* eggs after treatment with 5 g/L *Camellia* seed cake extract. The egg treated with 5 g/L extract at 72 h (A) and the egg treated with water control (B). Bars: 10 μm.

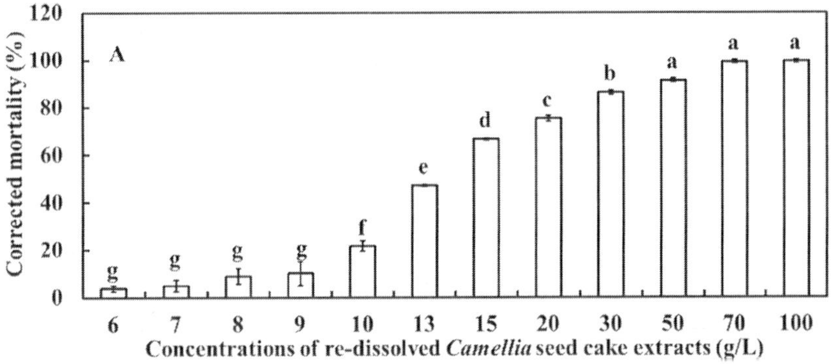

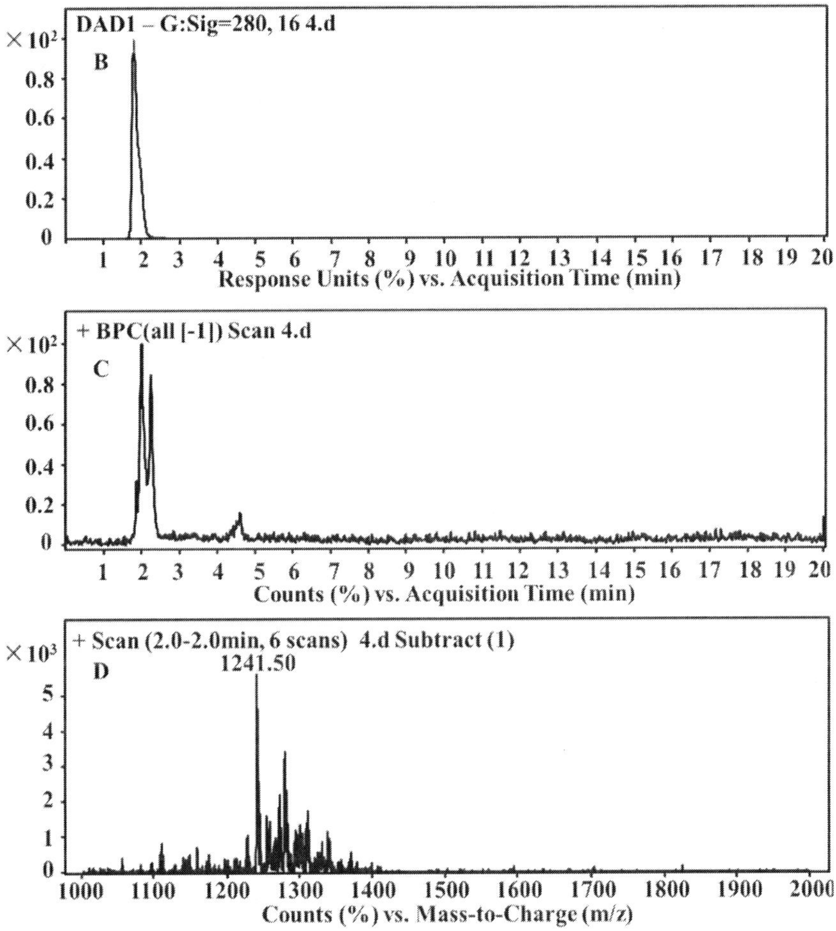

Figure 5. The effects of the re-dissolved extracts on the number of *M. javanica* J2s (A) and saponin detected by HPLC-MS (include HPLC profile (B), total ion chromatogram (C) and ESI-MS identification (D)).DAD: Diode-Array Detector; BPC: Base Peak Chromatogram.

Nematicidal activity of the volatile compounds in the plate

The effects of VOCs from *Camellia* seed cake extract on the juvenile population of *M. javanica*are presented in Fig. 6A. Nearly all the nematodes were killed from 50 g/L to 200 g/L. From 50 g/L to 10 g/L, the nematicidal rate showed a continual decline to approximately 7.56%. In addition, the nematicidal rate was nearly 50% at 20 g/L and 7.56% at 10 g/L (Fig. 6A).

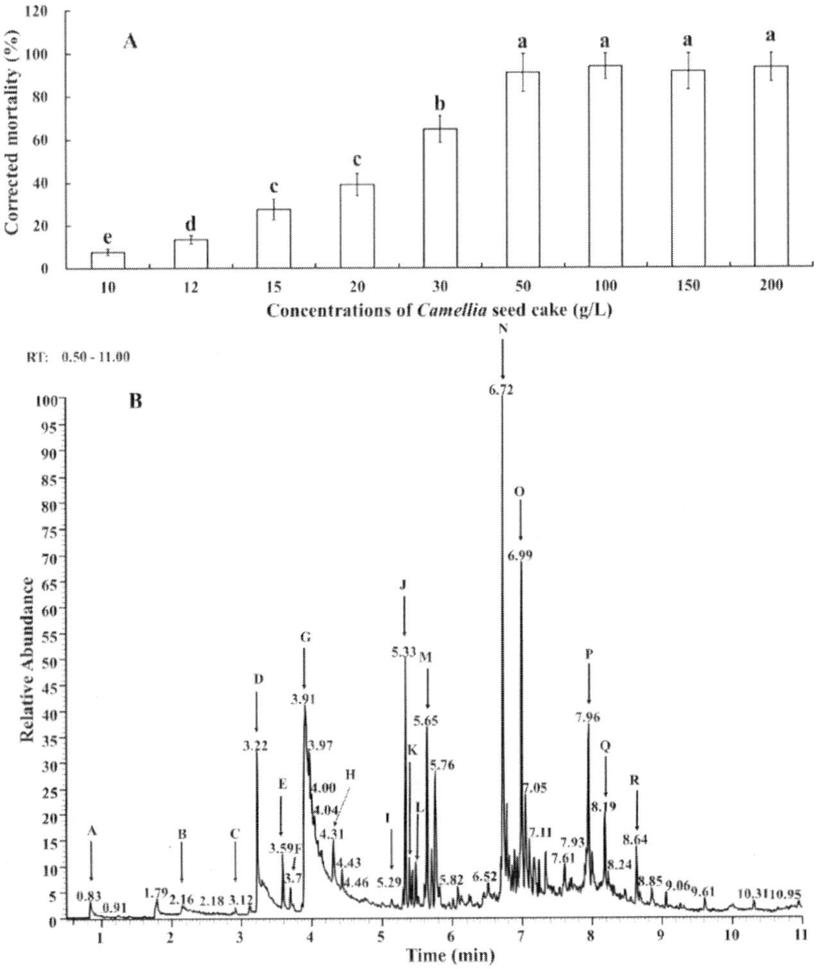

Figure 6. Effects of volatile components (VOCs) from *Camellia* seed cake extract on the *M. javanica* J2s (A) and the GC profiles for VOCs produced by the *Camellia*seed cake (B).The compounds corresponding to the retention times 0.83, 2.16, 3.12, 3.22, 3.59, 3.7, 3.91, 4.31, 5.29, 5.33, 5.39, 5.48, 5.65, 6.72, 6.99, 7.96, 8.19, 8.64 were designated as products A, B, C, D, E, F, G, H, I, J, K, L, M, N, O, P, Q and R, respectively.

VOCs were further identified by GC-MS. The retention times of the products are summarized in Table 1. According to the NIST library, products A, B, C, D, E, F, G, H, I, J, K, L, M, N, O, P, Q and R shown in Fig. 6B were identified as

1-butanol, 2-methyl-butanoic acid, 1-octen-3-ol, Butylesterbutanoic acid, 2-methylbutanebutyl, 3,3-dimethyloctane, 4-methylphenol, 2-methyl-2-butenoate butyl, 2-bromo dodecane, 2,7,10-trimelthyldodecane, Bcosane, 2,6, 10,15-tetramethylheptadecane, 2,6,11-trimethyldodecane, Hepta cosane, 2-methyltetradecane, 2,6,10,15-tetramethylheptadecane, 2-methy lnanadecane and 3,7,11,15-tetramethyl-2-hexadecen-1-ol, respectively (Table 1). We acquired 8 of the 18 pure compounds; the other 10 were not available for purchase. The effects of the 8 pure compounds on the juveniles of *M. javanica* are shown in Table 2. The highest efficiency was observed for 4-methylphenol, which yielded a corrected mortality of 9.50% with an added amount of 3 mg/L.

Table 1. Compounds from *Camellia* seed cake identified by GC-MS.

Metabolite	Rt(min)	Chemical name
A	0.83	1-butanol
B	2.16	2-methyl-butanoic acid
C	3.12	1-octen-3-ol
D	3.22	Butylester butanoic acid
E	3.59	2-methylbutanebutyl
F	3.7	3,3-dimethyloctane
G	3.91	4-methylphenol
H	4.31	2-methyl-2-butenoate butyl
I	5.29	2-bromo dodecane
J	5.33	2,7,10-trimelthyldodecane
K	5.39	Bcosane
L	5.48	2,6,10,15-tetramethylheptadecane
M	5.65	2,6,11-trimethyldodecane
N	6.72	Heptacosane
O	6.99	2-methyltetradecane
P	7.96	2,6,10,15-tetramethylheptadecane
Q	8.19	2-methylnanadecane
R	8.64	3,7,11,15-tetramethyl-2-hexadecen-1-ol

Table 2. Nematicidal compounds from *Camellia* seed cake identified by GC-MS.

Metabolite	Rt(min)	Chemical name	3000 mg/L	300 mg/L	30 mg/L	3 mg/L
A	0.83	1-butanol	24.26±0.37g	21.50±0.64g	7.18±1.31g	0g
B	2.16	2-methyl-butanoic acid	39.39±0.87e	33.97±0.91d	36.07±5.56c	4.10±0.32b
C	3.12	1-octen-3-ol	100a	45.58±0.18c	40.75±1.07b	4.08±0.11b
D	3.22	Butylester butanoic acid	64.58±2.95b	44.64±0.74c	26.01±1.43d	2.54±0.06de
E	3.59	2-methylbutanebutyl	39.39±0.87e	33.97±1.29d	26.28±0.42d	2.65±0.05d
F	3.7	3,3-dimethyloctane	66.55±0.61b	52.44±1.43b	32.21±0.82c	3.15±0.07c
G	3.91	4-methylphenol	100a	100a	100a	9.50±0.23a
H	5.29	2-bromo dodecane	33.99±2.70f	23.88±1.14f	22.08±1.43de	2.26±0.03e

All values are the mean of five replicates. Numbers following "±" represent the standard errors (SE). Different letters in the same column indicate statistically significant differences at the 0.05 probability level according to the Duncan test.

Pot experiments

For the first-season, sixty days after transplanting the banana seedlings, the outcomes of each treatment were compared to the control (CK). Treatment A showed the significant differences from CK by increasing the plant height, stem diameter, and the fresh weights of shoots and roots with 51.74%, 31.72%, 36.77% and 18.37%, respectively (Table 3). Treatment A also had the significant differences from CK by decreasing the eggmasses per plant and the eggmasses per gram root with 42.57% and 82.07%, respectively, compared with CK. Treatment B showed the significant differences from CK by increasing the plant height and stem diameter with 27.64% and 24.27%, respectively and decreasing the eggmasses per plant and the eggmasses per gram root with 31.08% and 57.30%, respectively. No significant differences were observed for other growth parameters in Treatment B. Treatment C only increased the stem diameter by 17.19% and decreased the eggmasses per plant by 33.27%. These results show that the application of 5 g/kg of *Camellia* seed cake to the soil effectively promoted banana growth and inhibited *M. javanica* infection.

Table 3. Effects of application of *Camellia* seed cake on biomass and egg masses densities of banana plants 60 days after transplantation in the pot experiments (first-season).

Treatment	Plant height (cm)	Stem diameter (mm)	Fresh weight of shoots (g)	Fresh weight of roots (g)	Eggmasses per plant	Eggmasses per gram root
CK	18.67±1.53c	25.13±0.26d	158.95±11.95b	113.42±5.50bc	197.33±13.05a	147.61±36.94a
A	28.33±1.53a	33.1±0.2a	217.47±12.90a	134.25±10.46a	113.33±26.31b	26.46±4.34c
B	23.83±3.40ab	31.23±0.15b	161.17±10.92b	122.32±4.75b	136±9.85b	63.03±21.44bc
C	20.4±3.44c	29.45±0.39c	160.82±13.56b	109.24±6.32c	131.67±7.02b	111.29±28.06ab

All values are the mean of five replicates. Numbers following "±" represent the standard errors (SE). Different letters in the same column indicate statistically significant differences at the 0.05 probability level according to the Duncan test.

For the second-season, sixty days after transplanting the banana seedlings, the outcomes of each treatment were compared to the control (CK). Treatment A showed the significant differences from CK by increasing the plant height, stem diameter, the fresh and dry weights of shoots with 26.17%, 20.71%, 52.59% and 66.35%, respectively. Treatment A also had significant differences from CK by increasing the nematode density in soil with 60.15% and decreasing the nematode density in the roots with 28.23%. Treatment B increased the plant height, stem diameter and the dry weight of shoots by 13.73%, 16.79% and 15.48%, respectively. Treatment B also increased the nematode density in soil by 14.18% and decreased the nematode density in roots by 14.29%. Treatment C increased the stem diameter and the dry weight of shoots by 12.49% and 18.93%, respectively, and decreased the nematode density in roots by 9.52% (Table 4). In addition, after identifying the nematodes in the soil using morphology and feeding-habit based classification, the numbers of plant parasites in treatments A and B were significantly lower than in the control, while the numbers of fungivores and bacterivores were greater in treatments A and B than in the control (S1 Table). For the culturable microbes, the numbers of bacteria, fungi and actinomycetes in the three treatments were all significantly higher than in the control (S2 Table).

Table 4. Effects of application of *Camellia* seed cake on biomass and nematode densities of banana plants 60 days after transplantation in the pot experiments (second-season).

Treatment	Plant height (cm)	Stem diameter (mm)	Fresh weight of shoots (g)	Dry weight of shoots (g)	Fresh weight of roots (g)	Nematode density in roots (individuals 10 g^{-1} root biomass)	Nematode density in soil (individuals 100 g^{-1} dry soil)
CK	20.9 ±1.55c	28.83±0.75d	160.71 ±20.69b	14.74±0.42c	118.91 ±14.12a	980±65.57a	870±52.92c
A	26.37 ±1.02a	34.80±0.36a	245.22±4.92a	24.52±0.49a	145.04 ±29.90a	703.33±65.06c	1393.33±30.55a
B	23.77 ±1.19b	33.67±0.45b	185.10 ±12.67b	18.51±1.27b	137.32 ±19.10a	840±26.46b	993.33±75.72b
C	21.5 ±0.61c	32.43±0.12c	174.27±5.78b	17.53±0.61b	117.41 ±18.15a	886.67±20.82b	923.33±41.63bc

All values are the mean of five replicates. Numbers following "±" represent the standard errors (SE). Different letters in the same column indicate statistically significant differences at the 0.05 probability level according to the Duncan test.

For the third-season, similar results for the growth parameters were observed as for the second-season. When compared with the control (CK), treatment A decreased the nematode density in roots by 31.84%. Moreover, treatment A increased the nematode density in soil by 73.30%, and treatment B by 21.87% (Table 5).

Table 5. Effects of application of *Camellia* seed cake on biomass and nematode densities of banana plants 60 days after transplantation in the pot experiments (third-season).

Treatment	Plant height (cm)	Stem diameter (mm)	Fresh weight of shoots (g)	Dry weight of shoots (g)	Fresh weight of roots (g)	Nematode density in roots (individuals 10 g^{-1} root biomass)	Nematode density in soil (individuals 100 g^{-1} dry soil)
CK	20.53 ±0.49c	27.53±0.7d	158.76 ±21.63b	14.9±0.24c	115.79 ±13.51a	975.67±48.02a	831.83±46.47c
A	27.77 ±0.45a	34.6±1.63a	240.18 ±14.33a	25.01±0.45a	135.67 ±13.83a	665±36.26c	1441.6±56.26a
B	24.87 ±1.24b	32.27±0.50b	175.27±6.50b	17.80±0.92b	127.58 ±13.87a	820±43.44b	1013.8±64.50b
C	20.80 ±0.72c	29.73±0.55c	163.28±8.31b	17.36±0.57b	118.83 ±18.14a	871±22.61b	918.62±43.38bc

All values are the mean of five replicates. Numbers following "±" represent the standard errors (SE). Different letters in the same column indicate statistically significant differences at the 0.05 probability level according to the Duncan test.

DISCUSSION

Root knot nematodes on bananas seriously damage its production and induce the complex diseases caused by the other plant pathogens, therefore exploring a high efficient nematicide with environmental security is urgently needed. In this study, the effects of *Camellia* seed cake on controlling *M. javanica* were evaluated and its possible mechanisms for nematicidal activities were elucidated. *M. javanica* has already been reported to be the dominant banana pathogen in Pakistan [27]. It has also been reported to be the dominant pathogen for almost all of the world's major crop plants, including papaya, potato, peanut, tomato, and rootstocks [28–31]. The populations isolated from banana roots in Hainan Province, China were mainly identified as *M. javanica* based on morphological and molecular characteristics, which was confirmed to be the dominant species of banana root-knot nematodes in our research.

Camellia seed cake acts as a non-conventional fertilizer and has been reported to effectively suppress plant-parasitic nematodes [32]. In this study, more than 50% of *M. javanica* J2s were killed at *Camellia* seed cake extract concentrations above 2 g/L, and at these concentrations, half of the *M. javanica* eggs could not be hatched. The results proved the potential nematicidal activity of *Camellia* seed cake, which is similar to those of Wang et al. [33], who reported that Camellia plant extracts had strong nematicidal activity on *B. xylophilus* and *M. incognita*. After treatment with *Camellia* seed cake extracts, the eggs of *M. javanica* were gradually dissolved, and the intestine of the juveniles gradually became indistinct in our observation. Similar symptoms have been reported that J2s and eggs of *M. incognita* were damaged after treatment with *Bacillus cereus* X5 [34] or 2, 4-diacetylphloroglucinol, the secondary metabolite of *Pseudomonas fluorescens* CHA0 [35]. These *in vitro* tests indicated the eggs and J2s of *M. javanica* could be controlled by the *Camellia* seed cake extract.

Several papers have reported that *Camellia* seed cake contained 15% camellia saponin. Therefore, the nematicidal effect and molecular weight of camellia saponin were investigated in this study. Saponins are steroid or triterpenoid glycosides, which are commonly observed in many plants [36]. Saponins from *Pulsatilla koreana* root or from *Medicago sativa* could control plant-parasitic nematodes [37]. Our results also confirmed that camellia saponin possesses a distinct nematicidal effect. After detection by HPLC-ESI-MS, a compound with a molecular weight of 1240.50 Da was observed. In previous reports, the molecular weight of triterpene saponin from *Lysimachia capillipes* is 1240.61 Da [24], from the roots of *Panax notoginseng*, is 1240.65 Da [25–26]. Therefore, we propose that compound detected in this study was confirmed as saponin.

After soaking, volatile gases produced by *Camellia* seed cake were observed to kill most of the root-knot nematodes. Therefore, the volatile gases produced by *Camellia* seed cake were collected using headspace-SPME and identified by GC-MS. The headspace-SPME represents an excellent, solventless analysis technique that has been applied to identify VOCs, e.g., in blood, viscera samples, urine, and food [38–40]. Through GC-MS, 18 compounds were identified, of which 8 standard compounds were reported for the first time to show nematicidal activities. Among them, 4-methylphenol showed the best efficiency. Our results are similar to those of Li et al. [41], who reported that the volatile compound methyl thiobutyrate has inhibiting effects on *Caenorhabditis elegans* and *M. incognita*, and can be further applied to prepare a vermifuge from its precursor substance. Most likely, the nematicidal effect of *Camellia* seed cake is due to the production of both volatile and non-volatile inhibitory compounds.

In our three seasons of pot experiments using sterilized or crude soils, the application of *Camellia* seed cake significantly suppressed *M. javanica* and promoted plant growth. We speculate that the growth promotion may be attributable to the suppression of the harmful nematode *M. javanica* and the nutrient content of the *Camellia* seed cake. Our results agree with previous reports of effective suppression of harmful nematodes by nematicidal agents from plant sources in pot and field experiments. Four medicinal plants, *Azadirachta indica*, *Calotropis procera*, *Datura stramonium* and *Tagetes erecta* had nematicidal effects against *M. incognita*[42], while the aqueous and ethanol extracts from leaves, stem, bark and fruit of *Eucalyptus* sp., against *M. javanica* [43]. The extracts of *Eucalyptus* sp not only reduced egg hatching and increased J2s mortality as exposure time increased, but also significantly increased the shoot length, shoot weight, root length and root weight [43], which showed the similar nematicidal activities and plant promotion as that of the *Camellia* seed cake.

After morphological observation at the end of the second-season pot experiment, the number of plant parasites in the *Camellia* seed cake treatments was lower than in the control, while the density of total nematodes and other functional groups showed the opposite tendency. This result may be caused by the fact that *Camellia* seed cake application increased the number of bacteria, fungi and actinomycetes, which in turn increased the number of total nematodes, especially bacterivores, fungivores and omnivores. Our results are similar to those of González [44], who showed that situ ingestion rates of fluorescently labeled bacteria could estimate the active bacterivores in natural aquatic systems. Remén [45] showed that ectomycorrhizal fungi constituted an important food source for fungivorous soil fauna and may be a factor regulating these faunal communities. Moreover, promotion of plant growth by bacterial-feeding nematodes has been previously reported [46–47]. The effects of bacterial-feeding nematodes may be another explanation of the growth-promotion effect observed in the pot experiments. Therefore, we proposed that the alteration of nematode populations and different functional groups is a new mechanism for promoting plant growth by the application of *Camellia* seed cake. However, further studies need to be provided to support this hypothesis.

CONCLUSIONS

The harmful root-knot nematode *M. javanica* was identified in this study and confirmed to be the dominant species on banana. The *Camellia* seed cake extracts effectively killed J2s of *M. javanica* and suppressed the hatching rate of eggs. Nematicidal compounds produced by *Camellia* seed cake, including saponins and VOCs, effectively killed J2s. In pot experiments of banana seedlings, application of *Camellia* seed cake not only suppressed *M. javanica* but also promoted plant growth. The direct biocontrol efficiency of *Camellia* seed cake in field experiments requires further study.

ACKNOWLEDGMENTS

We gratefully acknowledge Dr. Manqiang Liu of Nanjing Agricultural University for excellent assistance in nematode community analysis and thank Hainan Wanzhong Co., Ltd for huge help to us in banana planting.

REFERENCES

1. Wang B, Yuan J, Zhang J, Shen Z, Zhang M, Li R, et al. Effects of novel bioorganic fertilizer produced by *Bacillus amyloliquefaciens* W19 on antagonism of *Fusarium* wilt of banana. Biol Fertil Soils. 2013;49: 435–446.
2. Jaizme-Vega MC, Tenoury P, Pinochet J, Jaumot M. Interactions between the root-knot nematode *Meloidogyne incognita* and *Glomus mosseae* in banana. Plant Soil. 1997;196: 27–35.
3. Oka Y, Koltai H, Bar-Eyal M, Mor M, Sharon E, Chet I, et al. New strategies for the control of plant-parasitic nematodes. Pest Manag Sci. 2000;56: 983–88.
4. Gowen S, Quénéhervé P. Nematode parasites of bananas, plantains and abaca. In Plant Parasitic Nematodes in Subtropical and Tropical Agriculture, ed. Luc M, Sikora RA and Bridge J. CAB International, Institute of Parasitology, Wallingford, UK; 1990. Pp. 431–460.
5. Sikora RA, Schlosser E. Nematodes and fungi associated with root systems of bananas in a state of decline in Lebanon. Plant Dis Rep. 1973;57 (7): 615–618.

6. Oka Y, Shuker S, Tkachi N. Influence of soil environments on nematicidal activity of fluensulfone against *Meloidogyne javanica*. Pest Manag Sci. 2013;69: 1225–1234. pmid:23456999
7. Quénéhervé P, Valette C, Topart P, Montcel HT, Salmon F. Nematode resistance in bananas: screening results on some wild and cultivated accessions of *Musa* spp. Euphytica. 2009;165: 123–136.
8. Mogtahedi H, Santo GS, Hang AN, Wilson JH. Suppression of root-knot nematode populations with selected rapeseed cultivars as green manure. J Nematol. 1991;23 (2): 170–174. pmid:19283108
9. Wang KH, Hooks CRR. Survey of nematodes on banana in Hawai'i, and methods used for their control. Plant Disease. 2009;PD-69.
10. Chitwood DJ. Phytochemical based strategies for nematode control. Annu Rev Phytopathol. 2002;40: 221–249. pmid:12147760
11. Chen RS, Wang KL, Wu CY. Assessment of the camellia seed meal impact on loaches in paddy fields. Paddy Water Environ. 2012;10: 291–300.
12. Mansoor F, Sultana V, Ehteshamul-Haque S. Enhancement of biocontrol potential of *Pseudomonas aeruginose* and *Paecilomyces lilacinus* against root knot of mungbean by a medicinal plant *Launaea nudicalis* L. Pak J Bot. 2007;39(6), 2113–2119.
13. Long J, Cao F, Peng J. Effect of plant extracts on nematicidal activity of Bursaphelenchus xylophilus. North China Horticulture. 2010;22: 148–150.
14. Njuguna DG, Wanyoko JK, Kinyanjui T, Wachira FN. Mineral elements in the Kenyan tea seed oil cake. Int J Res Chem Environ. 2013;3 (1): 253–261.
15. Lee CP, Yen GC. Antioxidant activity and bioactive compounds of tea seed (*Camellia oleifera* Abel.) Oil. J Agric Food Chem. 2006;54 (3): 779–784. pmid:16448182
16. Adam MAM, Phillips MS, Blok VC. Molecular diagnostic key for identification of single juveniles of seven common and economically important species of root-knot nematode (Meloidogyne spp.). Plant Pathol. 2007;56: 190–197.
17. Li H, Trinh PQ, Waeyenberge L, Moens M. Bursaphelenchus chengi sp. n. (Nematoda: Parasitaphelenchidae) isolated at Nanjing, China, in packaging wood from Taiwan. Nematology. 2008;10 (3): 335–346.
18. Bybd DW, Kirkpatrick T, Barker KR. An improved technique for clearing and staining plant tissues for detection of nematodes. J Nematol. 1983;15 (1): 142–143. pmid:19295781
19. El-Rokiek KG, El-Nagdi WM. Dual effects of leaf extracts of *Eucalyptus citriodora* on controlling purslane and root-knot nematode in sunflower. J Plant Prot Res. 2011;51: 121–129.
20. Zhong S, He G, Zhang C, Hua H, Yan Q. RP-HPLC determination of tea saponin in tea-seed cake. PT CA (Part B: Chem Anal). 2008;44 (12): 1169–1175.
21. .Strobel GA, Dirkse E, Sears J, Markworth C. Volatile antimicrobials from *Muscodor albus*, a novel endophytic fungus. Microbiology. 2001;147: 2943–2950. pmid:11700345
22. Farag MA, Ryu CM, Sumner LW, Pare PW. GC-MS SPME profiling of rhizobacterial volatiles reveals prospective inducers of growth promotion and induced systemic resistance in plants. Phytochemistry. 2006;67: 2262–2268. pmid:16949113
23. Shen Z, Zhong S, Wang Y, Wang B, Mei X, Li R, et al. Induced soil microbial suppression of banana fusarium wilt disease using compost and biofertilizers to improve yield and quality. Eur J Soil Biol. 2013;57: 1–8.
24. Tian JK, Xu LZ, Zou ZM, Yang SL. Three novel triterpenoid saponins from *Lysimachia capillipes* and their cytotoxic activities. Chem Pharm Bull. 2006;54 (4): 567–569. pmid:16595968
25. Yoshikawa M, Murakami T, Ueno T, Yashiro K, Hirokawa N, Murakami N, et al. Bioactive saponins and glycosides. VIII. Notoginseng (1): new dammarane-type triterpene oligoglycosides, notoginsenosides-A,-B,-C, and-D, from the dried root of *Panax notoginseng* (Burk.) F.H. Chen. Chem Pharm Bull (Tokyo). 1997; 45(6):1039–1045. pmid:9214709

26. Sun HX, Qin F, Ye YP. Relationship between haemolytic and adjuvant activity and structure of protopanaxadiol-type saponins from the roots of *Panax notoginseng*. Vaccine. 2005;23: 5533–5542. pmid:16115706
27. Jabeen S, Bilqees FM, Khan A, Khatoon N. Pathogenicity of *Meloidogyne javanica* on banana in Pakistan. Proc Parasitol. 1996;21: 11–76.
28. Pradhan MAA, Rahaman MM, Paul SK, Ahamad MU, Goswami BK. Effect of BAU-biofungicide, neem oil and a nematicide on the root-knot (Meloidogyne javanica) of papaya (Carica papaya). Bangladesh J Agril Res. 2012;37 (2): 271–277.
29. Vovlas N, Mifsud D, Landa BB, Castillo P. Pathogenicity of the root-knot nematode *Meloidogyne javanica* on potato. Plant Pathol 2005;54: 657–664.
30. Tomaszewski EK, Khalil MAM, El-Deeb AA, Powers TO, Starr JL. *Meloidogyne javanica*parasitic on peanut. J Nematol. 1994;26 (4): 436–441. pmid:19279913
31. Rumbos CI, Khah EM, Sabir N. Response of local and commercial tomato cultivars and rootstocks to *Meloidogyne javanica* infestation. Aust J Crop Sci. 2011;5(11): 1388–1395.
32. Huang Y, Yang M, Li Y, Lin W, Hong S. Effects of *Camellia* cake organic fertilizer on growth of turf grass *Cynodon* spp. Rich. J Cent South Univ Forestry and Technol. 2013;4 (33): 1–6.
33. Wang L, Wan X, Hou R, Xu J, Wu H. Control of Camellia plant extracts and its preparation against plant parasitic nematodes. J Anhui Agri Univ. 2013;40 (4): 642–648.
34. Xiao TJ, Tan SY, Shen QR, Ran W. *Bacillus cereus* X5 suppresses root-knot nematode of tomato by colonizing in roots and soil. Afr J Microbiol Res. 2012;6 (10): 2321–2327.
35. Siddiqui IA, Shaukat SS. Suppression of root-knot disease by *Pseudomonas fluorescens* CHA0 in tomato: importance of bacterial secondary metabolite, 2, 4-diacetylphloroglucinol. Soil Biol Biochem. 2003;35: 1615–1623.
36. Dinda B, Debnath S, Mohanta BC, Harigaya Y. Naturally occurring triterpenoid saponins. Chem Biodivers. 2010;7: 2327–2580. pmid:20963775
37. D'Addabbo T, Carbonara T, Leonetti P, Radicci V, Tava A, Avato P. Control of plant parasitic nematodes with active saponins and biomass from *Medicago sativa*. Phytochem Rev. 2011;10: 503–519.
38. Tranthim-Fryer DJ, Hansson RC, Norman KW. Headspace/solid-phase microextraction/gas chromatography-mass spectrometry: a screening technique for the recovery and identification of volatile organic compounds (VOC's) in postmortem blood and viscera samples. J Forensic Sci. 2001;46 (4): 934–946. pmid:11451082
39. Silva CL, Passos M, Câmara JS. Investigation of urinary volatile organic metabolites as potential cancer biomarkers by solid-phase microextraction in combination with gas chromatography-mass spectrometry. Br J Cancer. 2011;105 (12): 1894–1904. pmid:22085842
40. Wardencki W, Michulec M, Curyło J. A review of theoretical and practical aspects of solid-phase microextraction in food analysis. Int J Food Sci Technol. 2004;39: 703–717.
41. Li G, Lu H, Zhang K. Volatile compound methyl thiobutyrate and application Faming Zhuanli Shenqing Gongkai Shuomingshu, CN103439426A. 2013. (in Chinese)
42. Hussain MA, Mukhtar T, Kayani MZ. Efficacy evaluation of Azadirachta indica, Calotropis procera, Datura stramonium and Tagetes erecta against root-knot nematodes Meloidogyne incognita. Pak J Bot. 2011;43: 197–204.
43. Dawar S, Younus SM, Zaki MJ. Use of *Eucalyptus* sp., in the control of *Meloidogyne javanica* root knot nematode. Pak J Bot. 2007;39 (6): 2209–2214.
44. González JM. Bacterivory rate estimates and fraction of active bacterivores in natural protist assemblages from aquatic systems. Appl Environ Microbiol. 1999;65 (4): 1463–1469. pmid:10103238

45. Remén C, Fransson P, Persson T. Population responses of oribatids and enchytraeids to ectomycorrhizal and saprotrophic fungi in plant-soil microcosms. Soil Biol Biochem. 2010;42:978–985.
46. Jiang Y, Wu Y, Xu W, Cheng Y, Chen J, Xu L, et al. IAA-producing bacteria and bacterial-feeding nematodes promote *Arabidopsis thaliana* root growth in natural soil. Eur J Soil Biol. 2012;52: 20–26.
47. Irshad U, Villenave C, Brauman A, Plassard C. Grazing by nematodes on rhizosphere bacteria enhances nitrate and phosphorus availability to *Pinus pinaster* seedlings. Soil Biol Biochem. 2011;43: 2121–2126.

CHAPTER 16

Effect of Pre-Harvest Application of Gibberellic Acid on Fruit Quality and Shelf Life of Mandarin (Citrus reticulata Blanco)

Pradeep Raj Rokaya[1], Dilli Ram Baral[1], Durga Mani Gautam[1], Arjun Kumar Shrestha[1], Krishna Prasad Paudyal[2]

[1]*Agriculture and Forestry University, Rampur, Chitwan, Nepal*

[2]*National Agricultural Research Institute (NARI), NARC, Khumaltar, Lalitpur, Nepal*

ABSTRACT

Mandarin (Citrus reticulata Blanco) is a promising fruit crop gaining popularity for its human nutrition and economic importance in Nepal. The qualitative losses during pre-harvest stage hinder the productivity and subsequently shorten the on-tree storability. An experiment was conducted to assess the effect of gibberellic acid on quality and shelf life of the mandarin fruit. GA_3 at 10, 20, and 30 ppm as against of control were evaluated. Observations on fruit weight (g), fruit firmness (kg/cm^2), rind colour (1 - 5 index), juice recovery (%), TSS/acid ratio, PLW (%), decay loss (%), and ascorbic acid (mg/100 ml) were recorded at three harvesting dates i.e. 20 Nov, 5 Dec, and 20 Dec and storage condition. It has been revealed that the fruits treated with GA_3 at 20 ppm retained higher fruit weight (128.6 g), more firmness (3.54 kg/cm^2), better juice recovery (57.75%), and greater TSS/acid ratio (21.24) at the end of study (20 December). The PLW was found less with GA_3 at 30 ppm in both ambient (5.17%) and cellar (6.69%) condition as against untreated fruits (9.52%

and 11.76%). Similarly, the decay loss was minimum in the fruits treated with GA_3 at 30 ppm both with ambient (1.02%) and cellar condition (8.21%) as against control with ambient (5.54%) and cellar (21.58%).

Keywords: Fruit Quality, Gibberellic Acid, Peel Puncture Resistance, Physiological Loss in Weight, Postharvest Shelf Life

1. INTRODUCTION

Mandarin (Citrus reticulata Blanco) is a promising fruit crop regarding its positive influence on human nutrition and economic importance in Nepal. Farmers in the mid hills of Nepal produce large amount of mandarin during its peak season i.e. November-December [1] . In the peak season, growers fetch lower price due to seasonal glutting, and in lean season, fruits are not available in the market. The consumers suffer due to unavailability of fruits beyond the peak harvesting and have to pay high price for cold stored coming all the way from India [2] . The fluctuation in the fruit supply and price can be minimized either by extending the harvesting period or by holding the fruits in storage [3] . Mandarin is a perishable fruit that cannot be kept on tree storage without any treatments.

Pre-harvest application of GA_3 has been reported to delay softening, delay rind colour, and minimize the fruits drops and puffiness [4] - [6] . Pre-harvest spray of GA_3 in mandarin delayed ripening and decay loss of fruits [7] . Kher et al. [8] found that GA_3 was the most effective chemicals in increasing the weight, specific gravity and decreasing the total acid content in guava fruits. Brahmachari and Rani [9] suggested that GA_3 and kinetin are beneficial in prolonging the shelf life and retaining the quality of guava fruits. Bhattacharya [10] reported that the fruits treated with GA_3 extended the storage life of lime fruits over control by delaying in ripening. Maturity and colour break stage were delayed with GA_3 50 ppm treatment in Satluj Purple plum [11] . There was a decrease in TA, increase in total juice volume, TSS/acid ratio, and along with delayed rind colour by GA_3 [12] . In citrus fruit, GA_3 application delayed the loss of chlorophyll [13] . Exogenous application of growth regulators significantly decreased fruit drop leading to increase in total number of fruits per plants, fruit weight, juice percentage, total soluble solids, acidity, and vitamin C [14] . Fruit skin colour advanced in GA_3 applied trees [15] . GA_3 delayed the maturity and reduced the postharvest losses in kinnow mandarin [16] . Considering perishability and market glut in peak season, this study was undertaken to minimize the pre-harvest losses and to get high price beyond peak season through increase in harvest span. Therefore, pre-harvest application of $GA_3$was investigated to evaluate its efficacy during on-tree storage and the consequent postharvest storage of mandarin.

2. MATERIALS AND METHODS

This investigation was undertaken on 15 years old uniform and healthy tree with spacing of 5m × 5m in the Lamjung district of Nepal in the year 2012 and 2013. On 16 October i.e. five weeks before normal harvesting (with fruits at colour break), tree were treated with GA_3 (10, 20, and 30 ppm) as a whole tree spray containing Tween-80 surfactant (0.02% v/v). The spray of surfactant served as control. The second spray was applied two weeks later. The experiment was laid out in randomized complete block design (RCBD) including four treatments with four replications. The fruits were harvested in November 20, December 5, and December 20 and analysis was done with the respective date of harvesting. 20 fruits in each replication were sampled for analysis. Fruit weight was taken in gram by top balance, fruit firmness in terms of peel puncture resistance by Effigy tester for citrus (model 011). The rind colour was evaluated as in index where colours were given numerical value: full green = 1, yellowish green = 2, greenish yellow = 3, yellowish orange = 4 and full orange = 5. Total Soluble Solids in juice were determined by hand refractometer in ^{0}Brix, acidity by titrating against 0.1 N titration and vitamin C as mentioned by AOAC [17] . Data were tabulated in excel and analysed by using standard procedures as proposed Gomez and Gomez [18] .

3. RESULTS AND DISCUSSION

3.1. Fruit Weight

Data presented in the Table 1 show that fruit weight was increased significantly in all the treatments of GA_3 with the advancement of the maturity as compared to control. Perusal on observations of fruit weight indicates that fruits treated with GA_3 was able to increase in weight. Maximum fruit weight was recorded with GA_3 at 20 ppm (128.6 g) against control (95.51 g) at the end of experiment (20 December). It was noted that GA_3 treated fruit had maximum increase on 5 December and beyond this date it was observed slow, stagnant and declined where as sharp decline in control fruits. Based on the findings, GA_3 with the concentration at 10 - 20 ppm was the most effective effect treatment for fruit growth in mandarin. The higher fruit growth with GA_3treated fruits might be due to mediating process for faster translocation and mobilization of photosynthates from source. These findings are in agreement with the reports of Singh et al. [19] in pear, and Pal et al. [20] in kinnow man-

Table 1. Periodical changes on fruit weight, rind colour, and peel puncture resistance in mandarin fruit during on-tree storage in 2012 and 2013.

Treatments (ppm)	Fruit weight (gram)			Rind colour index (1 - 5 scale)			Peel puncture resistance (kg/cm²)		
	A	B	C	A	B	C	A	B	C
GA_3 10	110.58	125.04	125.01	2.05	2.93	3.42	4.36	4.21	3.24
GA_3 20	113.50	127.46	128.60	1.92	2.84	3.16	4.19	3.80	3.54
GA_3 30	113.16	123.80	124.66	1.83	2.69	2.97	4.19	3.71	3.45
Control	98.75	98.26	95.51	3.37	4.63	5.00	3.75	2.80	2.73
LSD (0.05)	6.07	9.36	7.88	0.26	0.39	0.40	NS	0.27	NS

A = 20.11, B = 05.12, and C = 20.12.

darin who had observed GA_3 (10 ppm) has resulted in an increase in fruit diameter and Kumar et al. [21] in the fruit of strawberry who stated that gibberellic acid at 12.5 ppm resulted in the highest fruit length, breadth, and weight. Kaur et al. [11] has also reported that GA_3 at 25 and 50 ppm increased the fruit weight in plum and inconsonance with the observation of Bose et al. [22] who recorded three times increment on fruit weight in mandarin. The increment in fruit weight might be due to hormone directed to transportation and accumulation of phytosynthates which resulted in better fruit development and also acceleration of cell division, elongation, and enlargement. Similar observation was recorded by Daulta and Veniwal [23] in sweet orange who claimed maximum weight with GA_3 sprayed tree fruits.

3.2. Rind Colour

Rind colour is the perhaps most important and reliable index of mandarin fruit. GA_3 treated fruits resulted in delaying the rind colour development as well delay in the maturity of the mandarin fruits. Perusal from the observations in the Table 1 shows that GA_3 treated fruits resulted in significant delay in the rind colour development. On 20 November, the index value of 3.37 indicated orange colour in control as against index of less than 2.0 indicating greener fruits with GA_3treatments. Likewise at the end of the study (20 December), the colour index of GA_3 treated fruits was 2 - 3.5 indicating orange colour as against of 5.0 indicating orange yellow with over ripe fruits in control. These findings are in consonance with the findings of the Ladaniya [5] in mandarin who stated that GA_3 treatments significantly delayed the rind colour development in Nagpur mandarin. Kaur et al. [11] also observed that colour development of the fruits was delayed by gibberellin treatments in plum. Gibberellin has been reported to delay chlorophyll degradation and the senescence in the fruits [24] . Colour development is associated with a loss of texture, increasing sugar content and decreasing acidity [25] .

3.3. Peel Puncture Resistance

Peel puncture resistance serves as fruit quality determinant. Peel puncture resistance was significantly higher in the GA_3 treated fruits as compared to control. Table 1 shows that on 20 November harvesting, the firmness of the fruits was non-significant although GA_3 treated fruits had appeared higher index value (>4.0 kg/cm^2) than control (3.75 kg/cm^2). On 5 December, this value was decreased in all treatments, however, GA_3 treated fruits had significantly much higher (3.71 - 4.21 kg/cm^2) than control (2.8 kg/cm^2). On 20 December, the resistance sharply declined and observed as non-significant among the treatments. However, fruits treated with GA_3 at 20 ppm had higher firmness (3.54 kg/cm^2) than the untreated fruits (2.73 kg/cm^2). These results are in conformity with the findings of the Ladaniya [5] in Nagpur mandarin who stated that GA_3 treated fruits had higher peel puncture resistance and Kaur et al. [11] observed the similar trends in the plum. In all the treatments the firmness was found decreasing with ripening advancement. This might be due to cell wall loosening of the fruit. The pro-pectin, which acts as a cementing material for binding the cellulose and hemicelluloses is converted to soluble pectin. As a result it loosens the cell wall's binding force during ripening [25] . Bakhshi and Chadha [13] claimed that in citrus, GA_3application given to fruits delayed the loss of chlorophyll in limes and lemons.

Table 2. Periodical changes on juice recovery, TSS/acid ratio, and Vitamin C in mandarin Vitamin C of GA_3 treated mandarin fruit during on-tree storage in 2012 and 2013.

Treatments (ppm)	Juice recovery (%)			TSS/acid ratio			Vitamin C (mg/100ml)		
	A	B	C	A	B	C	A	B	C
GA_3 10	52.68	55.08	55.68	11.86	15.00	20.04	31.03	27.5	24.75
GA_3 20	53.72	56.23	57.75	12.35	15.48	21.24	30.49	27.14	24.75
GA_3 30	53.80	54.85	55.21	12.16	15.83	17.77	29.86	28.52	23.17
Control	48.90	47.11	45.06	12.66	15.99	17.83	28.71	25.97	21.95
LSD (0.05)	NS	NS	6.77	NS	NS	2.53	NS	NS	NS

A = 20.11, B = 05.12, and C = 20.12.

3.4. Juice Recovery

Perusal on data as indicated in Table 2 shows that the juice recovery percentage was significantly higher in fruits treated GA_3than non-treated fruits. At normal harvesting date (20 November), the juice percentage was non-significant but higher juice was obtained from the fruit treated with GA_3. On later stage i.e. 20 December the juice percentage was recorded highly significant (57.75%) in the fruits treated with GA_3 at 20 ppm as against of control (45.06%) because GA_3 treated fruits remained firm and non-treated fruits were overripe in their condition. Similar observations were recorded by Haribabu et al. [26] in kagzi lime, Bhat et al. [27] in lemon and Daulta and Veniwal [23] in sweet orange who have reported that GA_3 treatments increased the percentage of juice. Increased in juice percentage may also be explained by the fact that hormones play a regulating role in the mobilization of metabolites within a plant and it is well established fact that developing fruits are extremely active metabolic "sinks" which mobilize metabolites and direct their flow from vegetative structure.

3.5. TSS/Acid Ratio

Perusal on data as indicated in Table 2 shows that the TSS/acid ratio was not significant on 20 November and 5 December in all the treatments. However, non-treated fruits recorded higher TSS/acid ratio (12.66) than GA_3 treated fruits (11.86 - 12.16). The TSS/acid ratio was significantly higher in the GA_3 treated fruits (17.77 - 21.24) against to control (17.83). GA_3 at 20 ppm showed the highest TSS/acid ratio (21.24) followed by GA_3 at 10 ppm at the end of the study i.e. on 20 December whereas the lowest ratio was observed in control (17.83). These findings substantiate the earlier reports of Singh et al. [19] in pear and Kaur et al. [11] in plum and Ladaniya [5] in Nagpur mandarin who reported increase in TSS/acid ratio with GA_3 treatments. It was obvious that the TSS content increased while juice acidity decreased resulting in an increase in the TSS/acid ratio during holding of fruit on the tree. The increased TSS content could have been due to further synthesis and accumulation of photosynthates in the fruits on the tree [28] . The loss in acidity was reported to be due to normal respiration and conversion to other metabolities and loss was similar for fruit stored on the tree or in cold room [24] .

3.6. Vitamin C Content

Higher vitamin C content imparts higher nutritive value to fruits. Data in Table 2 shows that the vitamin C content decreased with the advancement of storage of fruits on tree in all treatments. However, more pronounced decrease was found in the non-treated fruits. A decrease in ascorbic acid could be due to enzymatic loss of L-ascorbic acid where it is converted to 2-3-dioxy?L-gluconic acid [29] . Vitamin C content in the GA_3 treated and non-treated fruit was non-significant during the study (20 November to 20 December). However, the content of vitamin C was recorded higher in the GA_3 treated fruits than in non-treated. The maximum vitamin C content was obtained with GA_3 at10 (24.75 mg/100 ml) and 20 ppm (24.75 mg/100 ml) followed by GA_3 at 30 ppm (23.17 mg/100) as against control (21.95 mg/100 ml). These results were in line with the findings of Sindhu and Sighrot [30] who reported maximum ascorbic acid content in GA_3 treated fruits.

Table 3. Effect of GA_3 on weight loss and decay loss of mandarin fruit during storage at ambient for 15 days (14°C-18°C and 45% - 73% RH) and cellar condition for 90 days (10°C-12°C and 85% - 90% RH).

Treatments (ppm)	Ambient		Cellar	
	Weight loss (%)	Decay loss (%)	Weight loss (%)	Decay loss (%)
GA_3 10	6.01	2.12	8.24	13.24
GA_3 20	5.61	1.18	6.77	10.27
GA_3 30	5.17	1.02	6.69	8.21
Control	9.52	5.54	11.76	21.58
LSD (0.05)	0.59	NS	1.63	3.78

3.7. Physiological Loss in Weight (PLW)

The perusal in Table 3, indicated the increase in PLW with the increase in storage period both at ambient and under cellar condition. The PLW was found significant among the treatments. The minimum loss in weight was observed in the fruits treated with GA_3 at 30 ppm (5.17%) followed by GA_3 at 20 ppm (5.61%), and GA_3 at 10 ppm (6.01%) and maximum loss in weight was recorded in control (9.52%) stored at ambient condition for 15 days. Under the cellar condition, the maximum PLW was recorded with the fruit treated with GA_3 at 30 ppm (6.69%) followed by GA_3 at 20 ppm (6.77%), and GA_3 at 10 ppm (824%) as against control (11.76%) at the end of the 90 days. The possible reason for reduction in weight loss by GA_3 could be due to its ability to retain more water against force of evaporation. GA_3 acts an antisenescent and antirespirant, which inhibited catabolic activities and consequently reduced the weight loss during storage [9] . These findings are conformity with the reports of Sindhu and Singhrot [30] in lemon, (Brahmachari and Rani [9] in guava, Ladaniya [5] in Nagpur mandarin, and Ahlawat et al. [30] in kinnow mandarin who reported decrease in weight loss of fruits during storage.

3.8. Decay Loss

Table 3 shows that the decay loss increased with the increase in period of storage in all the treatments. The decay loss among the treatments was not significant at ambient condition. However, the minimum decay loss was found in the fruits with GA_3 30 at ppm (1.02%) followed by GA_3 at 20 ppm (1.18%) and GA_3 at 10 ppm (2.12%) as against control (5.54%). On the other hand, the minimum decay loss was observed in the fruits treated with GA_3 at 30 ppm (8.21%).The lesser decay loss percentage in GA_3 treated fruits could be due to antisenesecent and antirespirant properties of the gibberellins which prevent the cellar disintegration with enhancing resistant ability in the fruit. Similar results were reported by Brahmachari and Rani [9] in guava, Sindhu and Singhrot [30] in lemon, Ladaniya [5] in Nagpur mandarin and Ahlawat et al. [31] in kinnow mandarin.

4. CONCLUSION

On the basis of the above experimental findings, we concluded that GA_3 at 20 ppm was more effective to increase fruit weight and overall fruit quality whereas GA_3 at 30 ppm was the best for the prolonging the shelf life and minimising the physiological loss in weight (PLW) and decay loss of mandarin in both ambient and cellar condition. GA_3 at 20 ppm retained higher fruit weight (128.6 g), higher peel puncture resistance (3.54 kg/cm^2), higher juice recovery (57.75%), and higher TSS/acid ratio (21.24) during the storage. In overall, GA_3 was proved to delay the ripening of the fruits for one month through delaying the chlorophyll development and reducing the loss of texture, decreasing sugar content and increasing acidity.

ACKNOWLEDGEMENTS

The authors are thankful to Director General of Department of Food Technology and Quality Control (DFTQC) of Ministry of Agricultural Development, and the chief of postharvest research unit of NARC, Khumaltar, Lalitpur for providing laboratory facility during the experiment.

REFERENCES

1. Subedi, P.P. (2001) Postharvest and Market Assessment of Citrus Fruits. In: Subedi, P.P., Dhakal, D.D. and Khanal, N.P., Eds., Integrated Management for Citrus Decline, Workshop, Damauli, 46-50.
2. Gautam, D. and Adhikari, R.R. (1989) Study on Modified Local Methods of Orange Storage. Journal of the Institute of Agriculture and Animal Science, 10, 91-95.

3. Bastakoti, P. and Gautam, D.M. (2007) Effect of Maturity Stages and Postharvest Treatments on Shelf Life and Quality of Mandarin Oranges in Modified Cellar Store. Journal of Institute of Agriculture and Animal Science, 28, 65-74.
4. Kawase, K., Susuki, K. and Hirose, K. (1981) Use of Growth Regulators to Control Rind Puffing of Satsuma Mandarin Fruit. Proceedings of International Society of Citriculture, 1, 237-239.
5. Ladaniya, M.S. (1997) Response of Nagpur Mandarin Fruit to Preharvest Sprays of GA3 and Carbendazim. Indian Journal of Horticulture, 54, 205-212.
6. Ladaniya, M.S. (2001) Post-Harvest Technology of Fresh Citrus Fruits. In: Singh, S. and Naqvi, S.A.M.H., Eds., Citrus, International Book Distributing Co., Lakhnow, 459-473.
7. Chapman, J.C., Book, G. and Peacook, B.C. (1979) Effect of Pre-Harvest GA Sprays. Ellendale Mandarin Fruits. Queensland Journal of Agriculture and Animal Science, 36, 53.
8. Kher, R., Bhat, S. and Wali, V.K. (2005) Effect of Foliar Application of GA3, NAA and CCC on Physic-Chemical Characteristics of Guava CV. Sardar. Haryana Journal of Horticultural Sciences, 34, 31-32.
9. Brahmachari, V.S. and Rani, R. (2005) Shelf Life Extension in Guava with Pre-Harvest Spray of Certain Growth Substances and Polyethylene Wrapping. Haryana Journal of Horticulture Science, 34, 36-38.
10. Bhattacharya, S.C. (1991) Storage Studies in Lime (Citrus aurantifolia). Master Thesis, Bidhan Chandra Krishi Vishowvidhaylaya, West Bengal, 27-26.
11. Kaur, H., Singh, A., Gupta, M. and Randhawa, J.S. (2008) Effect of NAA and Gibberellic Acid on Pre-Harvest Fruit Drop and Quality of Satluj Purple Plum. Haryana Journal of Horticultural Sciences, 37, 31-32.
12. Thapa, B.B. and Gautam, D.M. (2002) Pre-Harvest Application of Gibberellic Acid and Maturity Stage on Fruit Quality and Shelf Life of Mandarin. Journal of the Institute of Agriculture and Animal Science, 23, 41-47.
13. Bakhshi, J.S. and Chadha, K.L. (1968) Gibberellins and Plant Growth—A Review. The Panjab Horticultural Journal, 8, 78-113.
14. Nawaz, M.A., Ahmad, W., Ahamad, S. and Khan, M. (2008) Role of Growth Regulators in Pre-Harvest Fruit Drop, Yield and Quality in Kinnow Mandarin. Pakistan Journal of Botany, 40, 1971-1981.
15. Sen, A., Meyvaci, K.B., Can, H.Z. and Teksur, P.K. (2012) Effect of Preharvest Gibberellic and Calcium Applications On-Tree Storage of Satsuma Mandarins. Proceedings of the VII International Postharvest Symposium, 25-29 June 2012, 233-239. http://www.actahort.org/books/1012/1012_26
16. Amir, M., Chaturvedi, O.P. and Tripathi, V.K. (2003) Effect of Pre-Harvest Application of Gebberellic Acid and Calcium Nitrate on the Fruit Maturity and Storage Quality of Kinnow Mandarin. Farm Science Journal, 12, 148-149.
17. AOAC (1995) Official Methods of Analysis. 16th Edition, Association of Official Analytical Chemists, Washington DC.
18. Gomez, A.K. and Gomez, A.A. (1984) Statistical Procedures for Agricultural Research. John Wiley and Sons Inc., New York.
19. Singh, A. K., Singh, R. and Mann, S.S. (2003) Effect of Plant Regulators and Nutrients on Fruit Set, Yield and Quality of Pear CV. Le-Conte. Indian Journal of Horticulture, 60, 34-39.
20. Pal, R.N., Singh, R., Vij, Y.K. and Munshi, S.K. (1977) Effect of 2, 4-D, GA3 and Cycocel, and Fruit Growth Pattern of Kinnow Mandarin. Indian Journal of Horticulture, 34, 4-7.
21. Kumar, J., Rana, S.S., Verma, H.S. and Jindal, K.K. (1996) Effect of Various Growth Regulators on Growth, Yield and Fruit Quality of Strawberry CV. Tioga. Haryana Journal of Horticultural Sciences, 25, 168-171.
22. Bose, T.K., Hussain, T., Mitra, S.K. and Roy, A. (1988) Control of Pre-Mature Fruit Drop in Mandarin Orange. Haryana Journal of Horticultural Sciences, 17, 140-143.

23. Daulta, B.S. and Beniwal, V.S. (1983) Effect of Time of Application, Chemicals and Concentration of Plant Growth Regulators on Composition of Fruit in Sweet Orange (Citrus sinensis Obseck) CV. Campbell Valencia. Haryana Journal of Horticultural Sciences, 12, 168-172.
24. El-Otmani, M. and Coggins Jr., C.W. (1991) Growth Regulator Effect on Retention of Quality of Stored Citrus Fruits. Scientia Horticulturae, 45, 261-272. http://dx.doi.org/10.1016/0304-4238(91)90072-7
25. Rana, M.K. (2006) Ripening Changes in Fruits and Vegetables—A Review. Haryana Journal of Horticultural Sciences, 35, 271-279.
26. Sri Hari Babu, R., Rajput, C.B.S. and Rath, S. (1982) Effect of Zinc, 2, 4-D and GA3 in Kagzi Lime (Citrus aurantifolia Swingle) IV Fruit Quality. Haryana Journal of Horticultural Sciences, 11, 59-65.
27. Bhat, A., Kher, R. and Gupta, S.P. (2006) Effect of Growth Regulators and Nutrients on Fruit Cracking and Fruit Quality in Lemon. Haryana Journal of Horticulture Science, 35, 267.
28. Murata, T. (1977) Studies on the Postharvest Physiology and Storage of Citrus Fruit VIII. Metabolism of Citrate-14C and Glucose-14C in Satsuma Mandarin Fruit during Storage. Journal of the Japanese Society for Horticultural Science, 46, 375-379. http://dx.doi.org/10.2503/jjshs.46.375
29. Mapson, L.W. (1970) The Bio-Chemistry of Fruits and Their Products. In: Hulme, A.C., Ed., Vitamins in Fruits, Vol. 1, Academic Press, London, 369-384.
30. Sindhu, S.S. and Singhrot, R.S. (1993) Effect of Pre-Harvest Spray of Growth Regulator and Fungicides on the Shelf Life of Lemon CV. Baramasi. Haryana Journal of Horticulture Science, 22, 204-206.
31. Alhawat, V.P., Daulta, B.S. and Singh, J.P. (1984) Effect of Pre-Harvest Application of GA3 and Captan on Storage Behaviour of Kinnow—A Mandarin Hybrid. Haryana Journal of Horticulture Science, 13, 4-8.

CHAPTER 17

Chemical Composition and Antioxidant Activities of *Broussonetiapapyrifera* Fruits

Jie Sun, Shao-fang Liu, Chu-shu Zhang, Li-na Yu, Jie Bi, Feng Zhu, Qing-li Yang*

Shandong Peanut Research Institute, Qingdao, People's Republic of China

ABSTRACT

Fruits of *Broussonetiapapyrifera* from South China were analyzed for their total chemical composition, and antioxidant activities in ethanol and aqueous extracts. In the fruit of this plant, the crude protein, crude fat and carbohydrates was 7.08%, 3.72% and 64.73% of dry weight, respectively. The crude protein, crude fat and carbohydrates were 15.71%, 20.51% and 36.09% of dry weight, respectively. Fatty acid and amino acid composition of the fruit were analyzed. Unsaturated fatty acid concentration was 70.6% of the total fatty acids. The percentage of the essential amino acids (EAAs) was 40.60% of the total amino acids. Furthermore, *B. papyrifera*fruit are rich in many mineral elements and vitamins. Total phenolic content was assessed using the Folin-Ciocalteau assay, whereas antioxidant activities were assessed by measuring the ability of the two extracts to scavenge DPPH radicals, inhibit peroxidation, and chelate ferric ions. Their reducing power was also assessed. Results indicated that the aqueous extract of *B. papyrifera* was a more potent reducing agent and radical-scavenger than the ethanol extract. GC–MS analysis of the ethanol extract showed the presence of some acid-containing compounds. The changes in total phenolic content and antioxidant capacity in *B. papyrifera*from four different regions grown under normal conditions were assessed. The antioxidant activity of different extracts was positively associated with their total phenolic content. These results suggest that the fruit of *B. papyrifera* could be used in dietary supplement preparations, or as a food additive, for nutritional gain, or to prevent oxidation in food products.

INTRODUCTION

Plants have long been used as a source of traditional medicines to treat various diseases and conditions. Many of these medicinal plants are also excellent sources of phytochemicals, many of which have potent antioxidant activities. *Broussonetiapapyriferais* is a deciduous tree or shrub which grows naturally in the Asian and Pacific region in China, Thailand and the USA. The roots, leaves, bark and fruit are all used in traditional Chinese medicines. In China, the leaves have been used in folk medicine against chronic prostatitis, and the fruit has been used to treat impotence and ophthalmic disorders, the efficacy of which has been proved by pharmacological experiments [1]. Various compounds identified in this plant have been shown to inhibit lipid peroxidation [2], have antiplatelet effects [3], and inhibit the activities of the protein tyrosine phosphatase 1b (PTP1B) enzyme and aromatase [1], [4].

At present, there is an increasing interest both in industry and scientific research in useful plants which could be recognized as a nutritious food or dietary supplement as well as an important source of biologically active compounds of medicinal value [5], [6]. Abundant nutrient content of the plants help keep our body functioning well. Dietary antioxidants are important components because they protect against free radicals in the human body, such as reactive oxygen species [7]. In recent years, many studies have focused on the properties and composition of extracts obtained from the roots, bark and leaves of *B. papyriferais*. Rong Min Chen isolated three kinds of PTP1B inhibitors from its roots, and Son et al isolated a new prenylatedflavonol, papyriflavonol A, from its root bark [3], [8]. However, there are few studies on the chemical composition and antioxidant activities of this plant's fruit. The fruit of *B.papyrifera* is an important source of folk drugs in the Middle East [1], and further research on it would be of great interest. In view of the limited data on this fruit, we set out to: 1) study its chemical composition; 2) evaluate the antioxidant activities of different fruit extracts using several antioxidant assays; and 3) identify candidate constituents responsible for this activity.

MATERIALS AND METHODS

Ethics statement

In this study, no specific permits were required for the described field studies. The study is not privately-owned or protected in any way. The field studies did not involve endangered or protected species.

Materials

The principle reagents used, and their sources were as follows. 2, 20-diphenyl-1-picrylhydrazyl (DPPH), ferrozine, tocopherol, p-coumaric acid, and linoleic acid were purchased from Sigma-Aldrich (St Louis, MO, USA). Folin-Ciocalteau was sourced from AppliChem (GER). All other reagents were either of analytical grade, or of the highest quality available.

Plant materials

The ripe fruit of *B. papyrifera*, dark red in colour, were collected from the Zhongshan Arboretum, Qingdao Shandong Province, PR China, in September 2009. The fruits (200 g from 12 different trees) were dry heated in a hot air oven at 50°C for 24 h and the beans were allowed to cool at room temperature. Sample fruit were shelled and broken, before freeze-drying to constant weight, and then powdered to pass through a 40 mesh sieve. The *B. papyrifera* fruits from different regions (Hena Province, Hubei Province, Guangxi Province) were bought from traditional Chinese medicine pharmacy in Qingdao.

Proximate and mineral analyses

Proximate analysis, including moisture, crude fat, fiber, ash, and crude protein contents, was performed in three biological replicates, according to Association of Official Analytical Chemists AOAC (1995) [9] procedures. For mineral analyses, a Hitachi Z-8000 atomic absorption spectrometer procedure, described by AOAC (1995), was used. Phosphorus content was measured by phosphorus molybdenum blue spectrophotometry. Pb and Cd levels in samples were determined by a HGA graphite furnace, using argon as the inert gas. Other measurements were carried out in an air/acetylene flame. V_{B1}, V_{B2}, V_{B5} and V_{B6} contents were measured by spectrophotometer. V_C, ß- carotene, V_D, V_E and V_A levels in samples were measured by high performance liquid chromatography [10].

Fatty acid analysis

Samples (150 mg) were mixed with heptadecanoic acid methyl ester (internal standard) and extracted with chloroform/methanol (2:1) at 60°C for 1 h. The final extract was concentrated to 5 mL. Fatty acids in the extract were simultaneously hydrolyzed and derivatized as methyl esters with 1 mL of NaOH/methanol at 90°C for 10 min, and then a complete derivation was assured with 1 mL of BF_3 at 90°C for 10 min. The methyl esters were purified with 1 mL of hexane and 1 mL of water. Individual samples were passed through an anhydrous Na_2SO_4column, and then evaporated to dryness under a stream of nitrogen and re-dissolved in 100 μL of isooctane. The derivatized fatty acids were separated in a HP5890 Series II gas chromatograph equipped with a MS detector 5972 and a cross-linked column with a stationary phase of 5% phenyl methyl silicone [11].

Amino acid analysis

A modification of the AOAC (1995) method was used for amino acid analysis. Dry samples were hydrolyzed with 25 mL 6N HCl at 110°C for 24 h. Amino acid analysis was carried out by ion-exchange chromatography in an automatic amino acid analyzer (Hitachi L-8800).

Preparation of plant extracts

Fruit powders (50 g) were extracted by using a Soxhlet extractor for 1 h with 50 ml of extractant, including petroleum ether under reflux conditions. The drying residues (10 g) were then extracted with ethyl alcohol, at room temperature for 24 h with a mass to volume ratio of 1:20 (g/mL) and filtered through Whatman No. 4 filter paper. The extracts were evaporated to dryness on a rotary evaporator at 40°C. After ethyl alcohol extracted the residues were then treated with water under the same conditions. The ethyl alcohol and aqueous extracts were subsequently re-dissolved in the same solvent at a concentration of 5 mg/ml. Extracts were kept at 4°C until the bioassay analyses.

Total phenolic analysis

The Folin-Ciocalteau method was used for total phenolic analysis [12], [13]. Ten milliliters of 1:10 Folin-Ciocalteau reagent was added to 100 μL of sample (5 mg/ml) or standard (5 mg/ml). The mixture was mixed and incubated for 5 min before the addition of 2.5 mL of 2%Na_2CO_3. The resulting solution was incubated for a further 2 h before absorbance readings were taken at 760 nm. Gallic acid was used for the calibration curve and a range of concentrations from 100–1000 mg/L were prepared and analyzed as above. Results were expressed as mg gallic acid equivalents (GAE)/g dried plant material. All experiments were done in triplicate.

Antioxidant activity assays

Determination of 2, 20-diphenyl-1-picrylhydrazyl (DPPH) radical scavenging ability. The scavenging activity of the ethanol and aqueous extracts on DPPH radicals was measured according to the method of Tuberoso with some modifications **[14]**, **[15]**. 250 μL of extract, at different concentrations (0.5–5 mg/ml), were dissolved in 2.5 ml of DPPH, 0.1 mmol/l in methanol. Ethanol or water was used instead of the extract samples as a control. Spectrophotometric readings were carried out with a Cary 50 spectrophotometer at 517 nm, using a 10 mm quartz cuvette. Standards were prepared in a similar manner with ascorbic acid, tocopherol and 2, 6-Di-tert-butyl-4-methylphenol (BHT) at different concentrations, for comparison. All experiments were done in triplicate.

The scavenging activity (%) (SA) on DPPH radicals was calculated using the formula:

$$SA = (1 - \text{Abs in the presence of sample}/$$

$$\text{Abs in the absence of sample}) * 100$$

Chelation of metal ions (Fe^{2+}). The ability of the ethanol and aqueous extracts to chelate Fe^{2+} ions was evaluated using the method of Dinis, Madeira and Almeida (1994) with some modification **[16]**. 300 μL of sample, at various assay concentrations (0.5–5 mg/ml) was made up to 3.7 ml with deionized water. A solution of 2 mM ferrous chloride (0.1 ml) was added, and after 3 min the reaction was inhibited by the addition of 5 mMferrozine (0.2 ml). The mixture was shaken vigorously and left at room temperature for 10 min. Absorbance of the resulting solution was measured at 562 nm. A blank without the sample was prepared in a similar manner. Standards with ascorbic acid, tocopherol and 2, 6-Di-tert-butyl-4-methylphenol (BHT) at various concentrations were also made in a similar manner for comparison. All experiments were done in triplicate.

The chelating capacity was calculated as follows:

$$Fe^{2+} \text{ chelating activity}(\%) =$$

$$(\text{eabsorbance of blank} - \text{absorbance of sample})$$

$$/\text{absorbance of blank} * 100$$

Reducing power. The reducing power was measured following Oyaizu (1986) with some modification **[17]**. 250 μL of extract with different concentrations (0.5–5 mg/ml) were mixed with 1.75 ml of 0.2 M phosphate buffer (pH 6.6) and 1 ml of 1% potassium ferricyanide. The mixture was incubated at 50°C for 20 min followed by the addition of 1 ml 10% TCA. An aliquot (1 ml) from the incubation mixture was mixed in a test tube with 1 ml of distilled water and 0.2 ml of 0.1% ferric chloride. After 10 min the absorbance of the resulting solution was measured at 700 nm. An increased absorbance of the reaction mixture indicates increased reducing power **[18]**. A standard with ascorbic acid, tocopherol and 2, 6-Di-tert-butyl-4-methylphenol at various concentrations was also made in a similar manner for comparison. All experiments were done in triplicate.

Measurement of the lipid peroxidation inhibition activity in a linoleic acid emulsion system. The lipid peroxidation inhibition activity of the ethanol and aqueous extracts were measured in a linoleic acid emulsion system according to the methods of Qian, Jung, Byun, and Kim (2008) with some modification **[19]**. Briefly, 2.0 ml of extracts (0.5–5 mg/ml) was mixed with 2 ml of 2.5% linoleic acid dissolved in 95% ethanol. Then, 4 ml of 50 mM sodium phosphate buffer (pH 7.0) and 2 ml of distilled water were added. The mixture was incubated in a 20 ml centrifuge tube at 40±1°C in a dark room, and the degree of oxidation was evaluated by measuring the FeSCN values described below. The reaction solution (0.1 ml), incubated in the linoleic acid model system, was mixed at different intervals during the incubation period with 9.7 ml of 75% ethanol, 0.1 ml of 30% NH_4SCN, and 0.1 ml of 20 mM $FeCl_2$ solution in 3.5% HCl. After 3 min, the SCN value was measured by reading the absorbance at 500 nm. An equivalent volume of distilled water was used as a blank **[20]**. Ascorbic acid, tocopherol and 2,6-Di-tert-butyl-4-methylphenol (BHT) at 0.5–5 mg/ml was used as a control. All experiments were done in triplicate.

$$\text{The lipid peroxidation inhibition activity}(\%) =$$

$$\{1-(A_{s,t-X\,h}-A_{s,t-0\,h})*100\}/(A_{0,t-X\,h}-A_{0,t-0\,h})$$

Where: $A_{s,t\text{-}X\,h}$ and $A_{s,t\text{-}0\,h}$ are the absorbances of the sample at X h and 0 h, respectively; and $A_{0,t\text{-}X\,h}$ and $A_{0,t\text{-}0\,h}$ are the absorbances of the blank at X h and 0 h, respectively. Plots were made of scavenging activity against the concentration of the sample at 96 h and 168 h.

Antioxidant activity analysis of ethyl alcohol and aqueous extracts of *B. papyrifera* fruit from different regions. The antioxidant activity assays were performed on the extracts of *B. papyrifera* fruit from different regions (Hebei Province, Hubei Province, Jiangxi Province and Shandong Province). The ethyl alcohol and aqueous extracts of *B. papyrifera* fruit from different regions were obtained according to the methods described in 2.6.

The scavenging activity, Fe^{2+} chelating activity, reducing power and lipid peroxidation inhibition activity of ethyl alcohol and aqueous extracts from different regions were detected by the methods described in 2.8.1, 2.8.2, 2.8.3. All experiments were done in triplicate.

Measurements by high-performance liquid chromatography (HPLC)

0.01 g of dried ethanol and aqueous extracts were mixed with 2 mL of acetonitrile and 0.2 mL of 0.1 Nhydrochloricacid and stirred for 2 h at room temperature. The suspension was centrifuged at 10000 rcf for 1 min. The analyses were performed using a Waters Sun Fire C18 column (4.6×250 mm) and a 2998 diode matrix detector in a Waters e2695 HPLC system. The mobile phase was methanol (solvent A) and 5mlphosphate (solvent B). The gradient was 0–25 min, 20%–80% A, 80%–40% B. The mobile phase was 0.1% phosphate/methanol at a flow rate of 0.8 mL/min, and UV detection was at 273 nm. Phenolic acid standards, such as 7-hydroxycoumarin, ferulic acid, resveratrol, rutin, cinnamylate, meletin, protocatechuate acid, catechuic acid, epicatechin, protocatechuic acid and p-hydroxy phenyl ethyl ketone, were used for the identification of phenolic acids present in the *B. papyrifera* extracts. Sample compounds were identified on the basis of the retention times of standard materials and were quantified by comparing their peak areas with those of standard curves [21].

Statistical analyses

The SPSS 13.0 software (SPSS Inc., Chicago, IL, USA) was used to calculate the means and standard deviations in any experiments involving duplicate or triplicate analyses of any sample or condition. The statistical significance of any observed differences was evaluated by oneway analysis of variance (One-way ANOVA), using the Bonferroni Multiple Comparisons Test.

RESULTS AND DISCUSSION

Proximate and mineral analyses

The proximate composition of fruits of *B. papyrifera* is shown in Table 1, and the following data are expressed as average percentage dry weight (DW). The moisture content of the fruit was 3.75%. Carbohydrate (36.09%) formed the single largest component in comparison to other constituents detected. Crude fat content (20.51±0.21) was much higher compared to other fruit such as *Citrulluslanatusvar* (10.9%) [22] and white mahlab *Prunusmahaleb* (13.15%) ($P<0.05$) [23]. which indicated very clearly that the *B. papyrifera* formed a potential source of oils, and fats. However, it would be seen that, the fat content was lower than most Chinese conventional oil fruits (sunflower 34.4%, sesame 59.6% and groundnut 40.1%) ($P<0.05$) [22]. Crude protein content was 15.71%, which was less than the range of most conventional oil seeds, such as sesame, bean, and sunflower which contain about 18.3–25.3% protein [24].

Table 1. Moisture content and proximate composition of B. papyrifera fruits (g/100 g DW).

component	moisture	ash	crude fat	crude protein	carbohydrates	crude fiber
fruits	3.75±0.17	9.74±0.13	20.51±0.21	15.71±0.16	36.09±0.32	14.20±0.23

Table 2. Mineral composition of B. papyrifera fruits (mg/kg DW).

minerals	fruits
Ca mg/kg	17338.9±15.12
Cu mg/kg	19.75±0.43
Fe mg/kg	414.8±8.23
Mg mg/kg	3284.6±12.1
Zn mg/kg	29.8±1.01
Se µg/kg	<0.08±0.00
Cr mg/kg	3.48±0.11
K mg/kg	9098.7±14.18
Mn µg/kg	44.56±0.76
Cd mg/kg	0.009±0.00
Pb mg/kg	0.52±0.03
P mg/kg	4644.2±12.98
Hg µg/kg	4.91±0.79

The levels of trace elements in the fruit are shown in Table 2. *B. papyrifera* fruit were found to be a good source of minerals. Calcium content was 17338.9±15.12 mg/kgDW, whereas the magnesium content was 3284.6±12.1 mg/kgDW. Calcium is important for bone growth, and muscle and neurological function, whereas magnesium is linked with the prevention of ischemic heart disease. The *B. papyrifera* fruit had a high level of potassium (9098.7±14.18 mg/kgDW). Potassium is known to play an important role in regulating the permeation pressure of cells and takes part in the metabolism of sugar and protein in human cells in vitro. These fruit could theoretically provide other nutritionally useful minerals including copper, iron, magnesium and zinc (Table 2). The levels of iron and copper correspond with the recommended dietary allowances of NRC/NAS.

Table 3 shows the vitamin composition of the fruit analyzed. Of the nine vitamins analyzed, seven were found at various levels. V_A and V_D could not be detected in this species. V_E is a free radical scavenger, which can prevent damage to the body by free radicals, and was present in the fruit at a concentration of 7.29 mg/100 g.

Table 3. Vitamin composition of B. papyrifera fruits (mg/100 g).

vitamin	ß- carotene	V_E	V_D	V_A	V_{B1}	V_{B2}	V_{B5}	V_{B6}	V_C
fruits	0.09±0.00	7.29±0.054	-	-	0.32±0.023	0.12±0.01	1.85±0.017	1.21±0.09	0.01±0.00

Table 4 shows the fatty acid composition of the fruit. C18:2, C18:0, and C16:0 were the main fatty acid constituents. C18:2 is an essential fatty acid, which can reduce cholesterol levels in the blood. C18:2 constituted 52.5833% of the total fatty acids in the fruit. The C18:2 content was much higher than that of rapeseed, peanut and sesame [22]. These results suggest that the fruit of *B. papyrifera* could be a good source of C18:2 in the human diet. Other fatty acids, for example, C10:0, C12:0, C17:0 and C23:0 were found only in very small amounts.

Sixteen known amino acids were found in the fruit proteins of *B. papyrifera,* as shown in Table 5. The essential amino acids (EAAs) comprised 40.59% of the total, and the ratio of EAAs to non-EAAs in the fruit was 0.67, only slightly higher than the reference values of 40% and 0.6 recommended by FAO/WHO [25].

Total phenolic analysis

The total phenolic content of the two types of fruit extract were determined following a modified Folin-Ciocalteu reagent method and the results were expressed as gallic acid (GA) equivalents. The aqueous extract of the dried fruit had the highest phenolic content (5.79±0.12 mg GAE/g), followed by the ethanol extract (3.79±0.03 mg GAE/g).

DPPH radical scavenging activity

The DPPH radical inhibition assay is a widely used and comparably easy method to evaluate antioxidant activity, because DPPH is a stable free radical, which produces a violet solution in ethanol. It is reduced and decolorized in the presence of an antioxidant molecule [26]. Using this method, the radical scavenging activities of the fruit extracts of *B. papyrifera* were estimated by comparing the percentage inhibition of formation of DPPH radicals by the extracts compared to those of standards.

Table 4. Fatty acid composition (%) of B. papyrifera fruits.

fatty acids	fruits
C9:0	0.2316±0.0023
C10:0	0.0182±0.0009
C12:0	0.0766±0.0013
C 14:0	0.7742±0.0033
C 15:0	0.2375±0.0021
C 16:1 (cis - 9)	0.6219±0.0043
C 16:0	16.3625±0.0098
C 17:0	0.2374±0.0003
C 18:2 (cis,cis - 9,12) (omega 6)	52.5833±0.0148
C 18:1(cis - 9) (omega 9)	16.9175±0.0072
C 18:0	3.7695±0.0042
C 20:1 (cis - 11) (omega 9)	0.4953±0.0024
C 20:0	2.0615±0.0044
C 21:0	0.3348±0.0012
C 22:0	4.0033±0.0035
C 23:0	0.2011±0.0003
C 24:0	1.0757±0.0053

Table 5. Amino acid composition of B. papyrifera fruits (mg/100 mg of Protein).

amino acid	fruits
aspartate	1.61±0.23
threonine	1.67±0.43
serine	0.74±0.17
glutamate	2.46±0.21
glycine	1.07±0.15
alanine	0.61±0.12
cysteine	-----
valine	0.81±0.24
methionine	0.05±0.02
isoleucine	0.65±0.18
leucine	1.05±0.27
tyrosine	0.48±0.16
phenylalanine	0.76±0.16
lysine	0.71±0.17
histidine	0.34±0.12
arginine	0.71±0.11
proline	0.32±0.14

Fig. 1 shows a steady increase in the percentage inhibition of the absorbance of the DPPH radicals by the extracts up to a concentration of 5 mg dried extract/mL. This pattern of DPPH inhibition is commonly observed with plant extracts [27]. The ethanol and aqueous extracts of *B. papyrifera* were able to inhibit the formation of DPPH radicals with a percentage inhibition of 87.17±0.18, and 58.11±0.11% at the highest concentration, respectively (Fig. 1). When compared with standards, the DPPH radical scavenging capacity of the aqueous extract was weaker than ascorbic acid and tocopherol, and higher than BHT at the same concentration. DPPH radical scavenging capacity of the aqueous extract was weaker than all of standards we tested, at the same concentration.

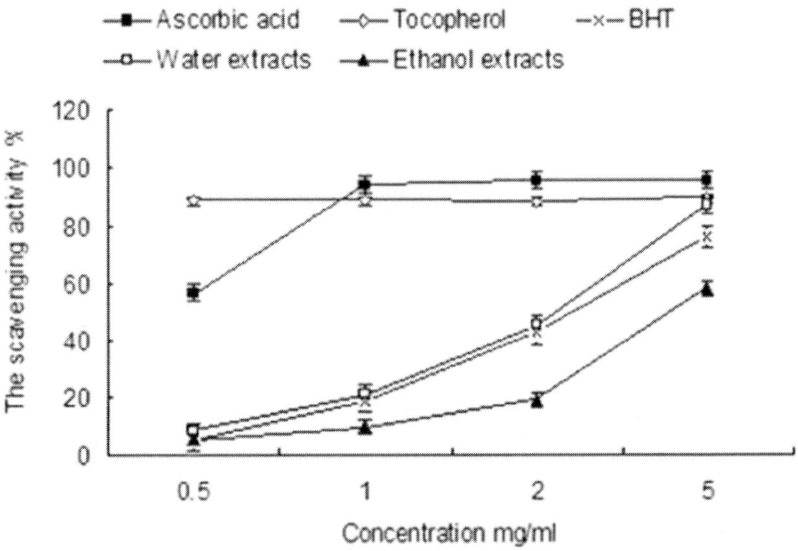

Figure 1. DPPH radical scavenging capacity of ethanol and aqueous extracts of *B. papyrifera*.The activity of different amounts (0.5–5 mg/ml) of ascorbic acid, tocopherol, BHT, ethanol extracts and aqueous extracts is expressed as per cent scavenging and the values represents the means of three independent experiments.

Chelation of metal ions (Fe^{2+})

In the present study, the Fe^{2+}-chelating activities of the different extracts were compared with that of Ascorbic acid, tocopherol and 2,6-Di-tert-butyl-4-methylphenol (BHT) (Fig. 2). The Fe^{2+}-chelating activities of the ethanol and aqueous extracts of *B. papyrifera* fruit were excellent and increased steadily with increasing concentration. With an Fe^{2+}-chelating activity of approximately 77.51%, the aqueous extract (5 mg/ml) showed significantly higher Fe^{2+}-chelating activity than the three standards, at the same concentration ($P<0.01$). The ethanol extract (5 mg/ml), showed a chelation capacity of 48.26%, and also showed a higher Fe^{2+}-chelating activity than the three standards ($P<0.05$). This result clearly shows that the lower molecular weight fractions contained some special components that can chelate metal ions such as Fe^{2+}.

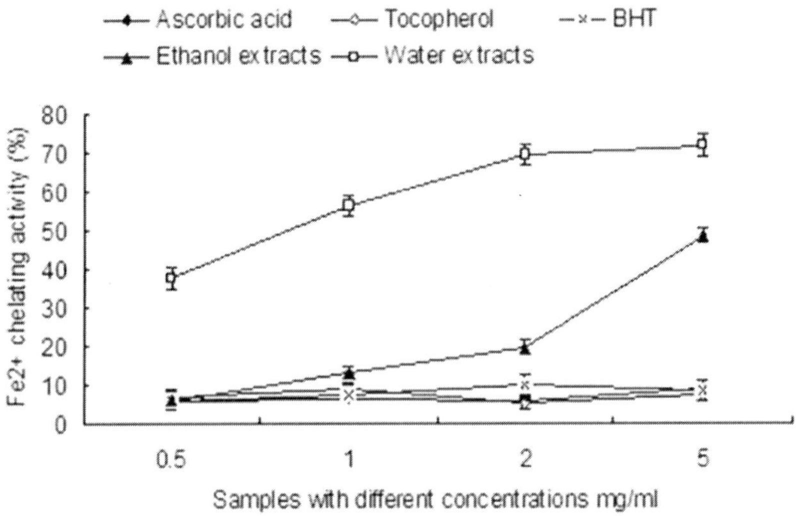

Figure 2. Fe^{2+}-chelating activity of ethanol and aqueous extracts of *B. papyrifera*.Determination of Fe^{2+}-chelating activity of ascorbic acid, tocopherol, BHT, and different concentrations (0.5–5 mg/ml) of ethanol extracts and aqueous extracts. The activities are expressed as % chelating and monitored in three biological replicates.

Reducing power

The reducing power of the ethanol and aqueous extracts are shown in Fig. 3. In this assay, the yellow color of the test solution changes to various shades of green and blue, depending on the reducing power of each compound. The presence of reducers (i.e., antioxidants) causes the reduction of the Fe^{3+} ferricyanide complex to the ferrous form. Therefore, measuring the formation of Perl's Prussian blue at 700 nm can monitor the Fe^{2+} concentration [21]. The ethanol and aqueous extracts showed lower reducing power when compared to the three standards. The aqueous extract showed higher reducing power than the ethanol extract at the same concentration. These findings indicate that the two kind of extracts contained minute quantities of compounds with high reducing power.

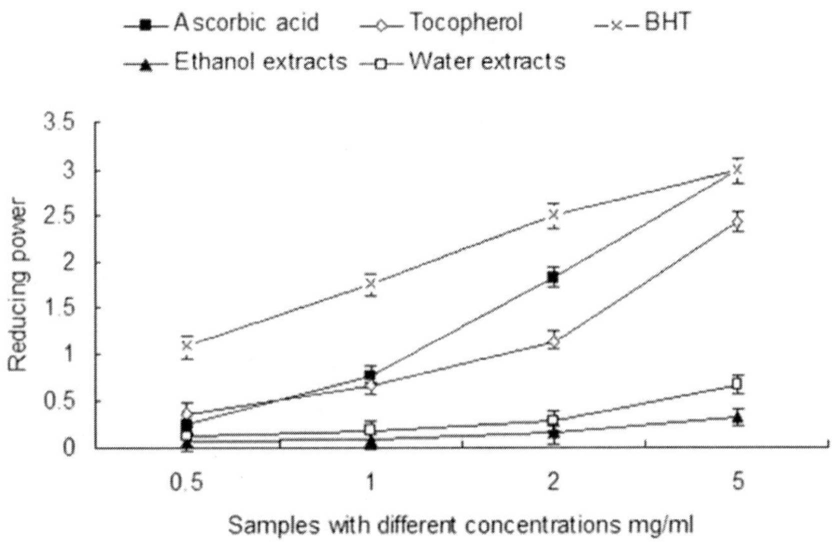

Figure 3. Reducing power capacity of ethanol and aqueous extracts of *B. papyrifera*. Ascorbic acid, tocopherol, BHT, ethanol and aqueous extracts at various concentrations (0.5–5 mg/ml) were used to detect the reducing power capacity. The reducing power capacities of different samples are monitored in three biological replicates.

Measurement of the lipid peroxidation inhibition activity in a linoleic acid emulsion system

The degree of lipid peroxidation inhibition of the ethanol and aqueous extracts at different times (96 h and 168 h) are shown in Fig. 4. Lipid peroxidation is thought to proceed via a radical-mediated abstraction of hydrogen atoms from methylene carbons in polyunsaturated fatty acids [28]. The amount of lipid peroxidation inhibition of the ethanol and aqueous extracts in the linoleic acid system was similar with that of tocopherol, and higher than ascorbic acid after 96 h. Nevertheless, the degree of lipid peroxidation inhibition of the two extracts was much higher than that of tocopherol and ascorbic acid after 168 h ($P<0.05$). That is to say that both the ethanol and aqueous extracts had a stronger lipid peroxidation inhibitory effect than tocopherol and ascorbic acid at the same molecular concentration, when the experiments were allowed to run for an extended period of time.

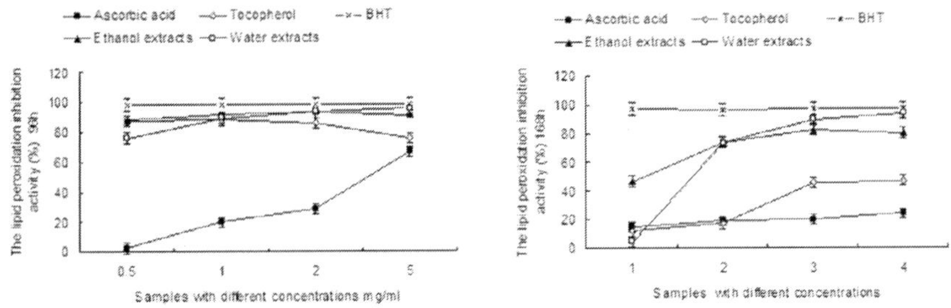

a b

Figure 4. The lipid peroxidation inhibitory activity of ethanol and water extracts of *B. papyrifera*.a, The lipid peroxidation inhibitory activity of ascorbic acid, tocopherol, BHT, ethanol and aqueous extracts at various concentrations(0.5–5 mg/ml) in 96 h. b, The lipid peroxidation inhibitory activity of ascorbic acid, tocopherol, BHT, ethanol and aqueous extracts at various concentrations(0.5–5 mg/ml) in 168 h.

Antioxidant activity analysis of ethyl alcohol and aqueous extracts of *B. papyrifera* fruit from different regions

The antioxidant capacity of ethanol and aqueous extracts of *B. papyrifera* fruit from different regions (Hubei,Hebei,Jiangxi and Shandong Province) was evaluated in order to establish how environmental influences could affect it, by using four different methods: DPPH, Fe^{2+} chelating activity, reducing power and lipid peroxidation inhibition activity tests (Fig. 5, 6). The total phenolic content of the all types of fruit extract were determined (Table 6). The cavenging activity and Fe^{2+} chelating activity of the eight samples were given in fig. 5. The four aqueous extracts or ethanol showed no significant differences ($P<0.05$) in the two analyzed tests. The ethanol and aqueous extracts from Hubei province showed higher levels of scavenging activity and Fe^{2+} chelating activity, while the extracts from Shangdong province showed lower levels of antioxidant activity. The reducing power and lipid peroxidation inhibition activity of ethyl alcohol and aqueous extracts from different regions were showed in fig. 6. The results showed that the reducing power and lipid peroxidation inhibition activity of Hubei and Jianhxi province were higher than those in the extracts derived from fruit originating from the other provinces. However, there are no significant differences ($p<0.05$) in all analyzed ethanol and aqueous extracts respectively. The results from antioxidant activity analysis of extracts from four different regions revealed that the different growing environment had certain effect on antioxidant activity, but the impact is not big. And the antioxidant activity of the extracts had positive correlation with the total phenolic content of the samples. The total phenolic content of different extracts showed a strong correlation their reducing power (R2=0.932, P<0.05) and lipid peroxidation inhibition activity (R2=0.96, P<0.01).

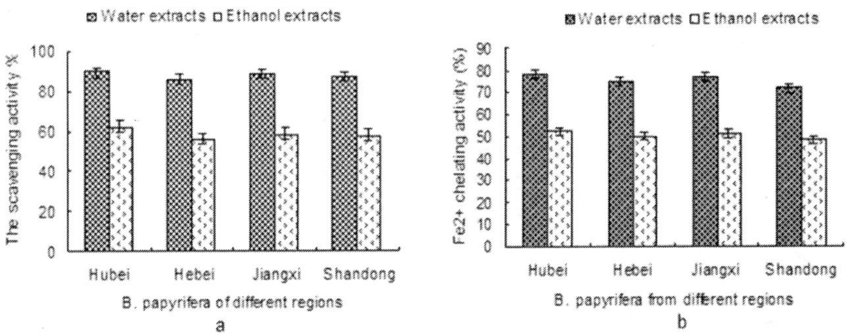

Figure 5. The scavenging activity and Fe^{2+} chelating activity of ethyl alcohol and aqueous extracts of *B. papyrifera* from different regions.a, DPPH radical scavenging capacity of ethanol and aqueous extracts (5 mg/ml) from different region (Hubei,Hebei,Jiangxi and Shandong Province). b, Fe^{2+} chelating activity of ethanol and aqueous extracts (5 mg/ml) from different region (Hubei,Hebei,Jiangxi and Shandong Province).

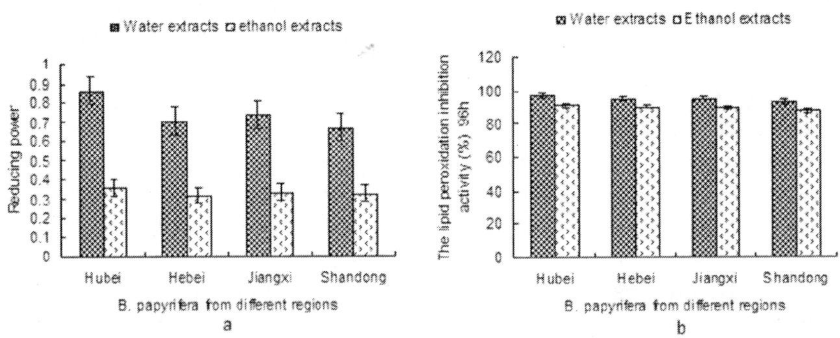

Figure 6. The reducing power and lipid peroxidation inhibition activity of ethyl alcohol and aqueous extracts from different regions.a, The reducing power of ethanol and aqueous extracts (5 mg/ml) from different region (Hubei,Hebei,Jiangxi and Shandong Province). b, Lipid peroxidation inhibition activity of ethanol and aqueous extracts (5 mg/ml) from different region (Hubei,Hebei,Jiangxi and Shandong Province).

Table 6. Total phenolic content of B. papyrifera fruits extracts from different regions (mg GAE/g).

regions	Hubei		Hebei		Jiangxi		Shandong	
extracts	water	ethanol	water	ethanol	water	ethanol	water	ethanol
total phenolic	6.13±0.17	4.18±0.11	5.94±0.13	4.02±0.51	5.91±0.21	3.841±0.32	5.79±0.1	3.79±0.03

High-performance liquid chromatography (HPLC) measurements

The antioxidant activities observed were associated with the total phenolic content, but the contribution of each phenolic compound to this activity should be evaluated. HPLC analysis was therefore performed on the two extracts from *B. papyrifera* fruit. The HPLC analysis is key to understanding the relationship between the compounds discovered and their activities.

In the present study, twelve components were analyzed as standards in the HPLC analysis, and four phenolic components were found in the ethanol and aqueous extracts. The four components were identified as 7-hydroxycoumarin, protocatechuate acid, protocatechuic acid and epicatechin. In addition, ferulic acid was found in the ethanol extract only. These results suggested that the ethanol and aqueous extracts had a similar phenolic compound composition, but at different concentrations. This might explain the differences in the antioxidant activities of the two extracts. Protocatechuate acid and protocatechuic acid have been found in many plants. These compounds possess strong antioxidative, antibacterial, and antimutagenic properties. 7-hydroxycoumarin has been found to have antioxidative properties in many plants such as *Fraxinusornus* and *Eysenhardtiasubcoriacea* [29], [30]. Hu confirmed that the 7-hydroxycoumarin group could contribute to the antioxidative activity of coumarin *via* an electron transfer mechanism [31]. Also, ferulic acid is found in many Chinese medicinal herbs and is an effective antioxidant and vasodilatory agent.

CONCLUSION

In conclusion, our data suggest that *B. papyrifera* fruit are an excellent source of fatty acids, essential in the human diet. They also contain considerable amounts of carbohydrates, proteins and a number of minerals including calcium, magnesium, and copper. These assays also show that these fruit contain very useful phytochemicals such as7-hydroxycoumarin, protocatechuate acid, ferulic acid, protocatechuic acid and epicatechin and have substantial antioxidant activity. Phenolic compounds and flavonoids appear to be the main components responsible for the antioxidant activity of the fruit extracts. The combination of bioactive compounds and rich nutritional composition of these fruit make them valuable as functional food or as nutritional supplements.

REFERENCES

1. Lee D, Bhat KPL, Fong HHS, Farnsworth NR, Pezzuto JM, et al. (2001) Aromatase inhibitors from *Broussonetiapapyrifera*. Journal of Natural Products 64: 1286–1293.D. LeeKPLBhatHHSFongNRFarnsworthJM Pezzuto2001Aromatase inhibitors from *Broussonetiapapyrifera*.Journal of Natural Products6412861293.
2. Ko HH, Yu SM, Ko FN, Teng CM, Lin CN (1997) Bioactive constituents of *Morusaustralis* and *Broussonetiapapyrifera*. Journal of Natural Products 60: 1008–1011.HH KoSMYuFNKoCMTengCN Lin1997Bioactive constituents of *Morusaustralis* and *Broussonetiapapyrifera*.Journal of Natural Products6010081011.
3. Lin CN, Lu CM, Lin HC, Fang SC, Shieh BJ, et al. (1996) Novel antiplatelet constituents from formosanmoraceous plants. Journal of Natural Products 59: 834–838.CN LinCMLuHCLinSCFangBJ Shieh1996Novel antiplatelet constituents from formosanmoraceousplants.Journal of Natural Products59834838.
4. Chen RM, Hu LH, An TY, Li J, Shen Q (2002) Natural PTP1B Inhibitors from *Broussonetiapapyrifera*. Bioorganic and Medicinal Chemistry Letters 12: 3387–3390.RM ChenLHHuTYAnJ. LiQ. Shen2002Natural PTP1B Inhibitors from *Broussonetiapapyrifera*.Bioorganic and Medicinal Chemistry Letters1233873390.
5. El-Adaway TA, Taha KM (2001) Characteristics and composition of different seed and flours. Food Chemistry 74: 47–54.TA El-AdawayKM Taha2001Characteristics and composition of different seed and flours.Food Chemistry744754.
6. Esenowo G, Ezeagu IE, Devi Sri, HaridasRao P, NathHarandra, et al. (2002) Prospects for incorporation of defatted Mucuna flour in biscuits formulation. Journal of Food Science and Technology 39(4): 435–438.G. EsenowoIEEzeaguSriDeviP. HaridasRaoHarandra Nath2002Prospects for incorporation of defatted Mucuna flour in biscuits formulation.Journal of Food Science and Technology394435438.
7. Irene D, Gian CT, Antonio D (2008) Chemical composition, nutritional value and antioxidant properties of Allium caepa L. Var. tropeana (red onion) seeds. Food Chemistry 107: 613–621.D. IreneCTGianD. Antonio2008Chemical composition, nutritional value and antioxidant properties of Allium caepa L. Var. tropeana (red onion) seeds.Food Chemistry107613621.
8. Son KH, Kwon SJ, Chang HW, Kim HP, Kang SS (2001) Papyriflavonol A, a new prenylatedflavonol from *Broussonetiapapyrifera*. Fitoterapia 72(4): 456–458.KH SonSJKwonHWChangHPKimSS Kang2001Papyriflavonol A, a new prenyl atedflavonol from *Broussonetia papyrifera*.Fitoterapia724456458.
9. AOAC (1995) Official methods of analysis, 15th edn.AOAC1995Official methods of analysis, 15th edn.Association of Official Analytical Chemists, Arlington.Association of Official Analytical Chemists, Arlington.
10. Balz M, Shulte E, Their HP (1992) Trennung von tocopherol und tocotrienolendurch HPLC. Fat SciTechnol 94: 209–213.M. BalzE.ShulteHP Their1992Trennung von tocopherol und tocotrienolendurchHPLC.FatSci Technol94209213.
11. AOCS (ed) (1993) Official methods and recommended practices of the American Oil Chemists' Society, 5th edn. Official method Cc 13b-45, reapproved (2000). AOCS (ed)1993Official methods and recommended practices of the American Oil

Chemists' Society, 5th edn. Official method Cc 13b-45, reapproved (2000).AOCS, Champaign, IL. AOCS, Champaign, IL.

12. Singleton VL, Rossi JA (1965) Colorimetry of total phenolics with phosphomolybdic–phosphotungstic acid reagents. American Journal of Enology and Viticulture 16: 144–158.VL SingletonJA Rossi1965Colorimetry of total phenolics with phosphomolybdic–phosphotungstic acid reagents.American Journal of Enology and Viticulture16144158.
13. Razali N, Razab R, Junit SM, Aziz AA (2008) Radical scavenging and reducing properties of extracts of cashew shoots (*Anacardiumoccidentale*). Food Chemistry 111: 38–44.N. RazaliR. RazabSMJunitAA Aziz2008Radical scavenging and reducing properties of extracts of cashew shoots (*Anacardiumoccidentale*).Food Chemistry1113844.
14. Tuberoso CIG, Melis M, Angioni A, Pala M, Cabras P (2007) Myrtle hydroalcoholic extracts obtained from different selections of *Myrtuscommunis L.*. Food Chemistry 101: 806–811.CIG TuberosoM. MelisA. AngioniM.PalaP. Cabras2007Myrtle hydroalcoholic extracts obtained from different selections of *MyrtuscommunisL.* .Food Chemistry101806811.
15. Tuberoso CIG, Rosa A, Bifulco E, Melis MP, Atzeri A, et al. (2010) Chemical composition and antioxidant activities of *MyrtuscommunisL.*berries extracts. Food Chemistry 123: 1242–1251.CIG TuberosoA. RosaE.BifulcoMPMelisA. Atzeri2010 Chemical composition and antioxidant activities of *MyrtuscommunisL.* berriesextracts.Food Chemistry12312421251.
16. Dinis TCP, Madeira VMC, Almeida LM (1994) Action of phenolic derivatives (acetaminophen, salicylate, and 5-aminosalicylate) as inhibitors of membrane lipid peroxidation and as peroxyl radical scavengers. Archives of Biochemistry and Biophysics 315: 161–169.TCP DinisVMCMadeiraLM Almeida1994Action of phenolic derivatives (acetaminophen, salicylate, and 5-aminosalicylate) as inhibitors of membrane lipid peroxidation and as peroxyl radical scavengers.Archives of Biochemistry and Biophysics315161169.
17. Oyaizu M (1986) Studies on products of browning reactions: Antioxidative activities of browning products of browning reaction prepared from glucosamine. Japanese Journal of Nutrition 44: 307–315.M. Oyaizu1986Studies on products of browning reactions: Antioxidative activities of browning products of browning reaction prepared from glucosamine.Japanese Journal of Nutrition44307315.
18. SabeenaFarvin KH, Baron CP, Nielsen NS, Jacobsen C (2010) Antioxidant activity of yoghurt peptides: Part 1-in vitro assays and evaluation in x-3 enriched milk. Food Chemistry 123: 1081–1089.KH SabeenaFarvinCPBaronNSNielsenC. Jacobsen 2010Antioxidant activity of yoghurt peptides: Part 1-in vitro assays and evaluation in x-3 enriched milk.Food Chemistry12310811089.
19. Qian ZJ, Jung WK, Byun HG, Kim SK (2008) Protective effect of an antioxidative peptide purified from gastrointestinal digests of oyster, Crassostreagigas against free radical induced DNA damage. BioresourceTechnology 99: 3365–3371.ZJ QianWKJungHGByunSK Kim2008Protective effect of an antioxidative peptide purified from gastrointestinal digests of oyster, Crassostreagigas against free radical induced DNA damage.BioresourceTechnology9933653371.
20. You L, Zhao M, M J (2011) Regenstein, JiaoyanRen.In vitro antioxidant activity and in vivo anti-fatigue effect of loach (*Misgurnusanguillicaudatus*) peptides prepared by papain digestion. Food Chemistry 124: 188–194.L. YouM. ZhaoJ. M2011Regenstein, JiaoyanRen. In vitro antioxidant activity and in vivo anti-fatigue effect of loach (*Misgurnusanguillicaudatus*) peptides prepared by papain digestion.Food Chemistry124188194.
21. Chen XH, Xia LX, Zhou HB, Qiu GZ (2010) Chemical Composition and Antioxidant Activities of Russulagriseocarnosa sp. nov.Journal of agricultural and food chemistry 58: 6966–6971.XH ChenLXXiaHBZhouGZ Qiu2010Chemical Composition and Antioxidant Activities of Russulagriseocarnosa sp. nov.Journal of agricultural and food chemistry5869666971.

22. AbdalbasitAM,Yousif MA, Bertrand M, Galaledin K, Abdeen S, et al. (2009) A Comparative Study of the Properties of Six Sudanese Cucurbit Seeds and Seed Oils. J Am Oil ChemSoc 86: 1181–1188.AM AbdalbasitMAYousifM. BertrandK.GalaledinS. Abdeen2009A Comparative Study of the Properties of Six Sudanese Cucurbit Seeds and Seed Oils.J Am Oil Chem Soc8611811188.
23. Abdalbasit AM, Kawthar MA, Ali AM, Siddig IAW (2009) Characterization of the Seed Oil and Meal from *Monechmaciliatum* and *Prunusmahaleb* Seeds. J Am Oil ChemSoc 86: 749–755.AM AbdalbasitMAKawtharAMAliIAW Siddig2009Characterization of the Seed Oil and Meal from *Monechmaciliatum* and *Prunusmahaleb* Seeds.J Am Oil Chem Soc86749755.
24. Ensminger ME, Oldfield JE, Heinemann WW (1990) Feeds and nutrition, 2nd edn. ME EnsmingerJEOldfieldWW Heinemann1990Feeds and nutrition, 2nd edn.Ensminger, Clovis.Ensminger, Clovis.
25. FAO/WHO (1990) FAO/WHO1990Protein quality evaluation in Report of Joint FAO/WHO expert consultation; Food and Agricultural Organization of the United nations: Rome, 23. Protein quality evaluation in Report of Joint FAO/WHO expert consultation; Food and Agricultural Organization of the United nations: Rome, 23.
26. Ko FN, Cheng ZJ, Lin CN, Teng CH (1998) Scavenger and antioxidant properties of prenylflavones isolated from *Artocarpusheterophyllus*. Free RadicBiol Med 25: 160–168.FN KoZJChengCNLinCH Teng1998Scavenger and antioxidant properties of prenylflavones isolated from *Artocarpusheterophyllus*.FreeRadicBiol Med25160168.
27. Kumaran A, Karunakaran RJ (2007) Antioxidant activity of Cassia auriculata flowers. Fitoterapia 78(1): 46–47.A. KumaranRJ Karunakaran2007Antioxidant activity of Cassia auriculata flowers.Fitoterapia7814647.
28. Rajapakse NE, Mendis WK, Jung JY, Kim SK (2005) Purification of a radical scavenging peptide from fermented mussel sauce and its antioxidant properties.Food Res Int 38: 175–182.NE RajapakseWKMendisJYJungSK Kim2005Purification of a radical scavenging peptide from fermented mussel sauce and its antioxidant properties.Food Res Int38175182.
29. Emma MM, Nedjalka VIY, Ivanka NK (1994) Antioxidative action of the ethanolic extract and some hydroxycoumarins of *Fraxinusornus* bark. Food Chemistry 51(2): 125–132.MM EmmaVIYNedjalkaNK Ivanka1994Antioxidative action of the ethanolic extract and some hydroxycoumarins of *Fraxinusornus* bark.Food Chemistry 512125132.
30. José M, Narváez M, Fernando N, Guillermo D (2008) Antioxidant aryl-prenylcoumarin, flavan-3-ols and flavonoids from *Eysenhardtiasubcoriacea*. Phytochemistry 69(2): 451–456.M. JoséM. NarváezN. Fernando D.Guillermo 2008Antioxidant aryl-prenylcoumarin, flavan-3-ols and flavonoids from *Eysenhardtia subcoriacea*.Phytochemistry692451456.
31. Hu XR, He JB, Wang Y, Zhu YW, Tian JJ (2011) Oxidative spectroelectrochemistry of two representative coumarins. ElectrochimicaActa 56(7): 2919–2925.XR HuJBHeY. WangYWZhuJJ Tian2011Oxidative spectroelectrochemistryof two representative coumarins.Electrochimica Acta56729192925.

CHAPTER 18

Liquid Sugar Extraction from Date Palm (Phoenix Dactylifera L.) Fruits

Gamal A El-Sharnouby[1*], Salah M Aleid[1,2] and Mutlag M Al-Otaibi[1]

[1]*Department of Food and Nutrition Sciences, College of Agricultural and Food Sciences, King Faisal University, Saudi Arabia*

[2]*Date Palm Research Center of Excellence, King Faisal University, Saudi Arabia*

ABSTRACT

In date syrup manufacturing, date fruits (Phoenix dactylifera L.) are mixed with a suitable amount of water with heating at a temperature greater than 70°C for about 1 h. These conditions are not satisfactory for extraction greater amount of liquid sugar (date syrup). In addition, overheating for an extended period could damage nutritious materials and changes product's color. In this study, pectinase/cellulase enzymes and sonication processes were used to achieve maximum syrup extraction from a date variety (Reziz) compared with classical method. The following variables were examined: fruit/water ratio, mixture of enzymes percent and ultrasonic power. Enzymes and Ultrasound were applied for improving the quantity and quality of extracted syrup. Date extract was filtered with filter system (Novox 200, filtrox) which could result entirely a clear syrup and then concentrated by a rotary evaporator until about 75% total solids to produce higher quality syrup. Physicochemical properties of each collected sample during process were evaluated. Results showed that sonication power 25% at ratio 1/3 D/W could lead to a higher extraction of RSS (Recovery of Soluble Solids) (74.30%) in a shorter time with a better physicochemical quality of syrup in comparison to mixture of enzymes 1.0% (66.70%) and classical methods of extraction (58.45%). Data indicated the possibility of employing ultrasound (25%) or enzymes processes (pectinase/ cellulase 0.1%) for producing greater syrup amount highly desirable for use in food product development.

Keywords: Date syrup; Reziz; Enzymes; Ultrasound; Extraction; Liquid sugar

INTRODUCTION

The date palm (*Phoenix dactylifera L.*) is one of the oldest fruit trees in the Arabian Peninsula and has played an important role in day-to-day life of the people of this region for the last 7000 years [1]. The production of date fruits in the Kingdom of Saudi Arabia reached about 1.07 million tons (annually). The global production of fruit from the date palm tree was estimated at 7.75 million tons worth US $3.82 billion in 2010 [2]. Saudi Arabia is ranked as the second largest date producer in the world after Egypt, with an annual production of 1.078 million tons of dates produced from over 24 million date palm trees [3].

The development of date fruits is divided into three stages, Khalal, Rutab and Tamr. Khalal stage dates are immature with hard texture, yellow or red or pink color, Total Soluble Solids (TSS) (A portion of the total solids which are dissolved in the water and pass through a filter prepared by depositing a layer of Kaolin paste into a standard filter paper) of 30-45 brix, astringent and edible; Rutab stage dates soften at the tip of the fruit, RSS of 55-60 °brix, free from astringency and edible; Tamr stage dates are fully ripped with RSS of 60–84 °Brix and edible. Dates are generally harvested at Tamr stage that is after the development of TSS of 60–70 °Brix that are edible at this stage. Date syrup is used in food preparations like sweets, snacks, confectionery, bakery products and health foods. Mature date fruits are also processed into products such as date bars, date syrup, etc. Dates are known to be rich in carbohydrates (80%) but quite low in protein (2-3%). Date fruits (Tamr) contained moisture ranged from 10 to 22%, total sugars 62 to 75%, protein 2.2 to 2.7%, fiber 5 to 8%, fat 0.4 to 0.7%, ash 3.5 to 4.2%, total acidity 0.06 to 0.20%, ascorbic acid 30.0 to 50.0 mg %,on dry weight basis. Date fruit as a nutritive product contains sugar substances about (70.6-76.3%), proteins (1.9-3%), fat (0.2-2.8%), minerals (1.3%) and vitamins [4]. The flesh of a fully ripe date (Tamr), consist of twothird sugars and one-quarter water, the rest being mainly cellulose, pectin, ash, and vitamins. The date is considered as a nutritious fruit as research has indicated the clear contribution of dates to human health when consumed with other food constituents. Dates contain sufficient quantities of minerals and vitamins that help to prevent deficiency [5]. Dates also have an emerging role of importance as a substitute for refined sugar. In food technology, date syrup as the main and general by-product of date is used for foodstuffs such as jams, marmalades, concentrated beverages, chocolates, ice cream, confectioneries, honey, etc [6-8]. As the Kabkab (Iranian Kabkab Dates) date has a high amount of wastes, it can be used for the date syrup production with economical advantageous [9]. Date syrup (dibs) is probably the most common derived date product. It is produced as an incidental by-product when bagged humid dates are heaped for several months, some syrups extracted by the force of their own weight. In addition, it is produced in the home and village by extraction and boiling down of juice, and on a semi and full industrial scale. As the Reziz date has a high amount of wastes, it can be used for the date syrup production with economical advantageous. In date's syrup industry the fruit are mixed with water and heated for around 1 h at 70°C, when the main components, sugars, are extracted. This method destroys some nutritive components and darkens the product's color.

Specific enzymes and enzyme combinations are needed to optimize the extraction of any particular fruit juice about yield and quality. The use of cellulases and pectinases has been an integral part of modern fruit processing technology, involving treatment of fruit mashes as they not only facilitate easy pressing and increase in juice recovery, but also ensure the highest possible quality of products. For example, pectinases can hydrolyze pectin and cause pectin-protein complexes to flocculate, so the resulting juice has a much lower amount of pectin and a lower viscosity, which is advantageous for the filtration process. These enzymes not only help in softening the plant tissue but also lead to the release of cell contents that may be recovered with high yield. Pectinases also used in sugar extraction process from date fruits. These enzymes break down complex polysaccharides of plant tissues into simpler molecules like galacturonic acids. Enzymes can be used economically if proper strategies are used for their optimization during clarification of fruit pulp.

A number of papers have dealt with the ultrasonically assisted extraction of different materials. High efficiency and shortening of extraction time are the main positive effects of such extractions. The most important proposed mechanism for these behaviors is the intensification of mass transfer and easier access of the solvent to the content of the cell. In addition, the collapse of cavity near the cell wall can lead to cell disruption and releases the cell contents into the extraction medium [10]. Ultrasound can also inactivate a range of microorganisms. Ultrasound in the range of 20-100 kHz frequency is mainly used in industrial processes and that those with higher frequency are mainly used for detection and analyses [11-13].

This present work focuses on the suitability of date extract (Syrup) as an alternative sweetener ingredient to replace expensive and imported sucrose in many food industries such as Candy, bakery products and soft drinks. In addition, the comparison of conventional, enzymes (pectinase and cellulase) and ultrasonically assisted extraction methods of liquid sugar from Reziz fruits to use in food product development.

MATERIALS AND METHODS

Materials

Date fruits: Date fruits (*Phoenix dactylifera* L.) at the tamr stage of maturity, i.e. Reziz, obtained from Al-Hassa Govemorate, Date Factory, Kingdom of Saudi Arabia were used in this work.

Date khalas syrup: Date syrup of Khalas variety at TSS (78%) was obtained from local market in Alhassa and used as control Sample for sensory evaluation.

Enzymes: Pectinase and cellulase enzymes: (in liquid state) were provided from Sigma company (USA) in form of aqueous solution. Activity of pectinase (≥ 3,800 units/mL) and cellulase (≥ 1000 units /g).

Ultrasound device: ÔMisonix' model XL-2020, 20 kHz, 500 W, horn type machine with 12.7 mm tip diameter.

Filtration system: Filtration system (Novox 200, filtrox, Swiss) for fine purification was used.

Chemicals: All chemicals and solvents were obtained from Sigma- Aldrich Co. (Dorset, UK), unless otherwise specified.

Methods

The date fruits were pitted and sliced to pieces with the approximate dimensions of 1 cm×1 cm×0.3 cm. Then the prepared date fruits were used for three extraction methods.

Extraction of date syrup with water: Date pulp (100 gm) in a typical expermints put in an erlenmeyer flask (1 L) and water were added at different ratios. The date pulp/water ratios (D/W) were 1:2, 1:3, 1:4 and 1:5 and samples were blending using a hand-held blender (Phillips, Holland). The pH was adjusted to 6.0 ± 0.2. Each sample, in triplicate, was placed in water bath at 70°C for 2 h. After heating the slurry was filtered through a cheese cloth with a hand press to remove large impurities and insoluble matters, then centrifuged (model RC 28 S, USA) at 8000 g for 10 min, and the supernatant was decanted and weighed.

Extraction of date syrup with pectinase/cellulase enzymes enzymes: The extraction procedure was also repeated at chosen ratio (1/3 D/W) with addition of enzymes mixture. Samples were blending using a hand-held blender (Phillips, Holland), then pectinase and cellulase preparation (1:1) at the rate of 0.25, 0.50 and 1.0% (v/w) were added and pH was adjusted to 6.0 ± 0.2 before the addition of enzyme preparations. Each sample, in triplicate, was placed in an water bath at 50°C for 4 h. Then the slurry was filtered through a cheese cloth with a hand press to remove large impurities and insoluble matters.They were then centrifuged and the supernatant was decanted and weighed.

Extraction of the date syrup with ultrasonic waves: The extraction procedure was also repeated at ratio 1:3 W/D ratio with using ultrasonic device (Hielscher, Model UP400S, 400 W, with 14 mm tip diameter, Germany) at different power of ultrasonic device (15, 20 and 25%). After blinding, each sample, in triplicate, was placed in an Erlenmeyer flask (1 L), and water bath at 40°C for 2 h and performed to ultrasonic for 60 min. as described by. Then the slurry was filtered through a cheese cloth with a hand press to remove large impurities and insoluble matters. They were then centrifuged and the supernatant was decanted and weighed.

Filtration method: Finelly, the obtained extract from different extraction methods was filtered in Filter system (Filtrox AG, CH 9000 St. Gallen, Switzerland) with an effective clarifying filtration area of 1.02 m^2 and 2.0 µm pore size, operated at a pressure of 1 bar for fine purification. The clear extract was concentrated to about 78 ºBrix using rotary evaporator (Rotavapor R-124, Buchi, Switzerland) at 70. The concentrated samples were packed in sealed glass bottles and stored at room temperature (20°C) until analyzed and evaluated.

Chemical and physical analyses: The date fruit samples were evaluated for moisture, ash, protein and pectin contents (as Ca pectate) and acidity (as citric acid) using standard methods [14]. Potassium and sodium were determined using Flame photometer. Calcium, magnesium and iron were detennined using Atomic absorption Spectrophotometer 2380 as described in [14].

(TSS %): Total soluble solids (TSS %) with main component of sugar in date's syrup were performed by Brix refractometry using the Abbé refractometer (Milton Roy, USA) as Entezari et al.,.

(RSS %) : the recovery of soluble solids (RSS %) was computed as described by Al-Hooti et al. as follows:

$$\mathbf{RSS} = \frac{Weight\ of\ extract \times \%\ TSS\ of\ extract}{Weight\ of date\ pulp\ taken} \times 100$$

Sugars: The identification and quantificatin of Sugars (glucose, fructose and sucrose was conducted using HPLC (Agilent Technologies, 1260 Model Infinity,USA) with refractive index detection as described by [15]. The separation was carried out using a purospher star NH2 column (4.6∈250 mm∈5 μm, Merck, Germany). The color of the diluted date syrup was measured as optical density (O.D) at 520 nm using a spectrophotometer (Shimadzu, Model UV-1800, Jpan). pH was determined by pH meter (Jenway, Model 3520, UK) [16].

Sensory evaluation of date syrup: The sensory characteristics of date syrups were conducted to determine the acceptapility of the product. Surup samples were presented in a sealed pouch coded with different numbers to six panalists of staff members at King Faisal University, Saudi Arabia, who were asked to rate each sensory attribute. Date surups samples were evaluated. Syrup samples were evaluated for color, taste, texture and overall quality on a 9-point hedonic scale [17]. However, evaluated date syrup was compared with the sugar cane syrup (Black Honey) from the local market (Alahsa -Saudi Arabia).

Statistical analysis: Results were given as means ± standard deviation of three independent determinations (n=3). One way analysis of variance (ANOVA) was used to test of significancecomparison between means were tasted by Duncan's multiple range test. Differences were considered to be significant at $p<0.05$ according to [18]. All statistical analyses were performed using SAS (9.3) [19].

RESULTS AND DISCUSSIONS

Chemical and physical evaluation of Reziz date

The data of gross chemical composition are shown in **Figure 1**. It could be concluded that date samples contained predominant amount of sugars and moderate levels of crude protein and ash. Reziz date variety recorded a higher amount of sugars (79%), Moisture (11%) crude protein (2%), crude fat (2%), ash (3%) and pectin (2.5%). However, it was clear from the obtained data that the moisture content of dates was less than the recorded average of semi-dry varieties (15%). This might be reasoned to a higher loss in moisture content after harvesting during storage and marketing.

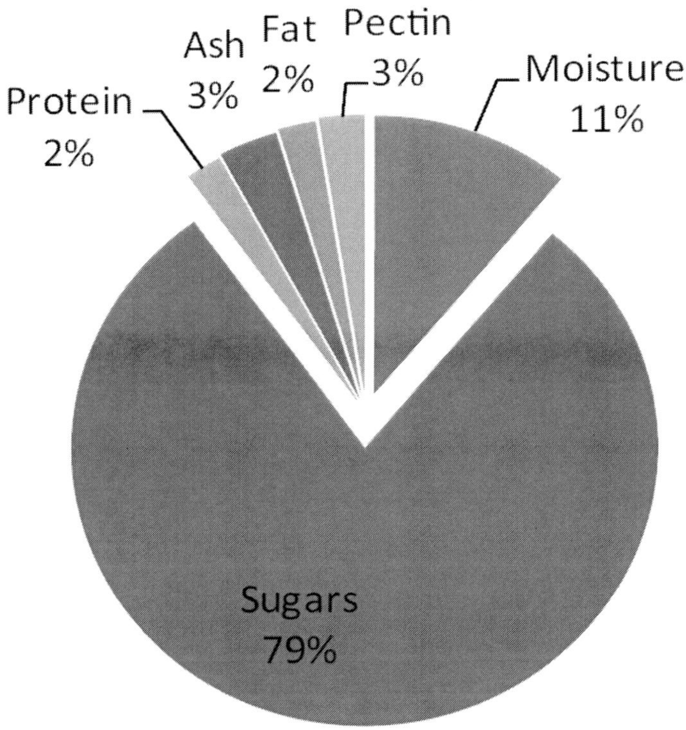

Figure 1: Chemical Composition of Riziz Date Fruits

Extraction of Reziz date syrup

Date syrup is a popular commodity produced by some date processing plants in Saudi Arabia. Moreover, considerable amounts of date syrup are produces at homes from local date varieties. Dates grown in Saudi Arabia and their sugar content should justify their use as a source of liquid sugar sautés to many food industries. Date syrups are composed of large amounts of reducing sugars. Fructose and glucose are the major sugar fractions in date syrup.

The effect of date pulp /water (D/W) ratio on the sugar extraction rate (RSS %) of date syrup was presented in (**Figure 2**) The data revealed that, increasing the D/W ratio up to 1/3 increased the RSS %. This positive relationship could be due to the increased rate of molecular diffusion, reduced liquid viscosity and better liquid solid contact. Also, it is evident from data that, RSS % was relatively affected by the moisture content of date pulp, whereas, date fruits (at ratio 1/3 D/W) recorded the highest RSS % (58.45%) compared with lowest ratio 1/1 D/W (52.55%). The water to date ratio was evidently a very important parameter in solids extraction, for technical and economic reasons. A high ratio will assist rapid and thorough extraction. The low RSS % of the extract is not easy to preserve on the industrial scale and will require very high energy to concentrate. At low D/W ratio will leave much of the unextracted soluble solids. The extraction of date syrup from dates usually gives a yield of 60%. It usually extracted by means of pressure or heat extraction (with 2.5 times their weight water), followed by filter press. Thus, this ratio (1/3 D/W) was chosen in further experimental.

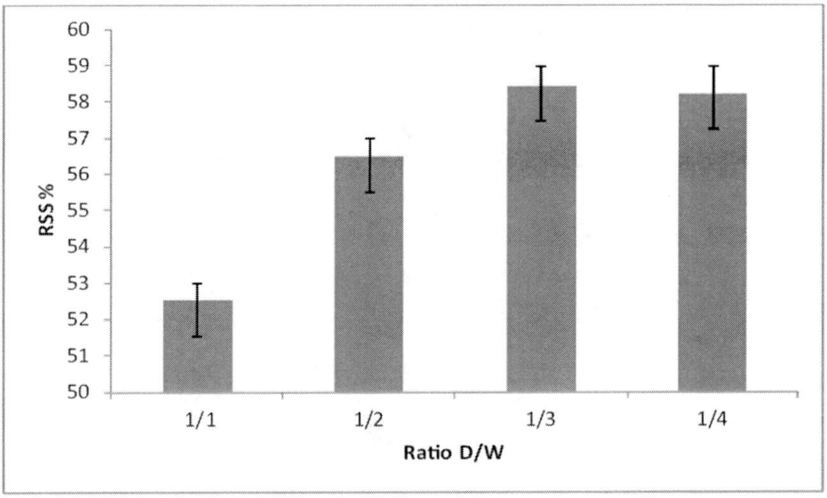

Figure 2: Effect of date / water ratio (D/W) on the extraction yield (RSS %) of soluble solids from Reziz date fruit.

Date syrup evaluation

Data concerning the physical characteristics of date syrup presented in **Table 1**. The total soluble solids (TSS %) and refractive index of the date syrup were higher (84.45% and 1.4885 respectively). This is principle due to the high level of sugars and relatively low moisture content. The pH value recorded 4.91 for date syrups. The optical density of 20% TSS of date syrups at 520 nm was 0.56. The data showed that, the prepared date syrup had high acidity 0.69% (as citric acid) , therefore the expected storage ability will be high.

Table 1: Characteristics of date syrup (at ratio 1/3 D/W)

Date Syrup	TSS (%)	Sugars %	Refractive index	Acidity %	pH	Color O.D
Reziz Syrup	84.25	81.88	1.4885	0.69	4.91	0.56

Effect of pectinase/cellulose enzyme preparations (ml/100 g) and ultrasonic process on the Recovery of Soluble Solids (RSS %) at ratio 1/3 D/W from Reziz date is shown in **Table 2**. Three treatments of pectinase / cellulose enzyme preparations used namely, 0.50, 0.75 and 1.0 (ml/100 g pulp date) as well as three treatments of ultrasound power (15, 20 and 25%) have been used for increasing the yield of syrup from fruits pulp at ratio 1/3 D/W. By using these methods of extraction the extraction yield of soluble solids for Reziz date variety was enhanced. However, The addition of enzyme preparations, even at the level of 1.0%, was found to increase the extraction yield of soluble solids from 62.40 to 66.70%. As the extraction yields of soluble solids from date pulps increased, this may open up the avenue for the use of such enzyme preparations for date fruit pulp processing for the development of sucrose substitutes based on date fruits. This causes considerable loss of soluble solids and the percentage of broken segments is also high. When pectinolytic enzymes are used deskinning is completed within 90 to 120 minutes. Pectinases are also used in sugar extraction process from date fruits. However, The use of sonication power at 25% was significantly different than the 15 and 20% level, so the use of 25% sonication power was found to be optimal for the maximum extraction yields of soluble solids from date fruit pulps.

The effect of different treatments (at ratio 1/3 D/W) on sugars content is shwon in **Figure 3** and **Table 3**. From such data, it was observed that the sugars content of date syrup samples arranged between 81.88 and 83.56%. However, ultrasound tratments resulted in increses in sugar content followed by enzymes treatments (83.56 and 82.85%, respectively). Meanwhile, the effect of different treatments (at ratio 1/3 D/W) on chemical composition of date syrup is shown in **Table 3**. The moisture content of date syrup samples was 17.20-18.50%, protein content was 1.16-1.22%, fat was 2.04-2.22% and ash ranged between 1.08-1.82%. On the contrary, pectin content not detected in date syrup samples. This is due to the impact of using of enzymes mixture and filter system (Novox 200, filtrox) which could result entirely a clear juice.

Table 2: Effect of pectinase/cellulose enzyme preparations (ml/100 g) and Sonication Power on the extraction yield (RSS %) at ratio (1/3 D/W) of soluble solids from Reziz date

Extraction yieldRSS %	**Concentration (ml/100g) of pectinase and cellulose enzyme**		
	0. 50	0. 75	1.0
RSS**	62.40	64.40	66.70
	Sonication Power		
Power setup	15%	20%	25%
RSS**	68.35	70.58	74.30

**Recovery of soluble solids

Table 3: Effect of pectinase/cellulase (1.0 ml/100 g) and Sonication Power 25% on chemical composition % of date syrup, on dry weight basis (at ratio 1:3 D/W)

Samples	**Moisture**	**Sugars**	**protein**	**Fat**	**Ash**	**Pectin**
1*	18.50	81.88	1.16	2.04	1.82	0.00
2**	17.50	82.85	1.22	2.22	1.06	0.00
3***	17.20	83.56	1.20	2.20	1.08	0.00

*Date Syrup ** Date Syrup with enzymes 1.0% *** Date Syrup with Sonication 25%

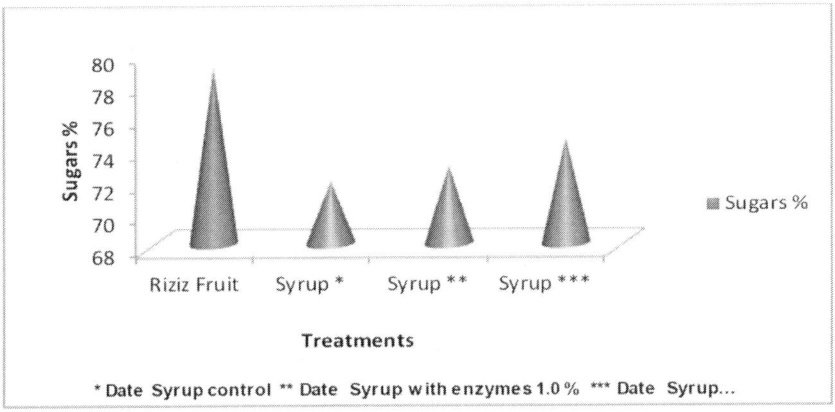

Figure 3: Effect of date / water ratio (D/W) on the extraction Sugars % from Reziz date fruit

Minerals composition of concentrated date syrup prepared from Reziz pulp date (at ratio 1/3 D/W) were tabulated in **Table 4**. Date syrups are good source of minerals. However, the mineral composition of date syrup shows high amount of K, fair amount of Na and low amounts of Ca, and Mg as well as trace amounts of Fe and Mn. Ultrasound tratments resulted in increses in minerals content followed by enzymes treatments. It could be concluded from the obtained results that, the slight amounts of minerals content were found after addition of enzymes mixture and using of ultsound process. This is due to the effect of enzyme mixture and destroying effect of ultrasound waves on date cell and more extraction of total soluble solids take place and the fact that elements content increased after drying as a result of concentration effect. Also, the different in element contents during different drying methods is due to the differences in moisture content of samples.

Physico-chemical compositions of concentrated date syrups before and after treatments with addition of pectinase/cellulose enzyme and ultrasound process have been presented in **Table 5**. Clear increase was observed in TSS % (83.25 to 86.00%), slight increase in acidity (0.69 to 0.72%), and slight decrease in pH (4.91 to 4.83) while the treatable acidity (expressed as percentage citric acid) was increased from 0.69 to 0.72%. However, sugas content (glucose, fructose and sucrose) were analysed by HPLC apparatus as mentioned in materials and methods. The date syrup had 81.88% total sugars on dry weight basis (36% fructose, 34.50% glucose and 1.33% sucrose), and 83.25% total solids on dry weight basis. However, it is clear that fructose and glucose are the major fractions of sugars. It can be reported that Reziz date syrups contains about equal amount of glucose and fructose (36.00, 34.50%), (37.05, 34.60%) and (38.23, 35.33%) for date syrup, syrup with enzymes mixture (1.0%) and syrup with sonication 25%, respectively. Date syrups contained more fructose than glucose. In general, Saudi dates varieties contain about 70% reducing sugars with almost equal quantities of glucose and fructose.

Table 4: Effect of pectinase/cellulase and Sonication Power on Mineral composition of date syrup, on dry basis (at ratio 1/3 D/W, on dry basis)

Date Syrup	Macro-elements					Micro-elements			
	Ca	P	Na	K	Mg	Fe	Zn	Cu	Mn
Reziz fruits	72.30	166.13	655.70	960.80	49.70	0.93	0.71	0.82	0.36
1*	66.30	168.21	630.51	949.21	42.13	1.00	0.66	0.62	0.22
2**	71.50	170.20	650.20	960.30	45.20	1.10	0.70	0.70	0.25
3***	73.40	173.15	668.70	983.70	53.20	0.13	0.75	0.78	0.32

*Date Syrup ** Date Syrup with enzymes 1.0% *** Date Syrup with Sonication 25%

Table 5: Effect of pectinase/cellulase and Sonication Power on Physicochemical composition of concentrated date syrup (at ratio 1/3 D/W, on dry basis)

DateSyrup	TSS %	pH	Acidity %	Sugars %	Fructose %	Glucose %	Sucrose %
1 *	83.25	4.91	0.69	81.88	36.00	34.50	1.33
2 **	84.16	4.85	0.72	82.85	37.05	34.60	1.20
3***	86.00	4.83	0.72	83.56	38.23	35.33	1.00

*Date Syrup ** Date Syrup with enzymes 1.0% *** Date Syrup with Sonication 25%

Organoleptic evaluation of date syrups

The prepared date syrups were palatability tested in terms of color, taste, consistency, and acceptability compared with date syrup from local market as control. The results are listed in **Table 6**, it was evident from these data that, syrup prepared at 25% of sonication power recorded the highest average scores for color, texture, taste, flavor and overall acceptability followed by syrup of 1.0% pectinase and cellulose enzyme mixture and date syrup (1/3 D/W). From the same data, it was clear that, the control syrup (from local market) recorded the lowest scores of all sensory evaluation terms. On the other hand, there were significant differences among all date syrups and control syrup in their organoleptic properties. The results stated that, prepared date syrups by sonication 25% followed by pectinase and cellulose enzyme are highly desirable and more acceptable.

Table 6: Analysis of variance for organoleptic evaluation of date syrup (at ratio 1/3 D/W)

Syrup	Color	Texture	Taste	Flavor	Overall quality
1 *	7.78 ±0.43^{c}	7.56±0.51^{b}	7.53±0.55^{c}	7.39±0.53^{c}	7.14±0.45^{c}
2 **	8.25±0.60^{b}	8.28±0.43^{a}	8.28±0.39^{b}	8.28±0.43^{b}	7.89±0.21^{b}
3***	8.64±0.38^{a}	8.53±0.32^{a}	8.64±0.33^{a}	8.56±0.34^{a}	8.61±0.32^{a}
Khalas control	7.06± 0.34^{d}	6.78±0.39^{c}	6.69±0.39^{d}	6.97±0.36^{d}	6.75±0.31^{d}

Within same column, means with the same letter are not significantly different *Date Syrup ** Date Syrup with enzymes 1.0% *** Date Syrup with Sonication 25%

CONCLUSION

The research data presented here indicate a strong possibility of producing concentrated date syrup for use in product development. Riziz date variety was found to be high in total sugar contents (about 83%). Among the various extraction procedures employed for producing date syrup, the use of date pulp to water ratio (D/W 1/3) with ultrasonic process at power 25% gave the highest recovery of total soluble solids (74.30%) compared with pectinase/cellulase enzymes at 1.0% (66.70%) and syrup control (58.45%). Reziz date syrups contains almost equal amount of glucose and fructose during different treatments. Glucose and fructose were the major sugars present in date syrup. The date syrup was found to be a good source of macro-elements like calcium, phosphorus, potassium, magnesium, but significantly low in sodium. The composition and sensory evaluation of Reziz date syrups confirmed that, it was rich in sugar, minerals and high acceptability. It can be concluded that there is good potential and a possibility to produce syrup from date fruits for use as a replacement of sucrose in food products. Further studies are suggested for the use of these ultrasonic and enzymes process for the production of syrup from date fruits at different temperature and times.

ACKNOWLEDGEMENTS

The investigators acknowledge gratefully the financial support from the Deanship of Scientific Research, King Faisal University, Saudi Arabia, material and moral support for this project.

REFERENCES

1. Ahmed A, Ahmed AW, Robinson RK(1995) Chemical composition of date varieties as influenced by the stage of ripening. Food Chem 5:305-309.
2. FAO (2012) Agriculture Data: FAOSTAT.
3. Ministry of Agriculture (2011) Ministry of Agriculture of Saudi Arabia. Indicators about Date industry in Saudi Arabia.
4. Lambiote B(1982)Some aspects of the role of dates in human nutrition. Proceeding of the First International Symposium on Date palm. King Faisal University, Saudi Arabia.
5. Al-Farsi M, Al-Asalvar C, Al-Abid M, Al-Shoaily K, Al-Amry M(2007) Compositional and functional characteristics of dates, syrups, and their by-products. Food Chem 104: 943-947.
6. El-Sharnouby GA, AL-Eid SM, Al-Otibi MM(2009) Utilization of enzymes in the production of liquid sugar from dates. African Journal of Biochemistry Research 3:41-47.
7. Al-Hooti SN, Sidhu JS, Al-Saqer MN, Al-Othman JM(2002) Chemical composition and quality of date syrup as effected by pectinase/cellulase enzyme treatment. FoodChem 79: 215-220.
8. Bahramian S, Azin M, Chamani M, Gerami A (2011) Optimization of Enzymatic Extraction of Sugars from Kabkab Date Fruit. Middle-East Journal of Scientific Research 7: 211-216.
9. Taper AR, Jain RK(2014) Pectinases: Enzymes for fruit processing industry. International Food Research Journal 21: 447- 453.
10. Entezari MH, HaghNazary S, Haddad Khodaparast MH (2004) The direct effect of ultrasound on the extraction of date syrup and its micro-organisms. UltrasonSonochem 11: 379-384.
11. Vaillant F, Millan A, Dornier M, Decloux M,Reynes M(2001) Strategy for economical optimisation of the clarification of puply fruit guices using cross flow microfiltration. Journal of Food Engineering 48:83-90.
12. Mason T, Zhao Y(1994) Enhanced extraction of Tea solids. Ultrasounics 32: 375-377.
13. Vinatoru M (2001) An overview of the ultrasonically assisted extraction of bioactive principles from herbs. UltrasonSonochem 8: 303-313.

14. AOAC(2000) Official methods of analysis. Washington, DC: Association of Official Analytical Chemists, Methods 37.1.12, 44.1.05, 2.4.03, 37.1.51, 37.1.34.
15. Yang C, Ling X, Zhang C, Shao W(2008)The isolation and characterization of polysaccarides from longan pulp. Separation and Purification Technology 63: 226-230.
16. Salman M, Alghamdi M, Bazaid A, Abdel- Hameed S(2011) Determination of Fructose, Glucose and Sucrose in Taif Grape using High Performance Liquid Chromatography and analysis of mineral salts. Archives of Applied Science Research 3: 488-496.
17. Hoda S, Jood S(2005) Organoleptic and nutritional evaluation of wheat biscuits supplemented with untreated and treated fenugreek flour. Food Chemistry 90: 427-435.
18. Steel R, Torrie J, Dickey D (1996) Principles and Procedures of Statistics: A Biometrical Approach. McGraw-Hill Higher Education.
19. SAS(2007) Statistical analysis system. User manual SAS/STAT 9.3 second edition. SAS institute Inc., NC, USA.

CHAPTER 19

Proteomic Analysis of Sauvignon Blanc Grape Skin, Pulp and Seed and Relative Quantification of Pathogenesis-Related Proteins

Bin Tian[1*], Roland Harrison[1], James Morton[1], Santanu Deb-Choudhury[2]

[1]*Department of Wine, Food and Molecular Biosciences, Lincoln University, Lincoln, 7647, Canterbury, New Zealand,*

[2]*Agresearch, Lincoln Research Centre, Christchurch, 8140, Canterbury, New Zealand*

ABSTRACT

Thaumatin-like proteins (TLPs) and chitinases are the main constituents of so-called protein hazes which can form in finished white wine and which is a great concern of winemakers. These soluble pathogenesis-related (PR) proteins are extracted from grape berries. However, their distribution in different grape tissues is not well documented. In this study, proteins were first separately extracted from the skin, pulp and seed of Sauvignon Blanc grapes, followed by trypsin digestion and analysis by liquid chromatography-electrospray ionization-tandem mass spectrometry (LC-ESI-MS/MS). Proteins identified included 75 proteins from Sauvignon Blanc grape skin, 63 from grape pulp and 35 from grape seed, mostly functionally classified as associated with metabolism and energy. Some were present exclusively in specific grape tissues; for example, proteins involved in photosynthesis were only detected in grape skin and proteins found in alcoholic fermentation were only detected in grape pulp. Moreover, proteins identified in grape seed were less diverse than those identified in grape skin and pulp. TLPs and chitinases were identified in both Sauvignon Blanc grape skin and pulp, but not in the seed. To relatively quantify the PR proteins, the protein extracts of grape tissues were seperated by HPLC first and then analysed by SDS-PAGE. The results showed that the protein fractions eluted at 9.3 min and 19.2 min under the chromatographic conditions of this study confirmed that these corresponded to TLPs and chitinases seperately. Thus, the relative quantification of TLPs and chitinases in protein extracts was carried out by comparing the area of corresponding peaks against the area of a thamautin standard. The results presented in this study clearly demonstrated the distribution of haze-forming PR proteins in grape berries, and the relative quantification of TLPs and chitinases could be applied in fast tracking of changes in PR proteins during grape growth and determination of PR proteins in berries at harvest.

Funding: This research was funded by the New Zealand Winegrowers. The funder had no role in study design, data collection and analysis, decision to publish, or preparation of the manuscript. The co-author, Santanu Deb-choudhury, is employed by a commercial company Agresearch, Lincoln Research Centre. Agresearch, Lincoln Research Centre provided support in the form of salary for author SD-C, but did not have any additional role in the study design, data collection and analysis, decision to publish, or preparation of the manuscript. The specific role of this author is articulated in the 'author contributions' section.

INTRODUCTION

Protein stabilization of white wine is of great concern to winemakers as denaturation of proteins in wine may cause haze formation, which is usually considered a wine fault. Pathogenesis-related (PR) proteins originally derived from grape berries are the major soluble proteins remaining in finished wine and they are mainly responsible for haze formation [1,2]. Pathogenesis-related proteins are a group of plant proteins induced in pathological or related situations [3]. They were first discovered in tobacco as a result of a hypersensitive reaction to tobacco mosaic virus (TMV) [4]. Pathogenesis-related proteins are typically acidic, of low molecular mass and highly resistant to proteolytic degradation and to low pH values. On the basis of similarities in amino acid sequences, serological relationship, and/or enzymatic or biological activity, eleven families have been recognized and classified for tobacco and tomato [5]. Some of these PR protein family members have also been found in grapevine. The two prominent soluble proteins accumulated in grapes during ripening have been identified as chitinases (PR-3 family) and thaumatin-like proteins (PR-5 family) [6,7]. However, in early studies, the β-1,3-glucanases (PR-2 family), a potential indicator of pathogen attack, were not found in grape juice and/or berry extracts [7–10]. With the accomplishment of grapevine genome sequencing programmes in 2007 [11,12] and the development of technology in protein analysis, proteomic analysis of grapevine has significantly improved knowledge of grape proteins and produced a better understanding of their characteristics [13]. These have identified other PR protein family members found in grapevine, such as osmotins (PR-5 family), β-1,3-glucanases (PR-2 family) and the PR-10 proteins [14–16].

Thaumatin-like proteins (TLPs) and chitinases are the two predominent PR protein families present in finished white wine [2,10,17] and they are usually removed by fining with bentonite, a clay material that has a strong affinity for proteins and other larger molecules [18]. However, the addition of bentonite may result in the loss of wine volume (5–20%) as lees and remove important aroma and flavour compounds [19,20]. Recent study showed that bentonite requirement to achieve wine protein stability is strongly correlated with concentration of PR proteins in wine, and specifically has a positive linear correlation with the concentration of chitinases [21]. Thus, a lower concentration of PR proteins in juice and wine, in particular the concentration of chitinases, could reduce the bentonite usage required in white wine protein stabilization. Since both TLPs and chitinases found in wine are derived from grape berries, the distribution and quantification of them in grape berries could be of great interest for winemakers to potentially reduce their concentrations in juice through managing the extraction during juice processing. Deytieux and co-workers have observed that the TLPs and chitinases are present in the skin of *Vitis vinifera* L. cv. Cabernet sauvignon, and their concentrations in grape skin increase during ripening [14]. A recent study on the effects of water stress on grapes [22] also shows the presence of chitinases in pericarp (skin and pulp). Proteomic studies since the completion of grapevine genome sequencing in 2007 have investigated the diversity of PR proteins [23] and protein changes during ripening [24], but there is little reported on the distribution and quantification of PR proteins in specific grape tissues, especially with focus on white wine haze formation related TLPs and chitinases. Therefore, in this study, the liquid chromatography-electrospray ionization-tandem mass spectrometry (LC-ESI-MS/MS) was carried out to investigate the distrubition of PR proteins in different grape tissues and provide some initial assessment towards quantification. The protein extracts of grape tissue were also analysed by high performance liquid chromatography (HPLC) and sodium dodecyl sulphate polyacrylamide gel electrophoresis (SDS-PAGE) to obtain the relative quantification of TLPs and chitinases in grape tissues.

MATERIALS AND METHODS

Sauvignon Blanc grapes and protein extraction

Sauvignon Blanc grapes were collected from the Dillons Point vineyard (with permission granted from the owner of vineyard) in Marlborough, New Zealand (41°30'52.9"S, 174°01'15.5"E) at harvest in 2012. The grape skin was obtained by hand-peeling 20 frozen grape berries and the pulp was accordingly obtained by removing the seeds. Protein extraction from the grape skin, pulp and seed was then carried out according to a method optimised for plant tissues rich in phenolics [25]. Approximately 2 g of skin, 2 g of pulp and 0.5 g of seed were ground to a fine powder in liquid nitrogen with mortar and pestle. The powder was vortexed in 5 mL of sucrose buffer (0.7 M sucrose, 0.5 M Tris-HCl pH 7.5, 50 mM EDTA, 0.1 M potassium chloride, 2 mM PMSF, 2% 2-ME and 1% PVP) and incubated for 30 min at 4°C. An equal volume of 1 M Tris-saturated phenol (pH 7.9) was added. The mixture was stored at -20°C for 30 min with vortexing every 10 min. The phases were separated by centrifugation (for 30 min at 0°C and at 3210 g). The upper phenol phase was collected and re-extracted twice with an equal volume of sucrose buffer. From 5 mL initially collected of the phenol phase, 2 mL was recovered after two re-extractions. Five volumes of 0.1 M w/v ammonium acetate in cold methanol were added to the phenol phase followed by incubation at -20°C overnight to precipitate proteins. The pellet was washed three times with 5 mL of cold 0.1 M ammonium acetate/methanol (w/v) and once with 5 mL of cold acetone before the trypsin digestion treatment. The protein pellet was further utilised either for the LC-ESI-MS/MS analysis or for the HPLC analysis.

Trypsin digestion and LC-ESI-MS/MS

The washed protein pellet was dissolved in 100 µL 50 mM ammonium bicarbonate by sonication for 5 minutes. The dissolved proteins were then reduced with 50 mM tris-(2-carboxyethyl) phosphine (TCEP), alkylated with 360 mM acrylamide and finally digested with sequencing grade trypsin (Promega). Nanoflow LC-MS/MS was performed on a Nano-Advance (Bruker) HPLC. Samples were loaded onto a C_{18} trap column and then switched in-line with an analytical column (Bruker, 0.1 x 150 mm Magic C_{18} AQ 3.0µm, 200Å). Elution was performed at 0.8 µL/min, using a tailored gradient from 0%-35% acetonitrile (with 0.1% formic acid) over 60 minutes and then from 35%-45% acetonitrile (with 0.1% formic acid) in 10 minutes. The column outlet was directly interfaced to an amaZon Speed ETD (Bruker) mass spectrometer. Automated information dependent acquisition (IDA) was performed using Hystar PP 3.2.44.0 software, with a

MS survey scan over the range m/z 350–1200 followed by three MS/MS spectra from 50–3000 m/z acquired during each cycle of 30 ms duration.

Analysis of MS/MS data

After each LC-MS/MS run, peak lists were queried against *Vitis vinifera* sequences in the Uniprot database using the Mascot search engine (v2.2.03, Matrix Science) maintained on an in-house server. The following Mascot search parameters were used: 'semitrypsin' was selected as the proteolytic enzyme with two missed cleavages permitted; error tolerance was set to 0.3 Da for MS and 0.6 Da for MS/MS. Search results were compiled and analysed using ProteinScape 3.1.0 (Bruker) using the ProteinExtractor function. Acceptance thresholds for peptide and protein scores were set at 20 and 80, respectively. The identification score for at least one peptide used for protein identification was calculated by the search engine. Results assessed as being true matches were used for further analysis.

Assignment of identified proteins to functional classes

Identified proteins were assigned a Gene Ontology (GO) term according to their molecular function. Protein NCBI accession numbers were used to perform batch retrieval on the Protein Information Resource website (http://pir.georgetown.edu/), and grouped into functional categories using the 'GO-MIPS funcat conversion table' (http://www. Geneontology .org/external2go/mips2go) set up at the Munich Information Center for Protein Sequences (MIPS Institute) [26]. In some cases, where the GO term assigned to the protein appeared too broad, proteins were assigned to MIPS funcats (http://www.mips.gsf.de/projects/funcat) according to their role described in the literature.

HPLC analysis of protein extracts

The washed protein pellet was dissolved in 1 mL 7 M urea. Protein extracts (50 μL) of grape tissues, Sauvignon Blanc juice (50 μL), and purified thaumatin-like proteins (TLPs) and chitinases which were prepared using the two step purification method decribed by Van Sluyter et al. [27], were loaded at 1 mL/min onto a C8 guard column (4.6x5 mm, Vydac 208GK54) which was equilibrated using a mixture of 83% (v/v) solvent B [0.1% trifluoroacetic acid (TFA) in 8% acetonitrile] and 17% solvent A [80% acetonitrile, 0.1% (v/v) TFA] at 35°C at a flow rate of I ml/min. HPLC analysis was performed using a C8 column (4.6x250 mm, Vydac 208TP54) equilibrated with the above solvent mix. A gradient elution of the proteins was performed; 17% solvent A to 49% solvent A in the first 7 min, from 49 to 57% in 7 to 15 min, from 57 to 65% in 15 to 16 min, from 65 to 81% in 16 to 30 min, and then held at 81% for 5 min before re-equilibrating the column in the starting conditions for an additional 6 min [28]. Elution was monitored using wavelengths at 210, 220, 260, 280, and 320 nm. Sauvignon Blanc juice proteins eluted at 9.2 min and 19.2 min were assigned to TLPs and chitinases respectively according to the previous studies [10,29,30]. For the protein extracts of grape tissues, since there was only a single peak at 19.2 min which was assigned to chitinases without further investigation. However, there were two possible peaks, eluting at 9.3 min and 10.6 min respectively, which might correspond to TLPs in this study. Thus, proteins from these two peaks were collected and labelled as fraction 1 (F1) and fraction 2 (F2) accordingly and further investigated by sodium dodecyl sulphate polyacrylamide gel electrophoresis (SDS-PAGE). Relative quantification of TLPs and chitinases in the protein extracts using HPLC was carried out by comparison of the corresponding peak areas against the area of a commercial thaumatin protein from *Thaumatococcus daniellii* (Sigma-Aldrich, New Zealand) used as a standard, and thus the protein concentration was expressed as thaumatin equivalent.

Protein composition analysis by SDS-PAGE

Fractions F1 and F2 of protein extracts and TLPs collected from Sauvignon Blanc juice were freeze-dried and re-dissolved in 30 μL of Tris-HCl buffer (pH 7.5). Aliquots of these concentrates (15 μL) were mixed with 5 μL of the NuPAGE LDS sample buffer (Novex, Life Technologies, US) and then denatured at 70°C for 10 minutes. Denatured protein samples were then loaded onto the NuPAGE Bis-Tris Mini Gel (Novex, Life Technologies, US). Electrophoresis was run in MOPS SDS running buffer with constant voltage mode (200 V) at room temperature for 50 min. The gel was stained using SimplyBlue SafeStain (Novex, Life Technologies, US) according to the manufacturer's protocol.

RESULTS AND DISCUSSION

Identification of PR proteins in grape tissues

Proteins extract from skin, pulp and seed tissues of Sauvignon Blanc grapes and analysed by LC-MS and are shown in Table 1. Hypothetical proteins, uncharacterised proteins and unnamed proteins are excluded from this table. Genome sequencing has greatly enhanced the analysis of the proteome of *Vitis vinifera* [12]. In this study, over 100 proteins were identified in specific Sauvignon Blanc grape tissues. This direct injection method provides rapid identification of the major proteins extracted from grape tissues, but high intensity peaks from liquid chromatography separation may eclipse lower intensity peaks due to the lack of pre-fractionation, so proteins with low concentration in protein mixture may be less easily encountered and identified in whole tissue proteomic studies. Among the identified proteins, 38 annotated proteins were present in both grape skin and pulp, 15 annotated proteins present in both grape pulp and seed, and 11 annotated proteins present in both grape skin and seed. However, some of the identified proteins were exclusively present in specific grape tissues: proteins involved in photosynthesis such as chlorophyll a-b binding protein (gi|225447576) and photosystem II 44 kDa protein (gi|91983988), were only identified in grape skin; and proteins found in alcoholic fermentation such as alcohol dehydrogenase 7 (gi|7264742) and pyruvate decarboxylase (gi|10732644), were only identified in grape pulp.

Table 1. Identified proteins and their distribution in Sauvignon Blanc grapes using LC-MS/MS.

Identified proteins	NCBI database accession	MW [kDa]	pI	skin	pulp	seed
01 Metabolism						
01.01 Amino acid metabolism						
5-methyltetrahydropteroyltriglutamate-homocysteine methyltransferase	225439223	84.90	6.08	-	-	+
serine hydroxymethyltransferase, mitochondrial	225429452	57.10	8.94	+	-	-
serine hydroxymethyltransferase 1	225433510	51.90	7.80	-	+	-
01.05 C-compound and carbohydrate metabolism						
1,3 beta glucanase	6273716	13.40	6.11	+	-	-
2,3-bisphosphoglycerate-independent phosphoglycerate mutase isoform 2	359480976	60.20	5.63	+	-	+
5'-methylthioadenosine/S-adenosylhomocysteine nucleosidase 1 isoform 2	359475059	28.30	4.59	+	-	-
adenosylhomocysteinase-like isoform 1	225456806	53.00	5.71	+	+	-
alpha-1,4 glucan phosphorylase L isozyme, chloroplastic/amyloplastic-like	359489019	111.30	5.05	+	+	-
beta-glucosidase 42-like	359495874	55.20	5.09	+	-	-
chitinase 5-like	225434076	43.40	6.51	+	+	-
class IV endochitinase	2306811	27.20	5.31	+	+	-
endochitinase-like	359497495	21.20	8.94	-	-	+
isocitrate lyase-like	225447308	64.60	7.07	-	-	+
phosphoglycerate kinase, chloroplastic	359494603	40.90	9.54	+	-	-
phosphoglycerate kinase, cytosolic	225464999	42.40	6.31	+	+	-
01.07 Metabolism of vitamins, cofactors, and prosthetic groups						
c-1-tetrahydrofolate synthase, cytoplasmic	359479954	32.70	7.12	+	-	-
01.20 Secondary metabolism						
4-coumarate—CoA ligase-like 7	225436506	59.50	9.59	+	-	-
chalcone—flavonone isomerase 2	225448801	25.10	5.13	-	-	+
Polyphenol oxidase, chloroplastic	1172587	67.30	6.28	+	-	-
02 Energy						
02.01 Glycolysis and gluconeogenesis						

probable galactinol—sucrose galactosyltransferase 2	225441787	84.80	5.43	+	+	-
Putative 2–3 biphosphoglycerate independant phosphoglycerate mutase	239056191	61.00	5.57	-	+	-
pyrophosphate—fructose 6-phosphate 1-phosphotransferase subunit alpha-like	225457674	67.30	8.78	+	+	-
ribulose bisphosphate carboxylase/oxygenase activase 2, chloroplastic isoform 2	359478916	48.60	5.78	+	-	-
sedoheptulose-1,7-bisphosphatase, chloroplastic	225466690	42.50	5.92	+	-	-
sucrose synthase 2	225437428	92.40	5.69	-	+	+
UDP-glucuronic acid decarboxylase 1	225449563	38.80	6.53	+	+	+
UTP—glucose-1-phosphate uridylyltransferase isoform 2	359476943	50.20	6.41	+	+	-
01.06 Lipid, fatty acid and isoprenoid metabolism						
9,10[9',10']carotenoid cleavage dioxygenase	61654494	61.10	6.04	+	-	-
acyl-[acyl-carrier-protein] desaturase, chloroplastic-like	359496940	45.00	7.87	+	-	-
annexin D1	225459318	35.20	7.82	+	+	-
lipid transfer protein isoform 4	28194086	11.70	10.40	+	-	-
lipoxygenase	268636245	101.60	6.06	+	-	-
non-specific lipid-transfer protein	225439679	11.60	10.52	+	-	-
non-specific lipid-transfer protein 2-like	359490972	9.70	9.18	+	-	-
non-specific lipid-transfer protein A-like	225446753	11.90	9.96	-	-	+
phospholipase D alpha	84620126	91.70	5.52	+	+	-
probable acetyl-CoA acetyltransferase, cytosolic 2	225447510	41.10	6.00	+	+	+
probable acetyl-CoA acetyltransferase, cytosolic 2-like	359497005	42.60	9.59	-	+	+
Enolase	225455555	48.10	6.16	-	+	-
enolase 1	225441000	47.80	5.60	+	+	-
glyceraldehyde-3-phosphate dehydrogenase A, chloroplastic-like	225451685	43.10	7.76	+	-	-
glyceraldehyde-3-phosphate dehydrogenase B, chloroplastic isoform 2	225457604	47.20	7.84	+	-	-
glyceraldehyde-3-phosphate dehydrogenase, cytosolic	359491599	36.70	8.72	+	+	+
glyceraldehyde-3-phosphate dehydrogenase-like	225425884	36.70	7.77	-	+	+
protein disulfide-isomerase	225459587	55.60	4.79	-	-	+
triosephosphate isomerase, cytosolic	225449541	27.10	6.42	+	+	-
02.10 Tricarboxylic-acid pathway						
malate dehydrogenase, chloroplastic-like	225452831	43.70	9.06	+	+	+
malate dehydrogenase, cytoplasmic	225438145	35.50	6.20	+	+	-
malate dehydrogenase, mitochondrial	225443845	36.80	9.51	+	+	+
NADP-dependent malic enzyme	1708924	65.20	6.07	+	+	-
02.11 Electron transport and membrane-associated energy conservation						
enoyl-[acyl-carrier-protein] reductase [NADH], chloroplastic-like	225441423	41.60	9.42	-	-	+
02.16 Fermentation						
alcohol dehydrogenase 7	7264742	39.40	5.95	-	+	-
pyruvate decarboxylase 1	10732644	62.20	6.10	-	+	-
pyruvate decarboxylase isozyme 1	225443847	62.40	5.59	+	+	-
02.30 Photosynthesis						
chlorophyll a-b binding protein 151, chloroplastic	225447576	28.40	5.61	+	-	-
chlorophyll a-b binding protein 40, chloroplastic isoform 1	225463428	27.90	4.97	+	-	-
photosystem II 44 kDa protein	91983988	51.80	6.76	+	-	-
02.45 Energy conversion and regeneration						
ADP, ATP carrier protein 1, mitochondrial-like	225450149	42.00	10.18	+	+	-
ATP synthase subunit beta, mitochondrial-like	225456079	59.10	5.86	+	+	-
ATP synthase subunit O, mitochondrial isoform 1	225450135	27.50	10.04	-	+	-
ATPase subunit 1	224365668	55.10	5.97	+	+	-
ATP-citrate synthase alpha chain protein 2 isoform 1	225450474	46.40	5.22	+	-	-
ATP-citrate synthase beta chain protein 2	225431960	65.90	7.10	-	+	+
V-type proton ATPase subunit B 1-like	225459744	54.20	4.85	-	+	-
11 Transcription						
11.02 RNA synthesis						
flavoprotein wrbA isoform 1	225461209	21.70	5.80	+	-	-
12 Protein synthesis						
12.01 Ribosome biogenesis						
40S ribosomal protein S16	225428853	16.40	10.21	+	-	-
40S ribosomal protein S23-like	225435203	15.60	11.02	+	+	-
40S ribosomal protein S5 isoform 2	225441583	23.10	10.19	+	-	-
60S ribosomal protein L35-like	225448819	14.30	11.39	-	+	-
putative 40S ribosomal protein S5, partial	37780996	16.80	11.26	-	+	+
12.04 Translation						
elongation factor 1-alpha	225439902	49.30	9.76	+	-	-
elongation factor 1-alpha-like	225435233	49.30	9.76	-	+	-
elongation factor 2-like isoform 1	225462164	93.90	5.74	+	+	-
seryl-tRNA synthetase	225450981	51.10	6.27	+	-	-

12.07 Translational control						
eukaryotic initiation factor 4A-2	225442221	46.90	5.34	-	+	-
14 Protein fate (folding, modification, destination)						
14.01 Protein folding and stabilization						
97 kDa heat shock protein-like	359482944	93.30	4.99	-	+	-
heat shock cognate 70 kDa protein isoform 2	359486799	71.10	5.02	-	+	-
heat shock cognate protein 80-like	359495606	80.00	4.84	+	+	-
luminal-binding protein 5-like	359490716	73.40	4.90	+	-	-
peptidyl-prolyl cis-trans isomerase CYP20-3, chloroplastic-like isoform 2	359480227	22.50	9.63	+	+	-
peptidyl-prolyl cis-trans isomerase isoform 1	225457957	17.90	9.87	+	+	+
14.07 Protein modification						
aspartic proteinase	144228219	20.80	4.48	-	+	-
aspartic proteinase nepenthesin-2	225455876	53.10	5.64	-	-	+
dolichyl-diphosphooligosaccharide—protein glycosyltransferase subunit 2-like	359480291	75.30	5.62	-	+	-
probable acetyl-CoA acetyltransferase, cytosolic 2-like	359497005	42.60	9.59	+	-	-
14.10 Assembly of protein complexes						
coatomer subunit gamma-2-like	359475304	98.60	4.94	+	+	+
ruBisCO large subunit-binding protein subunit beta, chloroplastic	225435794	64.60	5.71	+	+	-
14.13 Protein/peptide degradation						
probable mitochondrial-processing peptidase subunit beta	225452974	58.40	6.45	+	-	-
major allergen Pru av 1	225431844	17.10	5.96	+	-	-
36 Systemic interaction with the environment						
36.20 Plant/fungal specific systemic sensing and response						
vicilin-like antimicrobial peptides 2-1-like	359479651	63.90	7.84	-	-	+
Vitis vinifera Thaumatin-Like (VVTL) Proteins	410563154	21.30	4.76	+	+	-
16 Protein with binding function or cofactor requirement (structural or catalytic)						
16.07 Metal binding						
aconitate hydratase, cytoplasmic	225447278	107.40	7.84	-	-	+
aconitate hydratase 2, mitochondrial	225460961	110.00	6.71	+	+	-
16.21 Complex cofactor/cosubstrate/vitamine binding						
membrane steroid-binding protein 2	225470692	23.50	4.57	+	-	-
20 Cellular transport, transport facilitation and transport routes						
20.01 Transported compounds (substrates)						
importin subunit alpha-1	225431871	58.10	5.18	-	-	+
20.03 Transport facilities						
ABC transporter C family member 8-like	359482526	164.10	9.04	+	-	-
32 Cell rescue, defense and virulence						
32.01 Stress response						
catalase isozyme 1-like	359476986	56.90	6.77	+	+	-
peroxidase 4	225434381	34.00	9.56	+	-	-
32.05 Disease, virulence and defense						
major allergen Pru ar 1	225431840	17.30	5.10	+	-	-
34 Interaction with the environment						
34.11 Cellular sensing and response to external stimulus						
temperature-induced lipocalin	77744883	21.50	6.63	+	+	+

The functional distribution of identified proteins in specific grape tissues is shown in Fig 1. Most of the proteins extracted from Sauvignon Blanc grape tissues fell into the groups labelled as metabolism (41% for skin tissues, 29% for pulp tissues, and 38% for seed tissues), energy (24% for skin tissues, 33% for pulp tissues, and 27% for seed tissues), protein fate (11% for skin tissues, 17% for pulp tissues and 11% for seed tissues) and protein synthesis (9% for skin tissues, 11% for pulp tissues and 4% for seed tissues). The high frequency of proteins involved in metabolism and energy confirmed results of previous proteomic studies [24,31]. In addition, proteins identified in grape seed were less diverse than those identified in grape skin and pulp.

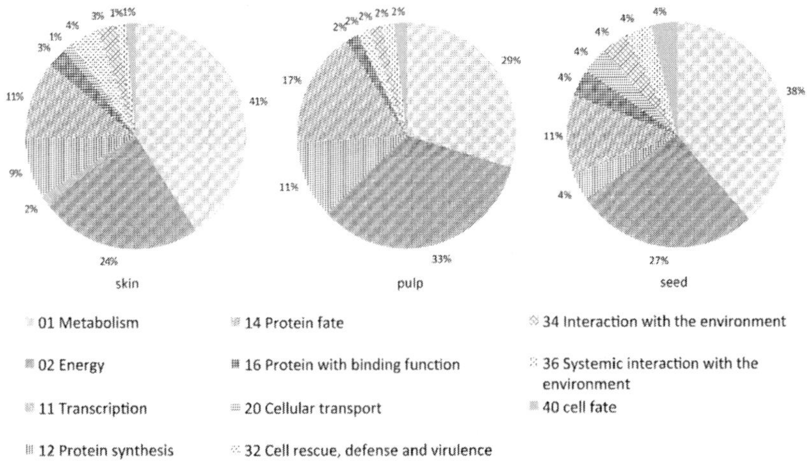

Figure 1. Functional distribution of the identified proteins in Sauvignon Blanc grape skin, pulp and seed.

The *Vitis vinifera* thaumatin-like proteins (VVTL1), one of TLPs isoforms [32], and chitinases were identified in both grape skin and pulp but not in the seed. The MS/MS analysis of a peptide from VVTL1 and chitinase are shown in Fig 2. In this study, more than one isoform of TLPs could be present but it was not possible to identify these because of high sequence homology between the isoforms thus reducing the probability of identifying unique peptides. These results are in agreement with previous studies in which the PR proteins were observed in grape skin and pulp of other grape cultivars [14,23,33]. The observation of PR proteins in grape skin suggests that grape processing techniques which facilitate skin extraction may result in increased PR protein extraction into juice, and may affect the final protein concentration in and bentonite requirement of wine. In this study, another PR protein, β-1,3- glucanase, was detected in Sauvignon Blanc grape skin. In a recent study, Wang et al [34] also observed β-1,3- glucanase in the skin of Sangiovese and Trebbiano. These results suggest that skin extraction can contribute to protein composition in wine and the β-1,3-glucanases observed in wine are likely to be derived from grape skin.

SDS-PAGE analysis of proteins seperated by HPLC

In order to identify and quantify the PR proteins from the protein extracts of grape tissues using HPLC, the retention times of all protein peaks from the protein extracts were compared with the retention times of purified TLPs and chitinases and with those previously reported [10,29,30]. For protein extracts from both grape skin and pulp, there were two adjacent peaks eluted at the retention time of 9.3 min (F1) and 10.6 min (F2), respectively, and there was a single peak eluted at retention time of 19.2 min (Fig 3). As described above, the fraction 1 (F1) from Sauvignon Blanc juice was assigned to TLPs and proteins eluted at 19.2 min were assigned to chitinases, and proteins eluted at 9.3 min and 10.6 min were collected and further analysed by SDS-PAGE. The analysis of these two protein fractions together with the TLPs separated from Sauvignon Blanc juice, for comparison, is shown in Fig 4. Although the molecular mass of TLPs determined by mass spectrometry was 21.3 kDa, the SDS-PAGE analysis result showed that the molecular weight of TLPs was about 18 kDa. In a previous study, Le Bourse et al. [35] also observed that the molecular weight of TLPs determined on electrophoresis gel was smaller than its theoretical value determined by mass spectrometer. The SDS-PAGE analysis showed that the main protein peak in F1 had a MW of 18 kDa which was

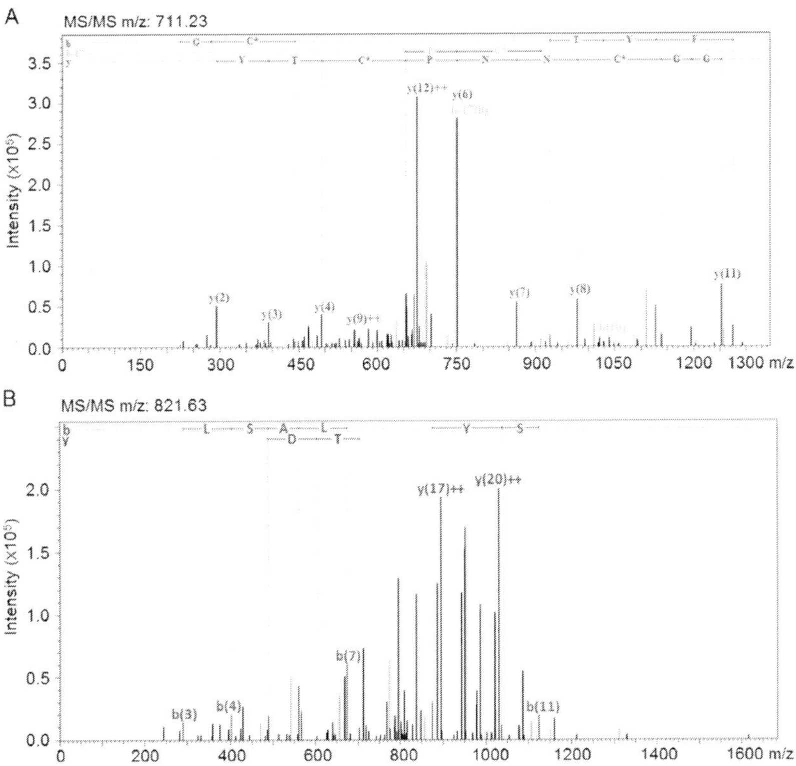

Figure 2. The MS/MS spectrum of 711.23 m/z (A) and 821.63 m/z (B), which identified the peptide unique to VVTL1 and chitinases from *Vitis vinifera*, respectively.

presumably TLP. Thus, the protein fraction eluting at 9.3 min was assigned to TLP and the area of this peak was used to relatively quantify TLP in protein extracts of grape tissues. In addition, the analysis of protein fraction 2 on SDS-PAGE showed that proteins with MW of 37 kDa were observed in protein extracts from both grape skin and pulp, and these proteins were presumably glucanases which were observed in wine by Sauvage et al. [36].

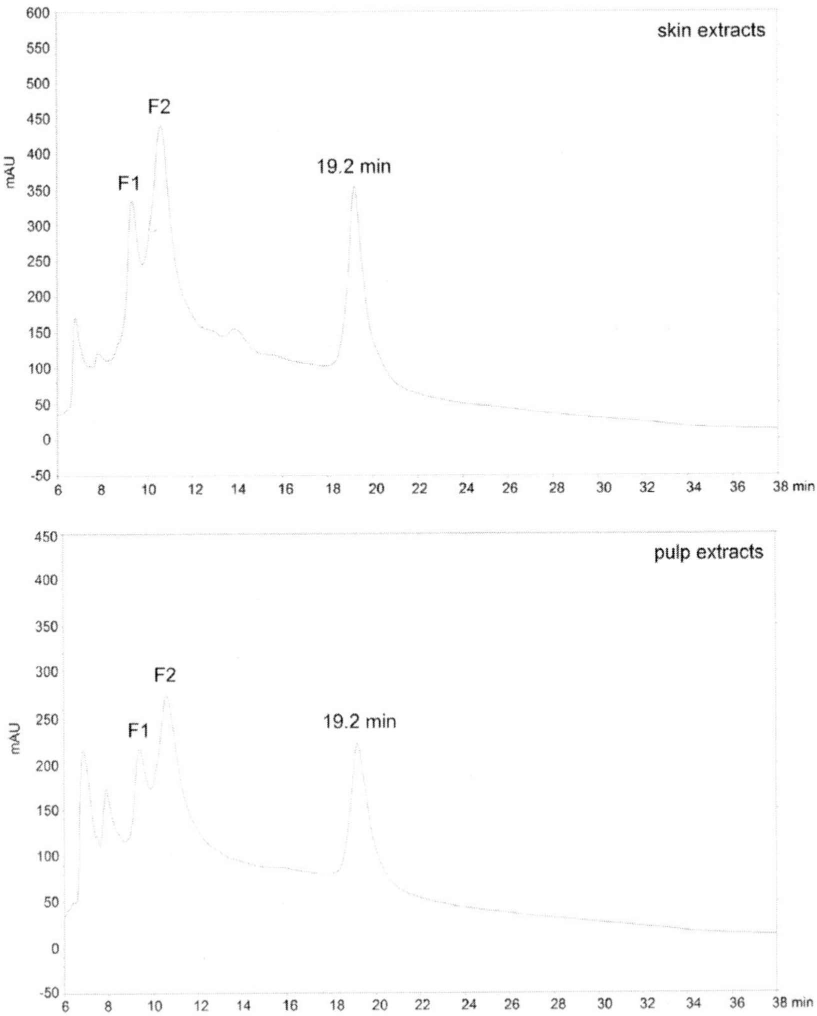

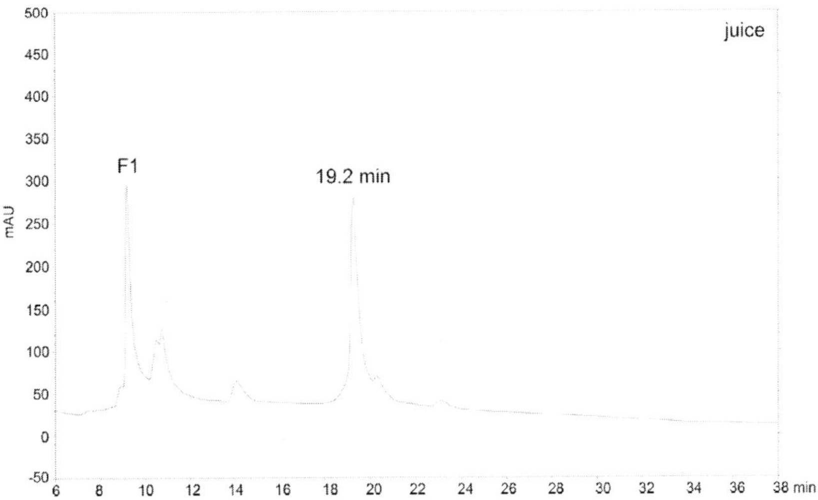

Figure 3. Comparison of chromatograms of protein extracts and juice from HPLC.

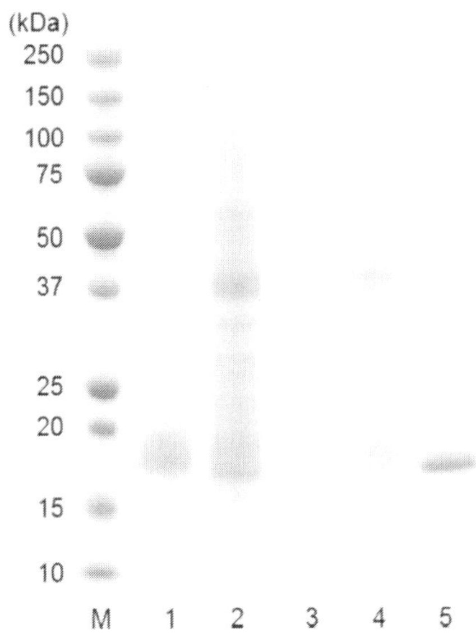

Figure 4. The SDS-PAGE analysis of protein fractions separated by HPLC: M, protein molecular weight marker; 1–2, F1 and F2 seperated from skin protein extracts and eluted at 9.3 min and 10.6 min respectively; 3–4, F1 and F2 seperated from pulp protein extracts and eluted at 9.3 min and 10.6 min respectively; 5, F1 (assigned to TLPs) collected from Sauvignon Blanc juice.

Relative quantification of PR proteins in grape skin and pulp

Relative quantification of TLPs and chitinases was carried out by comparing the area of the peaks eluted at 9.3 min and 19.2 min, respectively, against the area of a thaumatin standard. Using this method, the concentrations of TLPs and chitinases were determined in Sauvignon Blanc grape skin and pulp (Table 2). The results represented three measurements of TLPs and chitinases in specific grape tissues by HPLC. Comparison of the concentrations between two grape tissues showed that Sauvignon Blanc grape pulp contained similar concentration of TLPs and chitinases (275.1 mg/L and 248.2 mg/L, respectively). The concentrations of TLPs and chitinases determined in Sauvignon Blanc grape skin were 581.8 mg/L and 442.4 mg/L, respectively. In consideration of the weight ratio between skin and pulp in a single berry (approximately 1:8.8), the amount of these PR proteins in grape skin on per berry basis are actually much less than those in grape pulp. Furthermore, in this study the concentrations of TLPs and chitinases determined in grape skin and pulp were much higher than those reported in a previous study [37] in which the proteins were extracted by homogenizing the grape tissues in model grape juice, this is possibly because the protein extraction method used in current study was optimized for grape berries which have high sugar content and large amounts of polyphenols that can interfere the efficiency of protein extraction [25]. Furthermore, in this study, the observation of TLPs and chitinases in Sauvignon Blanc grape skin supported the conclusion that higher concentration of PR proteins in juice from mechanically harvested grapes coupled with long distance transport was likely caused by greater extraction of PR proteins from grape skin [37]. However, the concentration of PR proteins in juice is predominently determined by their concentration in grape pulp [38], the effect of skin extraction on enhancing the concentration of PR proteins in juice might be limited due to the relatively low amount of TLPs and chitinases in grape skin.

Table 2. Quantification of PR protein in Sauvignon Blanc grape skin and pulp (*n* = 3).

Tissue	TLPs (mg/L)*	Chitinases (mg/L)*
Skin	581.8 ± 17.2	442.4 ± 34.3
Pulp	275.1 ± 7.1	248.2 ± 14.1

* The concentration was determined using HPLC, and expressed as thaumatin equivalent.

CONCLUSIONS

Protein profiling of New Zealand Sauvignon Blanc was carried out in this study to provide additional information on the distribution and composition of proteins in specific grape tissues. Two major soluble haze-forming PR proteins, TLPs and chitinases, were identified in both Sauvignon Blanc grape skin and pulp. Furthermore, the relative quantification of TLPs and chitinases showed that their concentration on a per berry basis in grape pulp was mucher higher than those in grape skin. These results indicate that the predominent PR proteins in grape juice are likely coming from the pulp, but the potential for extraction of PR proteins from grape skin during juicing process may consequently increase the concentration of PR proteins in juice and thus require higher bentonite addition for resultant wine stabilization.

ACKNOWLEDGMENTS

The authors wish to acknowledge Jenny Zhao for her assistance with the RP-HPLC analysis and Robert Yu for his help in assignment of identified proteins.

REFERENCES

1. Ferreira RB, Picarra-Pereira MA, Monteiro S, Loureiro VB, Teixeira AR. The wine proteins. Trends in Food Science & Technology. 2002; 12: 230–239.
2. Waters EJ, Shirley NJ, Williams PJ. Nuisance proteins of wine are grape pathogenesis-related proteins. Journal of Agricultural and Food Chemistry. 1996; 44: 3–5.
3. Datta SK, Muthukrishnan S. Pathogenesis-related proteins in plants: CRC Press; 1999.
4. Van Loon LC, Van Kammen A. Polyacrylamide disc electrophoresis of the soluble leaf proteins from Nicotiana tabacum var. 'Samsun' and 'Samsun NN': II. Changes in protein constitution after infection with tobacco mosaic virus. Virology. 1970; 40: 199–211.
5. Van Loon LC, Pierpoint WS, Boller TH, Conejero V. Recommendations for naming plant pathogenesis-related proteins. Plant Molecular Biology Reporter. 1994; 12: 245–264.
6. Robinson SP, Davies C. Molecular biology of grape berry ripening. Australian Journal of Grape and Wine Research. 2000; 6: 175–188.
7. Robinson SP, Jacobs AK, Dry IB. A class IV chitinase is highly expressed in grape berries during ripening. Plant physiology. 1997; 114: 771–778. pmid:9232868
8. Tattersall DB, Van Heeswijck R, Hoj PB. Identification and characterization of a fruit-specific, thaumatin-like protein that accumulates at very high levels in conjunction

with the onset of sugar accumulation and berry softening in grapes. Plant physiology. 1997; 114: 759–769. pmid:9232867

9. Peng Z, Pocock KF, Waters EJ, Francis IL, Williams PJ. Taste properties of grape (Vitis vinifera) pathogenesis-related proteins isolated from wine. Journal of Agricultural and Food Chemistry. 1997; 45: 4639–4643.
10. Pocock KF, Hayasaka Y, McCarthy MG, Waters EJ. Thaumatin-like proteins and chitinases, the haze-forming proteins of wine, accumulate during ripening of grape (Vitis vinifera) berries and drought stress does not affect the final levels per berry at maturity. Journal of Agricultural and Food Chemistry. 2000; 48: 1637–1643. pmid:10820071
11. Velasco R, Zharkikh A, Troggio M, Cartwright DA, Cestaro A, Pruss D, et al. A high quality draft consensus sequence of the genome of a heterozygous grapevine variety. PLoS ONE. 2007; 2(12): e1326. pmid:18094749
12. Jaillon O, Aury JM, Noel B, Policriti A, Clepet C, Casagrande A, et al. The grapevine genome sequence suggests ancestral hexaploidization in major angiosperm phyla. Nature. 2007; 449: 463–467. pmid:17721507
13. Marzia G, Gabriella GM. Heard it through the grapevine: Proteomic perspective on grape and wine. Journal of Proteomics. 2010; 73: 1647–1655. pmid:20580953
14. Deytieux C, Geny L, Lapaillerie D, Claverol S, Bonneu M, Doneche B. Proteome analysis of grape skins during ripening. Journal of Experimental Botany. 2007; 58: 1851–1862. pmid:17426054
15. Jellouli N, Ben Jouira H, Skouri H, Ghorbel A, Gourgouri A, Mliki A. Proteomic analysis of Tunisian grapevine cultivar Razegui under salt stress. Journal of Plant Physiology. 2008; 165: 471–481. pmid:17942184
16. Okuda T, Fukui M, Takayanagi T, Yokotsuka K. Characterization of major stable proteins in Chardonnay wine. Food Science and Technology Research. 2006; 12: 131–136.
17. Waters EJ, Hayasaka Y, Tattersall DB, Adams KS, Williams PJ. Sequence analysis of grape (Vitis vinifera) berry chitinases that cause haze formation in wines. Journal of Agricultural and Food Chemistry. 1998; 46: 4950–4957.
18. Blade WH, Boulton R. Adsorption of protein by bentonite in a model wine solution. American Journal of Enology and Viticulture. 1988; 39: 193–199.
19. Lambri M, Dordoni R, Silva A, De Faveri DM. Effect of bentonite fining on odor-active compounds in two different white wine styles. American Journal of Enology and Viticulture. 2010; 61: 225–233.
20. Tattersall DB, Pocock KF, Hayasaka Y, Adams K, Van Heeswijck R, Waters EJ, et al. Pathogenesis related proteins—their accumulation in grapes during berry growth and their involvement in white wine heat instability. Current knowledge and future perspectives in relation to winemaking practices. Roubelakis-Angelakis KA, editor. Molecular Biology and Biotechnology of the Grapevine: Kluwer Academic. 2001. pp. 183–202.
21. Marangon M, Van Sluyter SC, Neilson KA, Chan C, Haynes PA, Waters EJ, et al. Roles of grape thaumatin-like protein and chitinase in white wine haze formation. Journal of Agricultural and Food Chemistry. 2011; 59: 733–740. pmid:21189017
22. Grimplet J, Wheatley MD, Jouira HB, Deluc LG, Cramer GR, Cushman JC. Proteomic and selected metabolite analysis of grape berry tissues under well-watered and water-deficit stress conditions. Proteomics. 2009; 9: 2503–2528. pmid:19343710
23. Monteiro S, Piçarra-Pereira MA, Loureiro VB, Teixeira AR, Ferreira RB. The diversity of pathogenesis-related proteins decreases during grape maturation. Phytochemistry. 2007; 68: 416–425. pmid:17188723
24. Giribaldi M, Perugini I, Sauvage FX, Schubert A. Analysis of protein changes during grape berry ripening by 2 DE and MALDI TOF. Proteomics. 2007; 7: 3154–3170. pmid:17683049
25. Vincent D, Wheatley MD, Cramer GR. Optimization of protein extraction and solubilization for mature grape berry clusters. Electrophoresis. 2006; 27: 1853–1865. pmid:16586412

26. Moser C, Segala C, Fontana P, Salakhudtinov I, Gatto P, Pindo M, et al. Comparative analysis of expressed sequence tags from different organs of Vitis vinifera L. Functional & Integrative Genomics. 2005; 5: 208–217.
27. Van Sluyter SC, Marangon M, Stranks SD, Neilson KA, Hayasaka Y, Haynes PA, et al. Two-Step Purification of Pathogenesis-Related Proteins from Grape Juice and Crystallization of Thaumatin-like Proteins. Journal of Agricultural and Food Chemistry. 2009; 57: 11376–11382. pmid:19886666
28. Marangon M, Van Sluyter SC, Haynes PA, Waters EJ. Grape and wine proteins: their fractionation by hydrophobic interaction chromatography and identification by chromatographic and proteomic analysis. Journal of Agricultural and Food Chemistry. 2009; 57: 4415–4425. pmid:19354294
29. Salazar FN, López F, Chiffelle I, López R, Peña-Neira A. Evaluation of pathogenesis-related protein content and protein instability of seven white grape (Vitis vinifera L.) clones from Casablanca Valley, Chile. European Food Research and Technology. 2012; 234: 509–515.
30. Pocock KF, Waters EJ. Protein haze in bottled white wines: How well do stability tests and bentonite fining trials predict haze formation during storage and transport? Australian Journal of Grape and Wine Research. 2006; 12: 212–220.
31. Sarry JE, Sommerer N, Sauvage FX, Bergoin A, Rossignol M, Albagnac G, et al. Grape berry biochemistry revisited upon proteomic analysis of the mesocarp. Proteomics. 2004; 4: 201–215. pmid:14730682
32. Marangon M, Van Sluyter SC, Waters EJ, Menz RI. Structure of Haze Forming Proteins in White Wines: Vitis vinifera Thaumatin-Like Proteins. PLoS ONE. 2014; 9(12): e113757. pmid:25463627
33. Colas S, Afoufa-Bastien D, Jacquens L, Clement C, Baillieul F, Mazeyrat-Gourbeyre F, et al. Expression and In Situ Localization of Two Major PR Proteins of Grapevine Berries during Development and after UV-C Exposition. PLoS ONE. 2012; 7(8): e43681. pmid:22937077
34. Wang W, Bianchi L, Scali M, Liu L, Bini L, Cresti M. Proteomic analysis of β-1, 3-glucanase in grape berry tissues. Acta Physiologiae Plantarum. 2009; 31: 597–604.
35. Le Bourse D, Conreux A, Villaume S, Lameiras P, Nuzillard JM, Jeandet P. Quantification of chitinase and thaumatin-like proteins in grape juices and wines. Analytical and Bioanalytical Chemistry. 2011; 401: 1541–1549. pmid:21465097
36. Sauvage FX, Bach B, Moutounet M, Vernhet A. Proteins in white wines: Thermo-sensitivity and differential adsorbtion by bentonite. Food Chemistry. 2010; 118: 26–34.
37. Pocock KF, Hayasaka Y, Peng Z, Williams PJ, Waters EJ. The effect of mechanical harvesting and long-distance transport on the concentration of haze-forming proteins in grape juice. Australian Journal of Grape and Wine Research. 1998; 4: 23–29.
38. Tian B, Harrison R, Jaspers M, Morton J. Influence of ultraviolet exclusion and of powdery mildew infection on Sauvignon Blanc grape composition and on extraction of pathogenesis-related proteins into juice. Australian Journal of Grape and Wine Research. 2015.

CHAPTER 20

Land Use History and Population Dynamics of Free-Standing Figs in a Maturing Forest

Larissa Albrecht[1], Robert F. Stallard[2,3], Elisabeth K. V. Kalko[1,2]

[1] *Institute of Experimental Ecology, University of Ulm, Ulm, Germany,*

[2] *Smithsonian Tropical Research Institute, Balboa, Panama,*

[3] *U.S. Geological Survey, Boulder, Colorado, United States of America*

ABSTRACT

Figs (*Ficus* sp.) are often considered as keystone resources which strongly influence tropical forest ecosystems. We used long-term tree-census data to track the population dynamics of two abundant free-standing fig species, *Ficus insipida* and *F. yoponensis*, on Barro Colorado Island (BCI), a 15.6-km^2 island in Lake Gatún, Panama. Vegetation cover on BCI consists of a mosaic of old growth (>400 years) and maturing (about 90–150 year old) secondary rainforest. Locations and conditions of fig trees have been mapped and monitored on BCI for more than 35 years (1973–2011), with a focus on the Lutz Catchment area (25 ha). The original distribution of the fig trees shortly after the construction of the Panama Canal was derived from an aerial photograph from 1927 and was compared with previous land use and forest status. The distribution of both fig species (~850 trees) is restricted to secondary forest. Of the original 119 trees observed in Lutz Catchment in 1973, >70% of *F. insipida* and >90% of *F. yoponensis*had died by 2011. Observations in other areas on BCI support the trend of declining free-standing figs. We interpret the decline of these figs on BCI as a natural process within a maturing tropical lowland forest. Senescence of the fig trees appears to have been accelerated by severe droughts such as the strong El Niño event in the year 1982/83. Because figs form such an important food resource for frugivores, this shift in resource availability is likely to have cascading effects on frugivore populations.

INTRODUCTION

For many decades, tropical primary forests have been increasingly affected by forest conversion and degradation. In response, research into the regeneration and restoration of deforested areas and the conservation of associated endangered animal species is receiving increased attention [1–3]. This requires development of an in-depth understanding of the underlying complex relations among organisms in tropical ecosystems, and more information is needed with regard to species abundances, metapopulations, landscape dynamics, dispersal, and animal movement among different habitats, each of which may have a wide range of land-use histories [3].

Secondary forests are characterized by high turnover rates and distinct successional stages, with initial plant populations that consist of mostly short-lived, light-demanding and fast-growing pioneer species which are gradually replaced by a mature forest having mainly long-lived, slow-growing, and shade-tolerant species [1, 4]. Plant and animal species richness, but not necessarily species composition, may resemble those of mature forests within a few decades after land abandonment. The exact trajectory of change strongly depends, however, on land-use intensity, closeness to old-growth forests, and availability of seed dispersers [3, 5, 6].

Climate change, in particular the increasing frequency and severity of droughts associated with El Niño Southern Oscillation (ENSO) events, is another factor that may influence forest dynamics and may have a detrimental effect on ecosystems [7, 8]. In this context, long-term studies focusing on the history and development of secondary forests, or the investigation of changes in species composition both in primary and secondary forests provide insight into forest dynamics that may be crucial in predicting and potentially mitigating effects of short- and long-term changes in forest ecosystems and their associated fauna (e.g., [1, 3–5, 9–11]).

Some pioneer trees may be particularly important for forest regeneration or restoration because they attract and maintain large populations of frugivorous seed dispersers [6]. Figs (*Ficus* sp.) play such a role in many tropical forests because they are often pioneer species and frequently share characteristics associated with keystone species such as considerable abundance, year-round fruiting phenology with large fruit crops, and a high nutritional value [12–15]. Terborgh [13] even proposed that loss of figs could result in a collapse of forest ecosystems. Therefore, knowledge about the population dynamics of figs is crucial for a better understanding of fruit-frugivore dynamics in tropical forests, but may also provide a tool in support of forest management and conservation.

Barro Colorado Island (BCI) in Panama constitutes one of the most intensely studied tropical forests in the world, with research going back to the early 1920's. It is at present covered by tall old-growth and secondary semideciduous rain forest at different successional states. Starting in the 1970's, during long-term studies on frugivorous bats, researchers observed that the previously very abundant free-standing fig trees were declining, and complete inventories of all figs on the island were conducted. To verify the early observations, in 2006, we collected all available data about fig trees on the island and conducted additional fig mapping. Trees were mapped and tagged during the first census, and all subsequent studies always used these tag numbers, so that the fate of individual trees can be traced.

Here, we present data of this long-term monitoring (> 35 years) documenting the distribution and population dynamics of two species of free-standing figs, *Ficus insipida* Willd. and *F. yoponensis* Desv., with a focus on the Lutz Catchment, a 25-ha area where fig census data, starting in 1973, are available [14, 16, 17]. Although the number of trees has declined profoundly, both species once occurred in some areas on BCI at rather high densities with about 5 trees per ha [14, 16, 17]. Distribution and current status of the *Ficus* populations can be explained by historical land use patterns, and we discuss likely reasons for the drastic population decline and its implications for frugivores, in particular bats.

Ficus insipida has one of the highest measured photosynthesis rates in plants [18], which explains the very fast growth rates observed in this species [19, 20]. These figs can reach 30 cm dbh (diameter breast height) within 19–39 years, and may gain 90% of their maximum size (130 cm dbh) in about 52 years [19]. In Corcovado National Park, Costa Rica, trees of *F. insipida* with an average 85 cm dbh are quite abundant in 27 year-old secondary forests growing on former cattle pastures [21]. Although *Ficus insipida* colonizes natural levees along rivers and hillslopes of abandoned agricultural lands, it rarely colonizes blow downs, because the saplings require especially high-light conditions for successful establishment [19, 22]. Accordingly, *F. insipida* and likely also *F. yoponensis* are good indicators of past human land use. Because free-standing figs are both a pioneer species and a keystone resource for many animals, our observations may be relevant for forest restoration and conservation in degraded landscapes.

MATERIALS AND METHODS

Study area and land use history

Our study was conducted on the 15.6-km^2 Barro Colorado Island (BCI) (9°09' N, 79°51' W), which was isolated in 1914 when the Chagres River was dammed to form Lake Gatún during the construction of the Panama Canal. Until 1914, BCI was a large hill at the lower Chagres River (Fig 1), surrounded by an extensive swampland fed by numerous small rivers through which the Chagres River meandered towards the Caribbean Sea from southeast to northwest.

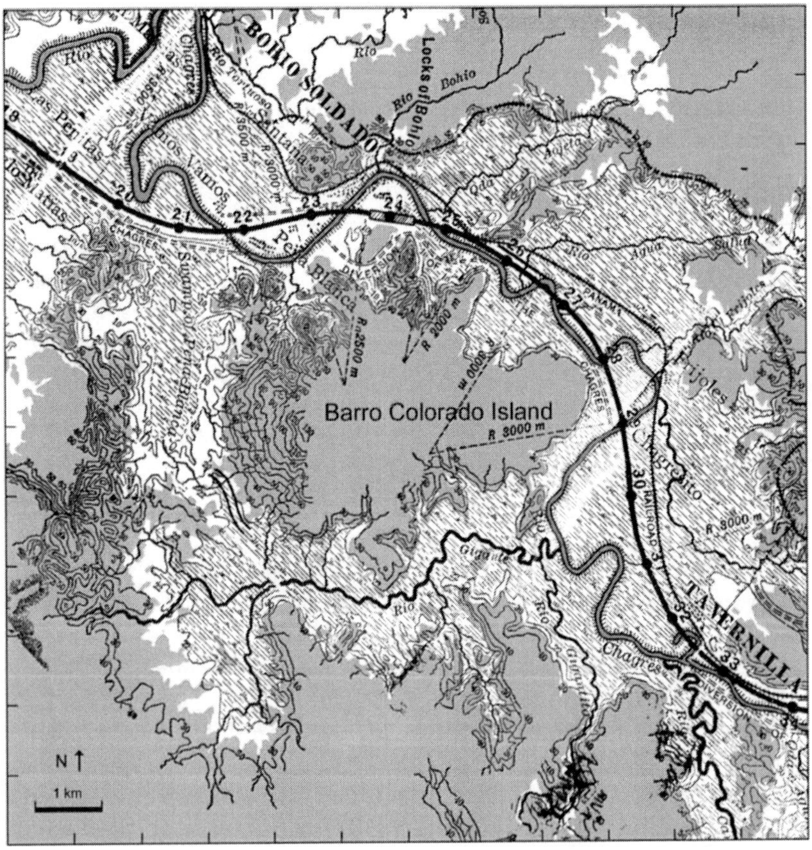

Figure 1. Map of the area around Barro Colorado Island before flooding (adapted from [38]).

This map depicts the natural course of the Chagres River (flowing southeast to northwest) and positions of former settlements. Land areas remaining after the filling of artificial Lake Gatún in 1914 are shaded in gray. The remaining white background represents the modern lake. Also indicated are the railway to the north (the black dash-dotted line) and the planned course of the Panama Canal (thick solid black line). Topographic contours, from the original map, in meters above sea level.

BCI receives about 2,600 mm rain per year with a pronounced dry season (less than 100 mm per month) from mid December to mid April, during which only about 10% of the annual rainfall occurs [23]. The island is covered by a semideciduous rain forest of mixed ages that reflect a complex history of human settlement. Some findings indicate that BCI has been subject to agriculture for at least 7,000 years [24, 25], and most areas have been deforested at some point in the last 1,000 years [23]. Pot shards and putative grinding stones found on the island are clear signs of human occupation prior to European arrival ([22, 26]; R. Stallard, personal observation).

After the founding of Panama City in 1519, the Chagres River was a major route of migration and commerce. In 1858 the first ocean-to-ocean railroad was completed along the course of the lower Chagres River from the City of Colon to Panama City. As a result, much of the lower Chagres Valley was deforested and transformed into an agricultural landscape and new settlements were founded (see Fig 1).

In the late 1880's, the French started excavations for a set of locks near Bohio Soldado [27], and were preparing to construct two major diversions of the Chagres alongside what is now BCI. This was presumably the time of greatest human population and impact along the north and east sides of BCI. The surrounding swamps and all the towns along the valley were flooded in 1914 by the creation of Lake Gatún.

In 1923, BCI became a natural reserve and a leading field station for tropical studies. Around that time the last agriculture land was abandoned [22, 28]. As a consequence of the pre-lake land use history, half of the island is currently covered with secondary forest mostly between 90 and 150 years old, but potentially up to 200 years old, and the south-western part is dominated by old-growth forest of least about 400–600 years old, and perhaps even over 2,000 years [22, 26].

Classification of forest age

The first reliable documentation of land cover on BCI dates from 1927. This documentation consists of a 1:10,000 topographic map [29] indicating forest, plantations, and non-forested land, and an aerial photo taken in early

June 1927 [30]. Although there was no active rain gauge on BCI in 1927, rainfall data from [31] for the Canal Basin indicate that 1927 was an especially wet year, and that May 1927 was a particularly wet month. Accordingly, it is reasonable to assume that after more than a month of rain, the canopy density in the June 1927 photograph reflects full leaf coverage. The original oblique photo was digitally flattened, aligned with the current shoreline of BCI, and georeferenced for further use in a geographic information system (GIS). The rectified photograph was digitally smoothed and each pixel received a value between 0 (black, dense canopy) and 255 (white, cleared). The color values were divided into five classes (posterized): (1) 0–31, (2) 31–95, (3) 95–160, (4) 160–220, and (5) 220–255, depicting different forest ages. Isolated patches (<100 m^2) were manually removed (Fig 2). Comparison with other BCI forest-age maps indicates that these categories correspond well to the forest disturbance categories distinguished on BCI [22, 30, 32]: old-growth (classes 1 and 2), tall-secondary (class 3), and low-secondary (classes 4 and 5) forest.

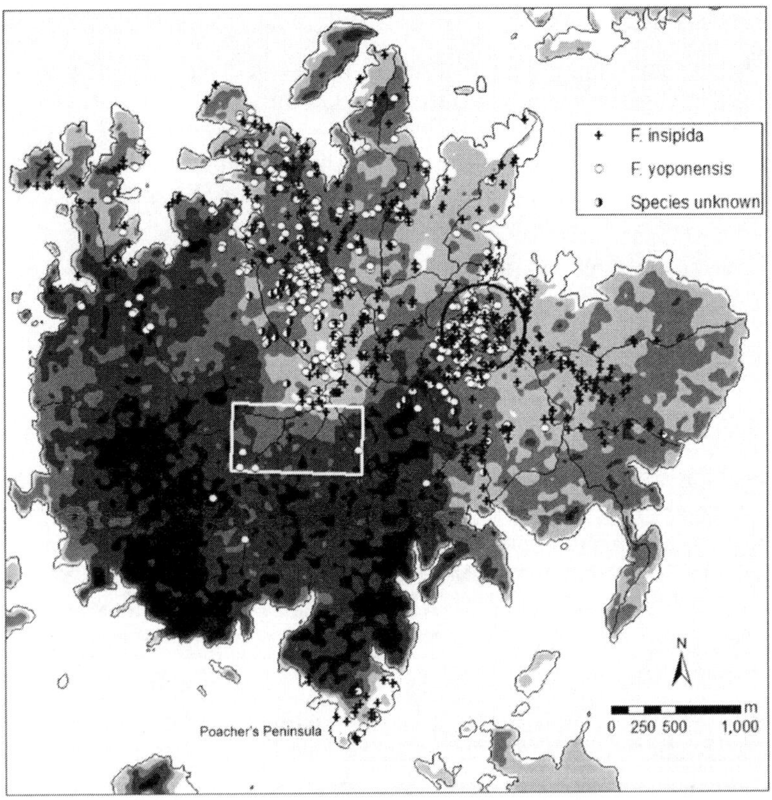

Figure 2. Distribution of *Ficus insipida* and *F. yoponensis* on Barro Colorado Island.

Depicted is the relative age of the forest from young (white) to old (black) forest as determined from an aerial photo from 1927 (compiled by R. Stallard and D. Kinner 2002). Secondary forest is about 90–150 years old, primary forest at least 400–600 years. The black circle marks the position of the Lutz Catchment, the white rectangle indicates the 50-ha forest-dynamics plot.

Early investigators mentioned that most of the secondary forest was probably about 50 years old when the reserve was founded [28, 32]. Based on the timeline and maps of the French activities, this would correspond to when the diversions were being constructed between the Chagres River and what is now BCI in the late 1880's. Although the secondary forest now appears to be fully grown (2011), the height of the canopy is still increasing. A canopy tower of 42 m in height was constructed in the Lutz Catchment in 1972. By 2001, the canopy height had increased sufficiently to compromise the meteorological observations done on the original tower, and in October 2001, the tower height was increased to 48 meters [33].

Data set

Adult fig trees were revisited and censused in the course of different studies over several decades to determine the fruiting phenology of figs in a selected area and to assess the fruit availability for fig-eating frugivores throughout the island. Although none of the fig censuses were performed to assess population dynamics of figs, condition of visited fig trees was generally noted and dead trees were always recorded. There are two data sets on figs available. In a 25-ha area called "Lutz Catchment", all figs were censused frequently during 38 years (1973–2011). The second set encompasses all fig trees found during a census of the whole island between 1985 and 1987, which were then partially recensused twice in the subsequent 25 years. Although the second data set is incomplete, it reveals that the decline of free-standing figs is an island-wide phenomenon and not restricted to the Lutz Catchment, providing a broader basis for discussion about possible reasons of the decline.

The fig census was approved by the Smithsonian Tropical Research Institute (STRI) who administrates BCI and whose permission is required to perform any research activities on the island.

Lutz Catchment census.

In 1973, a total of 142 fig trees encompassing six species were mapped in a 25-ha area called "Lutz Catchment" and tagged with individual numbers [16, 34]. Most of these trees (84%) belonged to the two free-standing fig species *Ficus insipida* (48 individuals) and *F. yoponensis*(71 individuals) (see also S1 Table). All individuals of these two free-standing fig species were mature and had a diameter of more than 50 cm. These trees were revisited at least once per census year (1973–1981, 1984, 1985, 1991, 1992, 1996, 2000, 2006–2011) in the course of different studies (e.g., [14, 16, 17]). In 2006 and 2007, we also searched systematically for new trees and saplings (> 50 cm in height) in the Lutz Catchment to determine whether there is any new recruitment of free-standing figs in this area. Between 2008 and 2010, all saplings were checked regularly and new saplings observed by chance were recorded.

Adult trees of *Ficus insipida* and *F. yoponensis* can be easily recognized from afar based on their smooth, grey trunks of up to 1.50 m in diameter with generally very conspicuous buttresses. Typically their distinctive leaves (dark, waxy green with bright yellow veins) can be found below the trees, so that even less striking trees are noticeable during systematic surveys. Saplings of both species have highly conspicuous, dark green leaves with striking yellow veins, about 20–40 cm in length and are thus easy to spot. We did not search for saplings smaller than 50 cm in height.

Island-wide census.

From 1985 to 1989 C.O. Handley Jr. mapped *Ficus* trees across the whole island. Handley described the condition of each tree (stem size, trunk angle, presence of vines, crown condition, trunk condition, slope) and marked their positions on hand-drawn maps. All trees in the census had a trunk diameter of at least 50 cm and up to about 150 cm above buttresses. Most trees had a diameter of about 90–120 cm. These trees were revisited in 1991/92 by E.K.V. Kalko and marked with numbered aluminum tags as well as with two labeled strips of distinctive pink flagging, one attached to the tag nail, the other to a nearby small tree or vine. Unfortunately, paper copies of these census periods archived on BCI were not complete for all areas of BCI, and some maps only depicted positions of fig trees without species identification and/or tag number (see 'Species unknown" in Fig 2). In total, data of about 140 *Ficus* trees tagged in 1991/92 (or later) were lost (most figs with the numbers 793–975). However, since Handley and others [14] noted that free-standing figs are scarce in the old-growth forest, we are confident that our general observations are still reliable despite missing data.

Most figs were then rechecked between 2006 and 2009 by L. Albrecht, and the accurate positions of all trees still present, or dead but identifiable, were geolocated using a high-resolution global positioning system (GPS) (Garmin 60CSx GPS, accuracy 5 m). Due to time constraints, about 280 of the original ~850 figs were not revisited in this last census. Positions of dead and decaying trees were often determined based on aluminum tags or the distinctive-pink flagging found on the ground or in the vegetation. Trees that were still alive were marked with new strips of labeled flagging on the tag and in the nearby vegetation. Trees without a tag (either new recruitment or where the original tag was overgrown) were marked with numerically consecutive aluminum tags following the census in 1991/92.

Analysis

Information from paper copies of fig census data and maps archived on BCI were entered into Excel and mapped in the GIS-Program ArcMap 9.2 (ESRI Inc., USA), using the hand-drawn maps of C.O. Handley as a basis for the new census combined with the GPS locations determined between 2006 and 2009 to estimate positions for now-absent trees. We estimated the positions of completely vanished dead or decayed trees based on relative distances depicted in the hand-drawn maps compared to accurate positions of geolocated trees. Based on these data, we can determine the distribution of free-standing figs throughout the island. However, as the data set of the whole island is incomplete, statistical analysis of tree mortality could only be performed for the Lutz Catchment.

Lutz Catchment.

Because of irregular census intervals and low sample size in the Lutz Catchment, we did not estimate mortality rates between census years and instead used the number of trees that were still alive for statistical analysis. In general, characterizations of time series should not make restrictive assumptions about the mathematical functions that are used to describe the data, because such assumptions can bias subsequent interpretation. Although we recognize that in a small population die-off is a series of steps, we assumed that data are smoothly varying. Spline functions are ideal computational tools for examining smoothly varying data, and we used relaxed cubic splines [35]. The relaxed cubic spline is a piecewise, continuous (connecting without gaps) set of cubic polynomials fit through the data, with one polynomial for each data segment; they are also piecewise continuous in their first derivative, used for rate calculations, and their second derivatives. With relaxed splines, the exact standard deviation is also specified. This standard deviation is chosen from other information, as we do below. The cubic spline regression gives the number of trees through time, $n(t)$, where t is time in years. The time derivative, $d\,n(t)\,d^{-1}\,t$, gives the rate of die-off, and the mortality rate, $\lambda(t)$, is $(d\,n(t)\,d^{-1}\,t)\bullet n(t)^{-1}$. The average annual mortality rate for a time interval was estimated from $\lambda_t = 100\bullet((\ln(n(t_0)) - \ln(n(t)))\bullet t^{-1})$, where $n(t_0)$ is the number of trees alive at the beginning of the census period and n(t) the number of trees alive after t years.

Because the derivative of the tree count through time gives the rate of tree death, we must take care as to how the data used in the spline regression are prepared. The algorithm has a zero first derivative for the beginning and end of the data series [35]. To get a derivative for a starting point, the most useful technique is to double the length of the series and use the first point, (t_0, n_0), in the time series as a center of symmetry, whereby a synthetic series is calculated by reflection through this point. The synthetic points are labeled from $-1 \geq -i \geq -n$, such that $t_{-i} = 2 \bullet t_0 - t_i$ and $n_{-i} = 2 \bullet n_0 - n_i$. With this, the spline curve passes exactly through (t_0, n_0) and has a slope in keeping with the general trend of the data approaching that point. We also assume that at some time in the future, we chose 25 years, there will be no fig trees and therefore a zero tree count and zero derivative. The main source of errors is not so much missing trees, but the fact that the censuses were not done at precisely the same time within a census year. We chose the error to be half of the average number of tree deaths per year for each type of fig. Accordingly, the *a priori* standard deviation of *F. yoponensis* was ±0.86 and for *F. insipida* it was ±0.45.

RESULTS

Lutz Catchment

In 1973, 119 individuals of the two free-standing fig species were identified in the Lutz Catchment, 48 *F. insipida* and 71 *F. yoponensis* (Table 1, Fig 3A). By 2011, only 20–21 trees, 14 *F. insipida* and 6–7 *F. yoponensis*, were still alive (Table 1, Fig 3B). One *F. yoponensis* tree was not accessible due to impassable treefalls. In 2010, it was in a very poor condition, and in 2013, this tree was dead. Thus, between 1973 and 2011, more than 70% of *F. insipida* and 90% of *F. yoponensis* died. This results in an average annual mortality rate of 3.2% per year for *F. insipida* and 6.1% per year for *F. yoponensis*. Almost all of the remaining trees showed distinct signs of age, in that the trunk and/or the roots exhibited slight to severe signs of rot and frequently major limbs were broken off (for details, see S2 Table).

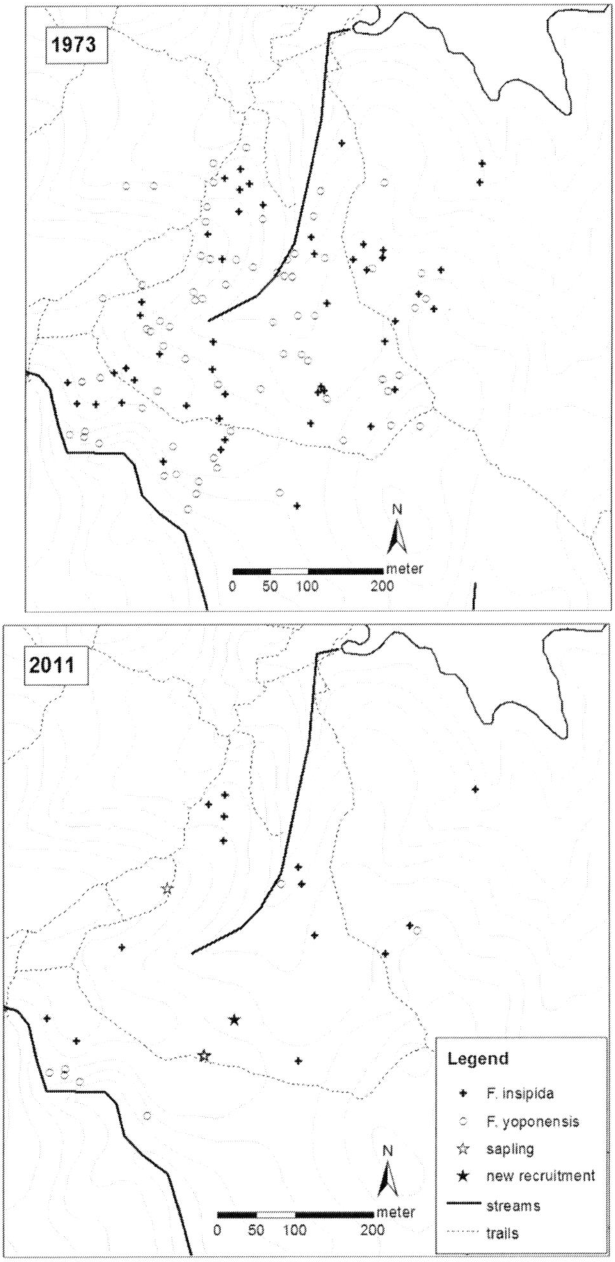

Figure 3. Results of the fig census in the Lutz Catchment. (a) Trees and saplings of *Ficus insipida* and *F. yoponensis* found alive in the Lutz Catchment in 1973 and (b) in 2011.

Table 1. Census data of each census year for *Ficus insipida* and *F. yoponensis*found in the Lutz Catchment.

census year	*Ficus insipida*		*Ficus yoponensis*	
	# alive	% dead	# alive	% dead
1973	48	—	71	—
1974	47	2.1	70	1.4
1975	47	2.1	70	1.4
1976	47	2.1	70	1.4
1977	47	2.1	70	1.4
1978	47	2.1	69	2.8
1979	46	4.2	66	7.0
1980	46	4.2	65	8.5
1981	46	4.2	65	8.5
1984	40	16.7	55	22.5
1985	39	18.8	47	33.8
1991	36	25.0	33	53.5
1992	35	27.1	28	60.6
1996	31	35.4	18	74.6
2000	28	41.7	12	83.1
2006	20	58.3	8	88.7
2007	20	58.3	8	88.7
2008	19	60.4	8	88.7
2009	18	62.5	8	88.7
2010	17	64.6	8	88.7
2011	14	70.8	7–6 [a]	90.1–91.5

[a] One tree not accessible in 2011 due to treefalls was reported dead in 2013

Given are the number of individuals of *Ficus insipida* and *F. yoponensis* still alive at each census and the percentage of individuals dead since 1973.

In 2007, we found a comparatively small *F. yoponensis* (approximately 10–15 m tall, 45 cm dbh) that lacked conspicuous buttresses and had produced a small crown (about 10 m in diameter). This was the only new recruitment we found after an extensive search in Lutz Catchment. This tree is not included in the subsequent statistical analysis on mortality rates.

Three saplings of *F. insipida* were found in large gaps close to trails during the systematic search of the whole 25-ha plot. One sapling found in 2006 was still vigorous and fast growing in 2010, while two other saplings observed in 2007 had died within two years (the positions of all saplings are shown in Fig 3B). Throughout the census period of the Lutz Catchment from 2006–2011, we did not find any saplings in the forest under close canopy.

Mortality of trees varied strongly between census years, with periods of high and low mortality (Fig 4). Between 1973 and 1981 only eight of the free-standing figs had died, two *F. insipida*(Fig 4, dark grey line) and six *F. yoponensis* (light grey). Four of these trees were uprooted in a severe

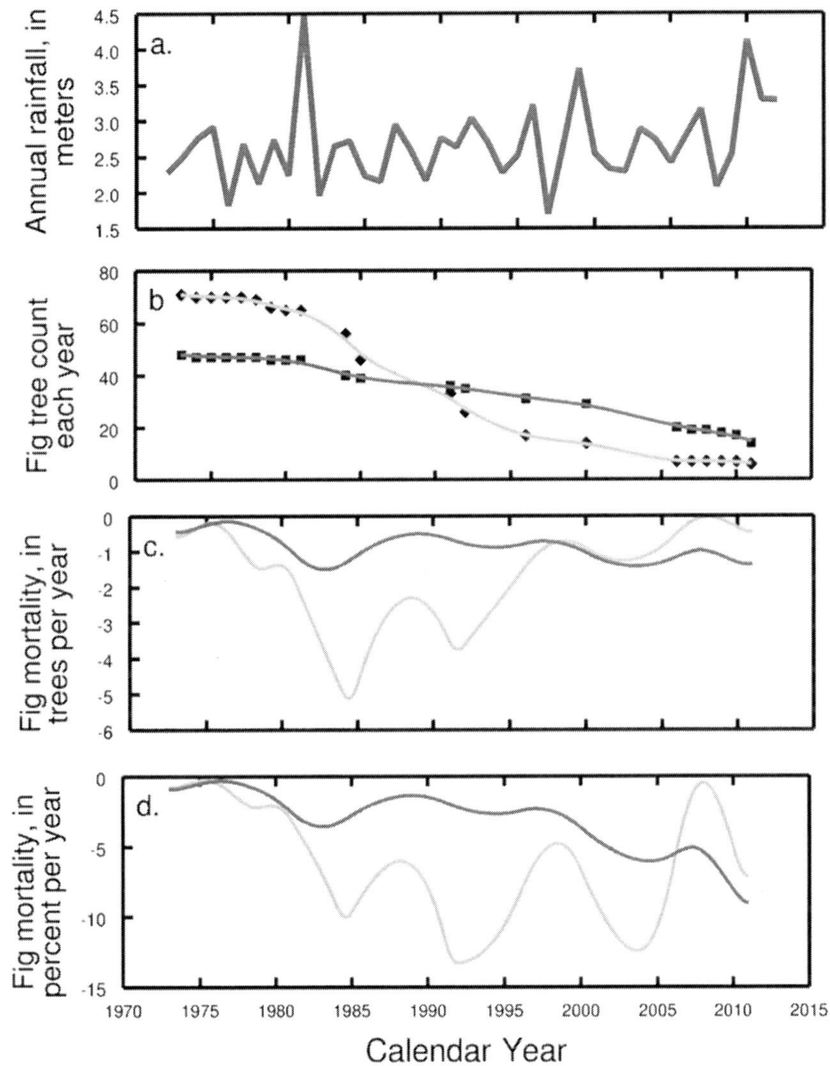

Figure 4. Mortality of fig trees in the Lutz Catchment. (a) Annual rainfall on BCI. Rainfall data are provided by the Environmental Studies Program (ESP). (b) Fig tree count per census year (light grey: *F. yoponensis*, dark grey: *F. insipida*). (c) Number of dead fig trees and (d) fig mortality in percent per year, estimated with relaxed cubic splines.

windstorm in 1979 [22]. High levels of mortality started in the early 1980's, particularly of *F. yoponensis*. Seven *F. insipida* and 18 *F. yoponensis* perished within four years (1981–1985). For *F. yoponensis*, mortality remained high in the 1990's, whereas the numbers of *F. insipida*started to decline more rapidly after year 2000. Interestingly, between 2006 and 2009, only two *F. insipida* and no *F. yoponensis* died, whereas between 2009 and 2011 four *F. insipida* and one *F. yoponensis* perished.

The relaxed-spline regression provides information about the rates and timing of the fig die-off. The mortality rate, $\lambda(t)$, tends to be less for *F. insipida*, with greater (more negative) mortality around 1982 and a pronounced increase in mortality in recent years. *Ficus yoponensis*demonstrates several episodes of increased mortality, around 1985, 1992, 2004, and the end of the sampling. Similarly, mortality was increased in *F. insipida* around 1983, 1994, 2004, and the end of the sampling (Fig 4C and 4D).

Island-wide census.

In total, almost 1,300 fig trees were mapped between 1985 and 1989 across the island. The census included 16 species: four free-standing fig species (subgenus *Pharmacosycea*) and 12 strangler fig species (subgenus *Urostigma*). The two most abundant species (>70% of all mapped trees) were *Ficus insipida* (>470 trees) and *F. yoponensis* (>365 trees) (see S2 Table). Most of the free-standing figs were growing in the secondary forest in the north-eastern part and the southernmost point of the island (Poacher's Peninsula), with almost none in mature forest (Fig 2).

As of 1991/92, about 90 individuals of *F. insipida* (19%) and almost 100 individuals of *F. yoponensis* (27%) had died. By 2009, about 220 individuals of *F. insipida* (45%) and 180 *F. yoponensis* (48%) were definitively dead or were at least not found even after an extensive search during the most recent census (2006–2009). However, about 280 of the original ~850 free-standing fig trees were not revisited due to time constraints. Of those trees that were alive in 1992 and revisited in the census of 2006–2009, 48.4% of *F. insipida* and 37.8% of *F. yoponensis* were still alive (see S2 Table). Based on these data, we estimate that of the about 470 individuals of *F. insipida* and about 370 individuals of *F. yoponensis* found in the first census in 1985 on BCI, about 290 individuals (61%) of *F. insipida* and 270 individuals (70%) of *F. yoponensis* were dead by 2011. Thus, fewer than 300 of the about 850 free-standing figs from the year 1985 were likely still alive in 2011. Many trees visited between 2006 and 2009 were in a poor condition, with rotting roots and/or trunks and broken limbs. There was no area visited in which not most of the original free-standing figs were dead, and in some areas no living tree was found. We located very few free-standing fig trees without tag number, and only three trees were small enough (< 60 cm dbh), two *F. yoponensis* and one *F. insipida*, to count as new recruitments. While searching for figs, we also looked out for saplings.

We observed nine saplings of *F. insipida* and one sapling of *F. yoponensis* of about 50 cm to 5 m in height.

DISCUSSION

Regeneration of deforested areas and its long-term implications on structure and composition of forests and the associated fauna have become important issues in tropical forest conservation worldwide (e.g. [3]). In this study on Barro Colorado Island, we demonstrate, using long-term census data sampled during different studies, that the transition from one successional state to another has the potential to transform the whole ecosystem by removing two keystone species, the free-standing figs *Ficus insipida* and *F. yoponensis*. Until a few decades ago, both figs were very abundant in the secondary forest of BCI, but their numbers have been declining rapidly since the early 1980's. Our observations on distribution and population dynamics of these two fig species can be linked to land use history and the normal transition from a secondary to a more mature forest.

With its long record of human settlement, the forest of BCI is a mosaic of different stages of natural succession, and thus can provide insights into ecological processes of a maturing Neotropical secondary forest. Most of this secondary forest is more than 90 years old and is approaching old-growth forest in general appearance and structure, yet some long-lived pioneer trees, such as *F. insipida* and *F. yoponensis*, remain as indicators of a secondary forest.

Population dynamics on BCI

In the Lutz Catchment, where a complete set of census data of 119 free-standing fig trees with frequent, though irregular census intervals are available, more than 70% of *Ficus insipida* and over 90% of *F. yoponensis* died between 1973 and 2011, while only one young *F. yoponensis*has successfully established (Fig 3). Accordingly, average annual mortality rate for the Lutz Catchment over 38 years (1973–2011) is very high, 3.2% per year for *F. insipida* and 6.1% per year for *F. yoponensis*. Thus, mortality rates for both species in Lutz Catchment are quite high and exceeded the normal long-term average of tree mortality rates in tropical forests of 0.5%-3.3% (e.g., [9, 10, 36, 37]). High mortality started in the early 1980's after the 1982/83 ENSO event that caused a severe drought on BCI. Between 1981 and 1985, 18 of 65 *F. yoponensis* (27.7%) and seven of 46 *F. insipida* (15.2%) had died (see Table 1).

The decline of these two free-standing fig species is not a local phenomenon of the Lutz Catchment, but has been documented for the whole island. For all of BCI, of the ~850 free-standing fig trees that were identified around 1985–87, at least 400–600 individuals (45–72%) have perished, which results in an average annual mortality rate of approximately 2.0% for *F. insipida* and 1.3% for *F. yoponensis*, which is within the normal tree mortality rates in tropical forests. However, the island-wide census started in 1985, several years after a period of very high mortality observed in Lutz Catchment in the early 1980's during and after the 1982/83 ENSO event, which particularly affected *F. yoponensis*. Because the drought affected the whole island [9], it is likely that many fig trees died before the census in 1985. Though repeat observations of many trees (28% of *F. insipida* and 39% of *F. yoponensis*) are lacking for the census in 2006–2009, in those areas revisited, many fig trees—sometimes all trees—were dead. Therefore, we conclude that the estimation that 61% of *F. insipida* and 70% of *F. yoponensis* have died since 1985 is justified. For *F. insipida*, the estimation is in the same range as observed in Lutz Catchment, where 64% of the trees died between 1985 and 2011. In Lutz Catchment, about 85% of *F. yoponensis* (40 of 47 trees) died since 1985, thus the estimation that 70% of *F. yoponensis* died on the whole island in the same time period may be conservative.

Overall, saplings were rare in the forest, and successful recruitment has been observed mainly along the lake shore or at the field-station lab clearing (L. Albrecht, personal observation). Altogether, very few young free-standing fig trees have been observed; thus there is virtually no replacement of dying fig trees. Free-standing figs need high-light condition for establishment, and presumably these trees rarely find suitable conditions to grow in older secondary or old-growth forest on BCI, and new recruitment is mainly restricted to forest edges. Our data indicate that *F. yoponensis* may be more successful in establishing in gaps, because several *F. yoponensis* trees, but no *F. insipida*, have been mapped in the old-growth forest areas (see Fig 2). Under the assumption that *F. yoponensis* exhibits similar growth rates as *F. insipida* [19], the only young tree (with 45 cm dbh) found in Lutz Catchment in 2007 might be about 20–40 years old.

With few exceptions, *F. insipida* and *F. yoponensis* were mostly in the secondary forest in the north-eastern part of BCI and on Poacher's Peninsula in the south. Free-standing figs are pioneer trees [19, 22], and distribution of both species is associated with the young-forest side of edges between older-forest patches and younger forest (Fig 2), indicating that these figs were mainly growing along edges of pastures or agriculture land when the area was gradually abandoned. In the 1920's, only a few trees in the deeper ravines in the secondary forest on BCI had stem sizes reaching 30 cm dbh [28]. Many free-standing figs in our study were located on top of hills or on flat areas, and some established on steep slopes. Mature fig trees have stem sizes of 60–150 cm dbh, thus most fig trees must have been very young in the 1920's. Consequently, most individuals found on BCI most likely were established about 90–150 years ago, shortly after the land was cleared between the French attempt to build the Panama Canal in the 1880's and the early 1920's. By the early 1920's, agriculture on BCI had been abandoned, and *Ficus* saplings had to establish before the then young secondary forest became too tall and dense. These pioneer free-standing fig trees remaining from that time are now old, exhibiting the typical signs of senescence such as rot on trunks and roots and broken main branches.

Before the creation of the Panama Canal, the Chagres River meandered through an alluvial floodplain [38] which probably was very similar in structure and forest composition to the floodplain area of the Rio Manu in Peru [4, 22]. In Cocha Cashu, lateral migration of meanders along the Rio Manu drives vegetation succession on recently deposited fluvial sediments, creating a chronosequence of different forest ages. In a transect of five 1-ha plots, from very young secondary forest close to the active river bank to an about 200 (±50) year-old mature forest, distinctive successional stages have been described [4, 19]. *Ficus insipida* and *Cedrela odorata* L. (Meliaceae) are saplings in the youngest hectare and dominate most of the secondary forest stages (hectares 2–4). In the oldest, about 200 year-old plot, *F. insipida* and *C. odorata* have almost been extirpated, and more shade-tolerant species dominate. Estimated mortality rates for *F. insipida* in the youngest to mid-successional stages in Cocha Cashu range from 1.4% to 5.6% per year [19]. Present-day populations of *F. insipida* and *F. yoponensis* on BCI correspond to the successional stages found in the fourth and fifth hectare in the Cocha Cashu plot, when long-lived pioneer trees such as free-standing figs are replaced by more shade-tolerant species.

It is likely that free-standing figs around BCI were naturally abundant in areas close to the meandering Chagres River before the Panama Canal was created, and some fig trees may also have been left as shade trees in the agricultural lands. These *Ficus* trees may have served as a seed source for the fast colonization of the abandoned agricultural land on BCI since the late 1880's to the early 1920's. Frugivorous bats, the main dispersers in early successional stages, usually carry fruits to a feeding roost up to 200 m away from the parent tree, creating a seed shadow of several hundred meters around the parent tree [6, 14]. The presence of such remnant fig trees in cleared areas may explain the often clumped distribution of figs on BCI (Fig 2).

Although the decline of free-standing figs observed on BCI is a normal process at this stage of succession, it does not necessarily happen smoothly.

Our data indicate that mortality may have been increased by some episodic events such as drought (see. Fig 4). Periods of increased mortality coincide in both species, although *F. yoponensis* was affected more strongly, and mortality was highest after the severe drought caused by the very strong El Niño Southern Oscillation (ENSO) event in 1982/83, which also triggered enhanced mortality of many species in the 50-ha forest-dynamics plot in the middle of the island [9, 10, 39].

During the 1970's fig mortality on BCI was low. Several trees died during a strong storm in 1979, which caused major localized blow downs [22]. High mortality rates started again in the early 1980's (Fig 4) after the 1982/83 ENSO drought. Our data from the Lutz Catchment indicate that many fig trees died during or shortly after this event. Our calculations indicate that the peak in mortality for *F. insipida* was in 1983, at the end or shortly after the drought, while the peak for *F. yoponensis* was two years later. Individuals of *F. yoponensis* were most strongly affected, resulting in mortality of over 50% of the trees by 1991. The second and third die-offs of the figs in 1992/94 and 2004 do not seem to be linked to particular droughts, and neither species demonstrated immediate effects of the ENSO-related drought of 1997/98. The year 2008 was very dry on BCI, which may account for the increased mortality between 2009 and 2011 of both fig species after a period of low mortality (Table 1, Fig 4).

Generally, free-standing figs are not drought tolerant, because high photosynthetic rates correlate with high stomatal conductance and a rapid hydraulic water transport [18, 20]. Moreover, *F. insipida* is already close to maximum heat tolerance threshold during normal dry seasons [40], and increased temperatures during droughts may result in further stress. After the severe drought in 1982, many fig trees were probably weakened and thus more susceptible to wind damage, disease, rot, or beetle infestation, leading to death of individuals several years after this event [8, 41].

Although droughts probably accelerated the death of fig trees on BCI, we interpret the decline of free-standing *Ficus* trees on BCI as a natural part of the maturation of a secondary forest. The similarity to successional stages observed in Cocha Cashu and observations of Gautier-Hion and Michaloud [42] regarding figs in African rain forests corroborate this assumption. They noted that terrestrial figs in any rain forest evolving from second-growth stage to older stages will not be renewed, thus fig density and fruit production will strongly decrease. Consequently, these figs will cease being a stable food source for many frugivores when the forest reaches the late stages of maturation [42].

Relevance for conservation

Free-standing figs are not only an integral part of many natural secondary forests, they may also serve as useful tools for forest management in fragmented landscapes. Trees of the genus *Ficus* have a very high impact on tropical forest ecosystems, particularly for the frugivore community (e.g., [13, 43, 44]). Overall, the abundance of figs is likely to be a main driver of abundance for many tropical frugivores (e.g., [44, 45]) which also has consequences on higher trophic levels. Although ripe crops appear to be widely scattered and unpredictable in time and space, the total fruit production per area is comparably high, as is the biomass of consumers [12, 14, 43, 46]. Because figs are fast-growing pioneer trees and very attractive for many seed-dispersing animals, planting of figs may aid in forest restoration in sites where natural regeneration is limited or needs to be accelerated.

Both *F. insipida* and *F. yoponensis* are very important food sources for frugivorous animals on BCI [14, 46, 47, 48]. They fulfill all requirements for keystone species: high abundance; year-round fruiting pattern, especially also during times of general fruit scarcity high fruit yield of up to 50,000 fruits per tree; and a high nutritional value (e.g., [14, 15, 17]). Thus, the disappearance of such a large number of figs will likely have an impact on the whole ecosystem on BCI.

There are many animal species feeding on figs on BCI, such as agoutis (*Dasyprocta punctata*), peccary (*Pecari tajacu*), kinkajous (*Potos flavus*), coati (*Nasua narica*), monkeys (e.g., *Alouatta palliata*, *Cebus capucinus*), frugivorous bats and frugivorous birds [14, 48]. Fig-eating bat species are among the most abundant bat species on BCI, likely due to the abundance of *Ficus*trees [14, 48, 49], with figs often comprising more than 70–80% of their diet [17, 50]. The most abundant bat species on BCI is the common fruit bat *Artibeus jamaicensis* (Leach, 1921) with an estimated population size of about 3,000 individuals on the island [14]. Although most fruit-eating bats consume quite a broad spectrum of fruits [14, 50], most of these fruits are only seasonally available, mainly during the rainy season (e.g., [14, 50]), and are usually eaten only when no figs are available [17]. Thus, in the future, if the population of *Ficus* trees becomes too small to produce reliable fruit crops during general fruit scarcity, the large populations of fig-eating bats on BCI will likely decrease due to food limitations [49, 50]. We assume that in 20–40 years, when the large free-standing *Ficus* trees have almost completely disappeared, the frugivore community on the island will have changed considerably, particularly the frugivorous bat guild.

There is no indication that other fig species can effectively replace *Ficus insipida* and *F. yoponensis* in terms of number or fruit availability. There are two other free-standing fig species known on BCI, *Ficus tonduzii* Standl.

and *F. maxima* Mill. Both are scarce on the island with 45 and four individuals recorded, respectively, in the fig census of the whole island. Of seven *F. tonduzii* observed in 1973 in Lutz Catchment, six were dead by 1985, indicating that this species may have a shorter lifespan than *F. insipida* or *F. yoponensis*.

Strangler figs, which recruit on standing trees and thus also in mature forest, cannot entirely substitute for free-standing figs on BCI, because the densities of mature trees, number of fruits per crop as well as nutritional value are generally lower than in free-standing figs (e.g., [14, 15, 23, 51]). About 30% of the mapped trees in the census of the whole island were strangler figs, though the actual number may be underestimated. Most strangler figs are small, inconspicuous epiphytes and hemiepiphytes and may therefore be easily overlooked. In 2006–2010, some new, and mostly small, strangler figs were recorded, while several large strangler figs observed in earlier census periods were dead and gone. Overall, it is probable that all twelve strangler-fig species on BCI together may play an important role for wide-ranging, mobile frugivores such as large bats, but they will not have the overall quantitative importance of the two free-standing fig species *F. insipida* and *F. yoponensis*. No other non-fig species has the abundance and year-round fruit availability to compensate the loss of these figs for the fig-eating bat guild on BCI, and few species are fruiting during the dry season, which is a period of general fruit scarcity [14, 50].

On the other hand, figs, particularly strangler figs, are quite abundant on many of the small islands in Lake Gatun and on the nearby mainland (L. Albrecht, personal observation). Many frugivorous bats are able to commute over longer distances to find fruits [14, 17, 52, 53]. Thus, it may be possible that larger frugivorous bats, like *Artibeus jamaicensis*, will be able to maintain their population size on BCI, by expanding their foraging ranges when the availability of non-fig fruits on BCI itself is not adequate. Thus, it will be interesting to observe whether and how the ecosystem on BCI changes over the coming years following the disappearance of this keystone resource.

CONCLUSIONS

Barro Colorado Island is an example of the regeneration of a Neotropical secondary rain forest over time following a long history of shifting agricultural land use. We found an obvious correlation between land-use history on BCI for the last 150 years and the distribution of the two abundant free-standing fig species *Ficus insipida* and *F. yoponensis*. Both fig species are a major food resource for many frugivores on BCI and are likely responsible for the high abundance of frugivorous bats on this island.

Now that the 90-to-150 year-old secondary forest on BCI is in a transitional stage between late secondary succession and a more mature forest, long-lived pioneer species, such as the abundant free-standing figs, are declining and shade-tolerant species are taking over. This transition, which affects the whole ecosystem, may be indicative for the future of many areas in the Neotropics that are recovering because of restoration associated with conservation programs. Generally, during secondary forest succession, tree composition changes drastically, often with a very high fruit resource density in middle to older successional stages due to the presence of free-standing figs and other pioneer fruit trees [19, 42]. Consequently, frugivore abundance will be highest in older secondary forests, and decrease together with the decline of figs when the forest matures (see also [21]). Overall, our observations on BCI may provide important information for long-term management and conservation of Neotropical forests and the associated fauna in times of increasing land-use pressure.

ACKNOWLEDGMENTS

We dedicate this study to the memory of Dr. Charles O. Handley Jr. († 2000), former curator of the Museum of Natural History in Washington, D.C., who initiated this study several decades ago, and to the memory of Prof. Dr. Elisabeth K.V. Kalko, whose death in September 2011 came completely unexpected. We thank the Smithsonian Tropical Research Institute (STRI) in Panama for logistic support and the use of their facilities. A particular thank goes to Katrin Heer, Verónica Albrecht-Borth and Marco Tschapka who provided useful comments on the first drafts of this manuscript. Thanks also to Thomas Hiller and Gloria Gessinger for completing data of the last fig census. Any use of trade, firm, or product names is for descriptive purposes only and does not imply endorsement by the U.S. Government.

REFERENCES

1. Guariguata MR, Ostertag R. Neotropical secondary forest succession: changes in structural and functional characteristics. Forest Ecol Manag. 2001;148: 185–206
2. Chazdon RL. Beyond deforestation: restoring forests and ecosystem services on degraded lands. Science. 2008;320: 1458–1460. pmid:18556551
3. Chazdon RL, Peres CA, Dent D, Sheil D, Lugo AE, Lamb D, et al. The potential for species conservation in tropical secondary forests. Conserv Biol. 2009;23: 1406–1417. pmid:20078641
4. Foster RB. Long-term change in the successional forest community of the Rio Manu floodplain. In: Gentry AH, editor. Four Neotropical forests. New Haven: Yale University Press; 1990. pp. 565–572

5. Dunn RE. Recovery of faunal communities during tropical forest regeneration. Conserv Biol. 2004;18: 302–309
6. Muscarella R, Fleming TH. The role of frugivorous bats in tropical forest succession. Biol Rev. 2007;82: 573–590. pmid:17944618
7. Harrison RD. Drought and the consequences of El Niño in Borneo: a case study of figs. Popul Ecol. 2001;43: 63–75
8. Allen CD, Macalady AK, Chenchouni H, Bachelet D, McDowell N, Vennetier M, et al. A global overview of drought and heat-induced tree-mortality reveals emerging climate change risks for forests. Forest Ecol Manag. 2010;259: 660–684
9. Condit R, Hubbell SP, Foster RB. Mortality rates of 205 Neotropical tree and shrub species and the impact of a severe drought. Ecol Monogr. 1995;65: 419–439
10. Condit R, Ashton PS, Manokaran N, LaFrankie JV, Hubbell SP, Foster RB. Dynamics of the forest communities at Pasoh and Barro Colorado: Comparing two 50-ha plots. Philos Trans R Soc Lond B Biol Sci. 1999;354: 1739–1748. pmid:11605618
11. .Chazdon RL, Redondo Brenes A, Vilchez Alvarado B. Effects of climate and stand age on annual tree dynamics in tropical second-growth rain forests. Ecology. 2005;86: 1808–1815
12. Janzen DH. How to be a fig. Ann Rev Ecol Syst. 1979;10: 13–51
13. Terborgh J. Keystone plant resources in the tropical forest. In Soulé ME, editor. The science of scarcity and diversity. Massachusetts: Sinauer Associates, Inc. Sunderland; 1986. pp. 330–344
14. Handley CO Jr, Gardner AL, Wilson DE. Demography and natural history of the common fruit bat, *Artibeus jamaicensis*, on Barro Colorado Island, Panamá. Smithson Contrib Zool. 1991;511
15. Wendeln MC, Runkle JR, Kalko EKV. Nutritional values of 14 fig species and bat feeding preferences in Panama. Biotropica. 2000;32: 489–501
16. .Morrison DW. The foraging behavior and feeding ecology of a Neotropical fruit bat, Artibeus jamaicensis. PhD Thesis, Cornell University, Ithaca, New York. 1975
17. Morrison DW. Foraging ecology and energetics of the frugivorous bat *Artibeus jamaicensis*. Ecology. 1978;59: 716–723.
18. Zotz G, Harris G, Königer M, Winter K. High rates of photosynthesis in the tropical pioneer tree, *Ficus insipida* Willd. Flora. 1995;190: 265–272
19. Terborgh J, Flores NC, Mueller P, Davenport L. Estimating the ages of successional stands of tropical trees from growth increments. J Trop Ecol. 1997;14: 833–856
20. Hao GY, Goldstein G, Sack L, Holbrook NM, Liu ZH, Wang AY, et al. Ecology of hemiepiphytism in fig species is based on evolutionary correlation of hydraulics and carbon economy. Ecology. 2011;92: 2117–2130. pmid:22164836
21. Weghorst JA. High population density of black-handed spider monkeys (*Ateles geoffroyi*) in Costa Rican lowland wet forest. Primates. 2007;48: 108–116. pmid:17265010
22. Foster RB, Brokaw NVL. Structure and history of the vegetation of Barro Colorado Island In: Leigh EG Jr, Rand AS, Windsor DM, editors. The ecology of a tropical forest: Seasonal rhythms and long term change. Washington D.C.: Smithsonian Institution Press; 1982. pp. 67–94
23. Leigh EG Jr. Tropical forest ecology–a view from Barro Colorado Island. Oxford University Press, UK; 1999
24. Bartlett AS, Barghoorn ES, Berger R. Fossil maize from Panama. Science. 1969;151: 642–643.
25. Bartlett AS, Barghoorn ES. Phytogeographic history of the Isthmus of Panama during the past 12,000 years. In: Grahm A, editor. Vegetation and Vegetational History of Northern Latin America. New York: Elsevier; 1973. pp. 203–299
26. Piperno DR. Fitolitos, Arqueología y cambios prehistóricos de la vegetación en un lote de cincuenta hectáreas de la Isla de Barro Colorado. In: Leigh EG Jr, Rand AS Windsor DM, editors. Ecología de un bosque tropical: Ciclos estacionales y cambios a largo plazo. Smithsonian Tropical Research Institute, Balboa, República de Panamá; 1990. pp. 153–156

27. McCullough D. The path between the Seas: The creation of the Panama Canal, 1870–1914. New York: Simon and Schuster; 1977
28. Kenoyer LA. General and successional ecology of the lower tropical rain-forest at Barro Colorado Island, Panama. Ecology. 1929;10: 201–222
29. .Miller F. Special map of Barro Colorado Island, Canal Zone, 1:10,000. Corozal, Canal Zone, Panama Canal Department, Office of the Department Engineer, Engineer Reproduction Plant; 1927
30. Svenning JC, Kinner DA, Stallard RF, Engelbrecht BMJ, Wright SJ. Ecological determinism in plant community structure across a tropical forest landscape. Ecology. 2004;85: 2526–2538
31. Matthew GE. Chagres River and Gatun Lake watershed and hydrological data: 1907–1948. The Panama Canal Section of Meteorology and Hydrology; 1949
32. Enders RK. Mammalian life histories of Barro Colorado Island, Panama. Bull Mus Comp Zool Harvard College. 1935;78: 385–502
33. Paton S. 2001 Meteorological and hydrological summary for Barro Colorado Island, Smithsonian Tropical Research Institute, Balboa, Republic of Panama; 2002
34. Thorington RW Jr. Tree map of Lutz watershed. In Rubinoff RW, editor. 1973 environmental monitoring and baseline data. Washington, D.C.: Smithsonian Institution; 1974. pp. 256–289
35. Reinsch CH. Smoothing by spline functions. Numer Math.1967;10: 177–183
36. Swaine MD, Lieberman D, Putz FE. The dynamics of tree populations in tropical forest: a review. J Trop Ecol. 1987;3: 359–366
37. Williamson GB, Laurance WF, Oliveira AA, Delamônica P, Gascon C, Lovejoy TE, et al. Amazonian tree mortality during the 1997 El Niño drought. Conserv Biol. 2000;14: 1538–1542.
38. Abbot HL. Problems of the Panama Canal. Macmillan, New York; 1907
39. Hubbell SP, Foster RB. Structure, dynamics, and equilibrium status of old-growth forest on Barro Colorado Island. In Gentry AH, editor. Four Neotropical forests. New Haven: Yale University Press; 1990. pp. 522–541
40. Krause GH, Winter K, Krause B, Jahns P, García M, Aranda J, et al. High-temperature tolerance of a tropical tree, *Ficus insipida*: methodological reassessment and climate change considerations. Funct Plant Biol. 2010;37: 890–900
41. McDowell N, Pockman WT, Allen CD, Breshears DD, Cobb N, Kolb T, et al. Mechanisms of plant survival and mortality during drought: why do some plants survive while others succumb to drought? Tansley Rev New Phytol. 2008;178: 719–739
42. .Gautier-Hion A, Michaloud G. Are figs always keystone resources for tropical frugivorous vertebrates? A test in Gabon. Ecology. 1989;70: 1826–1833
43. Shanahan M, So S, Compton SG, Corlett R. Fig-eating by vertebrate frugivores: a global review. Biol Rev. 2001;76: 529–572. pmid:11762492
44. Kissling WD, Rahbek C, Böhning-Gaese K. Food plant diversity as broad-scale determinant of avian frugivore richness. Proc R Soc B. 2007;274: 799–808. pmid:17251107
45. Hodgkison R, Balding ST, Zubaid A, Kunz TH. Temporal variation in the relative abundance of fruit bats (Megachiroptera: Pteropodidae) in relation to availability of food in a lowland Malaysian rain forest. Biotropica. 2004;36: 522–533
46. Milton K. The foraging strategy of howler monkeys: a study in Primate Economics. New York: Columbia University Press; 1980
47. Milton K, Morrison DW, Estribi MA:, Windsor DM. Fruiting phenologies of two neotropical *Ficus* species. Ecology 63;1982: 752–762
48. Kalko EKV, Herre EA, Handley CO Jr. Relation of fig fruit characteristics to fruit-eating bats in the New and Old World tropics. J Biogeogr. 1996a;23: 565–576
49. Kalko EKV, Handley CO Jr, Handley D. Organization, diversity, and long-term dynamics of a Neotropical bat community. In: Cody M, Smallwood J, editors. Long-term studies in vertebrate communities. Los Angeles, CA: Academic Press; 1996b. pp. 503–553.

50. Bonaccorso FJ. Foraging and reproductive ecology in a Panamanian bat community. Bull Florida State Mus Biol Sci. 1979;24: 359–408
51. Korine C, Kalko EKV, Herre EA. Fruit characteristics and factors affecting fruit removal in a Panamanian community of strangler figs. Oecologia. 2000;123: 560–568. pmid:28308765
52. .Albrecht L, Meyer CFJ, Kalko EKV. Differential mobility in two small phyllostomid bats, *Artibeus watsoni* and *Micronycteris microtis*, in a fragmented neotropical landscape. Acta Theriologica. 2007;52: 141–149
53. Kalko EKV, Estrada Villegas S, Schmidt M, Wegmann M, Meyer CFJ. Flying high–assessing the use of the aerosphere by bats. Integr Comp Biol. 2008;48: 60–73. pmid:2166977

Index